# Real-Time Diagnosis Algorithms in Biomedical Applications and Decision Support Tools

# Real-Time Diagnosis Algorithms in Biomedical Applications and Decision Support Tools

Editor

**Alfredo Rosado Muñoz**

MDPI

Basel • Beijing • Wuhan • Barcelona • Belgrade • Novi Sad • Cluj • Manchester

*Editor*
Alfredo Rosado Muñoz
University of Valencia
Burjassot
Spain

*Editorial Office*
MDPI
St. Alban-Anlage 66
4052 Basel, Switzerland

This is a reprint of articles from the Special Issue published online in the open access journal *Applied Sciences* (ISSN 2076-3417) (available at: https://www.mdpi.com/journal/applsci/special_issues/diagnosis_algorithms).

For citation purposes, cite each article independently as indicated on the article page online and as indicated below:

| |
|---|
| Lastname, A.A.; Lastname, B.B. Article Title. *Journal Name* **Year**, *Volume Number*, Page Range. |

**ISBN 978-3-7258-1299-8 (Hbk)**
**ISBN 978-3-7258-1300-1 (PDF)**
**doi.org/10.3390/books978-3-7258-1300-1**

Cover image courtesy of Alfredo Rosado Muñoz

# Contents

# About the Editor

**Alfredo Rosado Muñoz**

Alfredo Rosado Muñoz received his M.Sc. and Ph.D. degrees in physics from the Universitat de Valencia, Spain, in 1994 and 2000, respectively. He is a member of the International Federation of Automatic Control (IFAC). He is currently a Full Professor with the Department of Electronic Engineering, Universitat de Valencia. His research is related to digital hardware design (embedded systems) for digital signal processing, artificial intelligence and control systems, especially targeted for biomedical engineering, automation systems, and bio-inspired systems.

applied sciences

MDPI

*Editorial*

# Special Issue on Real-Time Diagnosis Algorithms in Biomedical Applications and Decision Support Tools

**Alfredo Rosado-Muñoz** [1]

Department of Electronics Engineering, University of Valencia, Burjassot, 46100 Valencia, Spain; alfredo.rosado@uv.es

## 1. Introduction

The use of automatic support tools in daily clicnical practice is increasing continuously. From family doctors to surgeons, specialists are using a wide range of devices and software, increasing the level of accuracy in their diagnoses. Deep learning algorithms and data analysis in general are providing new possibilities for doctors. Many doctors now use user-friendly tools and devices in their daily practice which contain an impressive level of research underneath. The algorithms and data processing required are hidden to doctors in order to allow them to concentrate on their main task: taking care of patients. Still, continuous research must be conducted in order to improve these algorithms.

Each of the ten papers published in this Special Issue is proof of such advances and continuously evolving developments. Different proposals from various fields show how new data analysis techniques can improve the daily tasks of doctors. This is especially important when those algorithms are included as part of the devices that doctors use, making it possible to provide doctors with important information to validate their diagnostics. In this sense, these algorithms must be not only be accurate, but they must also be able to be executed in real-time. That was one of the main goals of this special issue.

As a second main goal, decision support tools are an issue when dealing with the analysis of massive amounts of data from patients. These tools will help to provide relevant information to doctors, showing trends and variations in the information.

In both cases, research in this area is essential in order to provide the best care to patients.

**Citation:** Rosado-Muñoz, A. Special Issue on Real-Time Diagnosis Algorithms in Biomedical Applications and Decision Support Tools. *Appl. Sci.* **2023**, *13*, 13308. https://doi.org/10.3390/app132413308

Received: 21 November 2023
Revised: 11 December 2023
Accepted: 14 December 2023
Published: 16 December 2023



## 2. Real-Time Diagnosis Algorithms in Biomedical Applications

Five papers in this Special Issue mainly dealt with real-time issues:

- "Ventricular Fibrillation and Tachycardia Detection Using Features Derived from Topological Data Analysis [1]" by Azeddine Mjahad, Jose V. Frances-Villora, Manuel Bataller-Mompean, and Alfredo Rosado-Muñoz was published in July 2022 and dealt with the detection of an important cardiac pathology which can cause death if not adequately reverted in time.
- "Low-Cost, Compact, and Rapid Bio-Impedance Spectrometer with Real-Time Bode and Nyquist Plots" [2] by Didik R. Santoso, Bella Pitaloka, Chomsin S. Widodo, and Unggul P. Juswono, published in January 2020, presented a bio-impedance spectrometer with many possibilities for daily use by specialists, in addition to being accurate and compact.
- "Real-Time Localization of Epileptogenic Foci EEG Signals: An FPGA-Based Implementation" [3] by Jose V. Frances-Villora, Manuel Bataller-Mompean, Azeddine Mjahad, Alfredo Rosado-Muñoz, Antonio Gutierrez Martin, Vicent Teruel-Marti, Vicente Villanueva, Kevin G. Hampel, and Juan F. Guerrero-Martinez was published in 2020 and covered interesting research on epilepsia and proposed a method able to detect epileptic episodes in a short time, allowing the reversion of episodes as soon as possible.

- "Optimization of Physical Activity Recognition for Real-Time Wearable Systems: Effect of Window Length, Sampling Frequency and Number of Features" [4] by Ardo Allik, Kristjan Pilt, Deniss Karai, Ivo Fridolin, Mairo Leier and Gert Jervan, published in November 2019, provided important insight into wearable devices, commonly used nowadays.
- "A Prototype of a Portable Gas Analyzer for Exhaled Acetone Detection" [5] by Jakub Sorocki and Artur Rydosz in June 2019 described the use of a gas analyzer to detect exhaled acetone. Acetone is an important compound related to some diseases. This proposal showed an interesting approach to using the device and measurement method.

## 3. Decision Support Tools in Biomedical Applications

The other five papers were more related to decision support tools:

- "A Machine-Learning Model Based on Morphogeometric Parameters for RETICS Disease Classification and GUI Development" [6] by José M. Bolarín, F. Cavas, J.S. Velázquez, and J.L. Alió, published in March 2020, described a graphical user interface helping specialists to detect RETCS disease, assisted by artificial intelligence.
- "Wavelia Breast Imaging: The Optical Breast Contour Detection Subsystem" [7] by Julio Daniel Gil Cano, Angie Fasoula, Luc Duchesne, and Jean-Gael Bernard was published in February 2020 and described a tool to help specialists in analyzing breast images to detect anomalies and better establish a relationship between disease and deformities.
- "Analogy Study of Center-Of-Pressure and Acceleration Measurement for Evaluating Human Body Balance via Segmentalized Principal Component Analysis" [8] by Tian-Yau Wu and Ching-Ting Liou, published in Novemebr 2019, conducted an interesting analysis of human body balance by means of several data analysis tools, showing the results as an important parameter to evaluate certain human movement parameters in patients.
- "Dynamic Handwriting Analysis for Neurodegenerative Disease Assessment: A Literary Review" [9] by Gennaro Vessio, published in November 2019, proposed an approach on detecting neurodegenerative processes by means of the analysis of handwriting. These tests are very important to the specialist in order to achieve early detection.
- "Automatic Detection of a Standard Line for Brain Magnetic Resonance Imaging Using Deep Learning" [10] by Hiroyuki Sugimori and Masashi Kawakami appeared in September 2019 and showed how to develop and estimate accurate parameters from brain magnetic resonance as important markers for the specialist.

## 4. Conclusions

The goals of the Special Issue were fulfilled by the inclusion of research works related to relevant research topics in biomedical engineering, such as improvement in health care, therapy, and diagnosis. All of the contributions to this issue have a high social impact which, in turn, is what science and research is made for: improving human lives.

**Acknowledgments:** I want to express my gratitude to the authors for their careful research work and openness to suggestions, the reviewers for the detailed comments and constructive suggestions, the editors and proofreading team for the careful design and high quality of the published papers, both in the content of the research and the print quality.

**Conflicts of Interest:** The author declares no conflict of interest.

## References

1. Mjahad, A.; Frances-Villora, J.V.; Bataller-Mompean, M.; Rosado-Muñoz, A. Ventricular Fibrillation and Tachycardia Detection Using Features Derived from Topological Data Analysis. *Appl. Sci.* **2022**, *12*, 7248. [CrossRef]
2. Santoso, D.R.; Pitaloka, B.; Widodo, C.S.; Juswono, U.P. Low-Cost, Compact, and Rapid Bio-Impedance Spectrometer with Real-Time Bode and Nyquist Plots. *Appl. Sci.* **2020**, *10*, 878. [CrossRef]

3. Frances-Villora, J.V.; Bataller-Mompean, M.; Mjahad, A.; Rosado-Muñoz, A.; Gutierrez Martin, A.; Teruel-Marti, V.; Villanueva, V.; Hampel, K.G.; Guerrero-Martinez, J.F. Real-Time Localization of Epileptogenic Foci EEG Signals: An FPGA-Based Implementation. *Appl. Sci.* **2020**, *10*, 827. [CrossRef]
4. Allik, A.; Pilt, K.; Karai, D.; Fridolin, I.; Leier, M.; Jervan, G. Optimization of Physical Activity Recognition for Real-Time Wearable Systems: Effect of Window Length, Sampling Frequency and Number of Features. *Appl. Sci.* **2019**, *9*, 4833. [CrossRef]
5. Sorocki, J.; Rydosz, A. A Prototype of a Portable Gas Analyzer for Exhaled Acetone Detection. *Appl. Sci.* **2019**, *9*, 2605. [CrossRef]
6. Bolarín, J.M.; Cavas, F.; Velázquez, J.; Alió, J. A Machine-Learning Model Based on Morphogeometric Parameters for RETICS Disease Classification and GUI Development. *Appl. Sci.* **2020**, *10*, 1874. [CrossRef]
7. Gil Cano, J.D.; Fasoula, A.; Duchesne, L.; Bernard, J.G. Wavelia Breast Imaging: The Optical Breast Contour Detection Subsystem. *Appl. Sci.* **2020**, *10*, 1234. [CrossRef]
8. Wu, T.Y.; Liou, C.T. Analogy Study of Center-Of-Pressure and Acceleration Measurement for Evaluating Human Body Balance via Segmentalized Principal Component Analysis. *Appl. Sci.* **2019**, *9*, 4779. [CrossRef]
9. Vessio, G. Dynamic Handwriting Analysis for Neurodegenerative Disease Assessment: A Literary Review. *Appl. Sci.* **2019**, *9*, 4666. [CrossRef]
10. Sugimori, H.; Kawakami, M. Automatic Detection of a Standard Line for Brain Magnetic Resonance Imaging Using Deep Learning. *Appl. Sci.* **2019**, *9*, 3849. [CrossRef]

*applied sciences*

MDPI

*Article*

# Ventricular Fibrillation and Tachycardia Detection Using Features Derived from Topological Data Analysis

**Azeddine Mjahad *, Jose V. Frances-Villora, Manuel Bataller-Mompean and Alfredo Rosado-Muñoz**

Processing and Digital Design Group, Department of Electronic Engineering, University of Valencia, 46100 Burjassot, Spain; jose.v.frances@uv.es (J.V.F.-V.); manuel.bataller@uv.es (M.B.-M.); alfredo.rosado@uv.es (A.R.-M.)

* Correspondence: mjahad.azeddine@uv.es

**Featured Application: Automated External Defibrillation (AED) and Implantable Cardioverter Defibrillators (ICD) require accurate algorithms to detect arrhythmias and discriminate among them. This work proposes specific features for algorithms implemented in such devices.**

**Abstract:** A rapid and accurate detection of ventricular arrhythmias is essential to take appropriate therapeutic actions when cardiac arrhythmias occur. Furthermore, the accurate discrimination between arrhythmias is also important, provided that the required shocking therapy would not be the same. In this work, the main novelty is the use of the mathematical method known as Topological Data Analysis (TDA) to generate new types of features which can contribute to the improvement of the detection and classification performance of cardiac arrhythmias such as Ventricular Fibrillation (VF) and Ventricular Tachycardia (VT). The electrocardiographic (ECG) signals used for this evaluation were obtained from the standard MIT-BIH and AHA databases. Two input data to the classify are evaluated: TDA features, and Persistence Diagram Image (PDI). Using the reduced TDA-obtained features, a high average accuracy near 99% was observed when discriminating four types of rhythms (98.68% to VF; 99.05% to VT; 98.76% to normal sinus; and 99.09% to Other rhythms) with specificity values higher than 97.16% in all cases. In addition, a higher accuracy of 99.51% was obtained when discriminating between shockable (VT/VF) and non-shockable rhythms (99.03% sensitivity and 99.67% specificity). These results show that the use of TDA-derived geometric features, combined in this case this the k-Nearest Neighbor (kNN) classifier, raises the classification performance above results in previous works. Considering that these results have been achieved without preselection of ECG episodes, it can be concluded that these features may be successfully introduced in Automated External Defibrillation (AED) and Implantable Cardioverter Defibrillation (ICD) therapies.

**Keywords:** electrocardiography analysis; ventricular arrhythmia detection; ventricular fibrillation detection; ventricular tachycardia detection; ECG signal classification; Topological Data Analysis; representation of point cloud; persistent diagram representation; landscape representation; silhouette representation

**Citation:** Mjahad, A.; Frances-Villora, J.V.; Bataller-Mompean, M.; Rosado-Muñoz, A. Ventricular Fibrillation and Tachycardia Detection Using Features Derived from Topological Data Analysis. *Appl. Sci.* **2022**, *12*, 7248. https://doi.org/10.3390/app12147248

Academic Editor: Keun Ho Ryu

Received: 4 April 2022
Accepted: 10 July 2022
Published: 19 July 2022



## 1. Introduction

A rapid and accurate detection of ventricular arrhythmias is essential to taking appropriate therapeutic actions. These pathologies are very common, being considered one of the main causes of death in developed countries, given that even weak episodes of Ventricular Fibrillation (VF) eventually cause sudden death.

Although arrhythmias have different origins, they can be considered a consequence of changes in cellular electrophysiology of the heart. Moreover, in most cases of sudden cardiac death, arrhythmogenic cardiac disorders appear as the main causes of death without showing evidence of pathological abnormalities of the heart.

To revert VF, the current protocol is the electrical defibrillation of the heart using an Automatic External Defibrillator (AED) [1], which can be commonly found nowadays in public places such as airports, shopping centers, sports arenas, etc. This process involves an external application of a high-energy electrical shock through the chest wall of the patient to allow the reinstatement of the normal rhythm. Some studies [2–4] have established that defibrillation success is conversely proportional to the time interval between the start of the Ventricular Fibrillation episode and the time when the electrical discharge is applied.

However, similar pathologies exist, like Ventricular Tachycardia (VT), requiring a different treatment than VF. In these cases, the signal may share some characteristics (lack of organization, irregularity, etc.) with VF, but the administration of an electrical shock to a patient not suffering VF could result in serious injuries or even bring about VF itself. This is why an accurate detection and classification of ventricular arrhythmias is so relevant.

The electrocardiogram (ECG) is an inexpensive and noninvasive tool used in the diagnosis of cardiac conduction disorders. It enables the analysis of the heart rate and morphology of different cardiac electrical waves, which, in turn, may permit the identification of various types of heart diseases. Because of this, ECG signals are considered an important and reliable source of information [5,6].

Many statistical methods have been applied to detect VF or VT using ECG data. However, following these manual methods, it is difficult to make a feature extraction capable of capturing the deep characteristics of ventricular arrhythmias. This is the reason why machine learning techniques have been effectively applied for the recognition of cardiac arrhythmias. In this sense, Orozco et al. [7] used the Wavelet method to detect ECG arrhythmias with three types of episodes (Normal, VT, and VF). In [8], Pooyan et al. used an SVM with Gaussian Kernel to detect ventricular abnormalities with morphological features. Tripathy et al. [9] detected and classified shockable (VF/VT) arrhythmias using Variational Mode Decomposition with Random Forest (RF) decision trees. In [10], Jekova et al. used fixed thresholds to implement a real-time detection of shockable episodes (VF/VT). In addition, in the same manner, other works harnessed other machine learning techniques for the detection and recognition of ventricular arrhythmias, as in Mohanty et al. [11], who used a C4.5 classifier; Jothiramalingam et al. [12], who employed a k-Nearest Neighbor (kNN) classifier; Tang et al. [13], who used Bayesian decision; or Kuzilez et al. [14], who employed Independent Component Analysis (ICA) and Decision Trees.

Over the last few years, there has been a general surge in the use of algebraic topology to analyze statistical data. Using this method, complicated data shapes can be categorized. Specifically, a commonly used topological method very used to extract features from a Point Cloud (a set of data points in space) is the Topological Data Analysis (TDA). TDA employs tools from algebraic and combinational topology to draw out properties that express data shapes. It can be considered a key method in attempting to interpret and comprehend characteristics that are otherwise unattainable through the use of other practices due to noise, dimension, or incompletion. It is so unique in its nature that TDA bridges the way between geometry and topology.

Successful and remarkable applications have been made in a varied selection of fields, and the range of applications continues to expand. Some of these applications include neuroscience [15], materials science [16], detection and quantification of periodic patterns in data [17,18], analysis of turbulent flows [19], natural language processing [20], or even detection and classification of breast cancers [21]. However, it has been used in image processing [22], computer vision [23], or signal and time series analysis [24,25].

Specifically, over the past few years, researchers have also begun to use TDA along with Machine Learning methods [26,27].

Within TDA, there is an important method called Persistence Homology that can be considered the main tool of TDA. As well as being a modification of the representation of homology using Point Cloud data, this method computes the homological characteristics of datasets.

In addition, TDA uses Persistence Diagrams and Persistence Barcodes to represent the abundant homological information about the shape of data. However, note that the use of algorithms of Machine Learning along with Persistence Diagrams or Barcodes is an area of TDA under research, looking for a way to alter these diagrams to be adaptable and congruous with Machine Learning methods. An alternative approach to these two diagrams is Persistence Landscapes.

In this work, we hypothesized that using Topological Data Analysis (TDA), some geometric features condensing relevant information about the 'shape of data' can be very valuable for the detection and discrimination of VF and VT rhythms, even in noisy and complex signals. Extracted features can then be applied to machine learning classifiers.

Thus, the goal of this work is to assess the improvement of the classification performance to detect and discriminate VF and VT episodes, when incorporating a set of TDA-derived geometric features in the feature extraction and selection stage. Note that the main difference with previous works is that these kinds of features have been never applied before in the analysis and classification of ventricular arrhythmias.

The main contributions of this work are

- The proposal of a novel classification procedure using features derived from Topological Data Analysis (TDA).
- The application of the proposed classification procedure to the detection and discrimination of VF and VT. Specifically, an accuracy near 99% is obtained.
- The application of the proposed classification procedure to the detection of shockable (VF/VT) and non-shockable rhythms. In this case, a 99.5% accuracy is obtained, the highest in the bibliography.
- The evidence that features derived from Topological Data Analysis can overcome conventional feature selection limitations by providing information about the 'shape of data' to the classifier.
- The high performance obtained without preselection of episodes shows that geometric features are good candidates to be incorporated into Automated External Defibrillator (AED) and Implantable Cardioverter Defibrillation (ICD) devices.

The paper is organized as follows. Section 2 is dedicated to the description of fundamental TDA. Section 3 introduces the dataset, explains the proposed methodology and details the used classification procedure. The results of the analysis and a discussion of these are presented in Sections 4 and 5, respectively. Finally, Section 6 concludes the paper.

## 2. Fundamental Concepts of TDA

This section outlines a simplified description of the mathematics behind Homology and Persistent Homology (PH). In TDA, cloud data are frequently seen as a simplicial complex, which is a set of points, lines, segments, triangles, and its n-dimensional counterparts. This allows one to use the methods from simplicial homology to quantify the shape of the data in terms of connections [28] and enables us to make a topological feature extraction. The process of topological feature extraction using PH can be summarized in the following steps:

- Data Point Cloud $\chi \in \mathbb{R}^n$ is employed as an input.
- For each data point (or vertex) $v_i \in \chi$, make $B(v_i)$ a ball of radius $\epsilon$ centered at each $v_i$, where $\epsilon \in \mathbb{R}^+$.
- Raise the value of $\epsilon$.
- A simplicial complex is built for each $\epsilon$ using Vietoris Rips and filtration.
- Measure PH and take note of its appearance and disappearance.
- Plot the $(\epsilon_{birth}, \epsilon_{death})$ appearance and disappearance coordinates for each PH on an extended real plane $\mathbb{R}^2 \bigcup \{\pm\infty\}$. The Persistence Diagram comes as an output.
- Lastly, the topological features are extracted.

In terms of mathematics, the input to a PH are the Point Cloud data. In the case of ECG, the input data are the time series. Taken's Delay Embedding Theorem can be used in the conversion of time series data to point cloud data without losing topological properties. The approach consists of transforming a time series $x_t$, where $t \in \{1, 2, \ldots, T\}$, into its phase space representation. A point cloud or a set of points is obtained according to the following equation where $i = (1, 2, \ldots, T + n\tau)$ and $\tau$ is a delay parameter and $n$ specifies the dimension of the point cloud [29]:

$$v_i = x_i, x_{i+\tau}, \ldots, x_{i+n\tau} \tag{1}$$

Simplicial complexes are essential in the extraction of topological features from point cloud data. A single data point may define 0-Simplex. A line between two points denotes 1-Simplex. A triangle is a 2-Simplex. Tetrahedra represent 3-simplices (see Figure 1). Finally, a combination of simplices gives way to a Simplicial Complex called Vietoris Rips Complex [30–32].

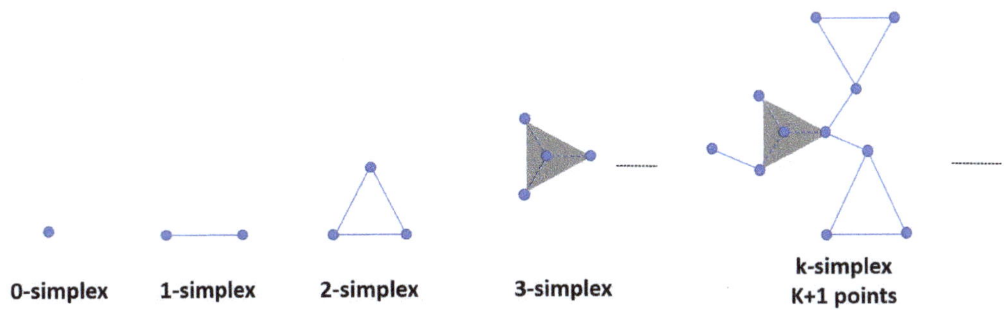


**Figure 1.** Gradual construction of various simplices (0-Simplex, 1-Simplex, 2-Simplex, ...) eventually gives way to a Simplicial Complex.

A simplicial complex can be taken from a dataset using the $Vietoris - Rips$ construction. Being $X = (x_1, \ldots, x_n)$ a point cloud in an euclidean space $\mathbb{R}_n$, for each distance $\epsilon > 0$, represented by $VR(X;\epsilon)$, there is a simplicial complex with vertex set in X where $x_0, x_1, \ldots, x_k$ spreads a $k - simplex$ if the reciprocal distance between any pair of its varieties is smaller than $\epsilon$, where $d(x, x) \leq \epsilon$, for all $0 \leq i, j \leq k$.

When building a Simplicial Complex with Point Cloud data, it is needed to follow a set of rules. Firstly, a circle should be drawn with radius $\epsilon$ for each point in a point cloud. Then, when two circles intersect with each other and the radius is increased, a line is drawn to link the two points, which can be seen in Figure 2.

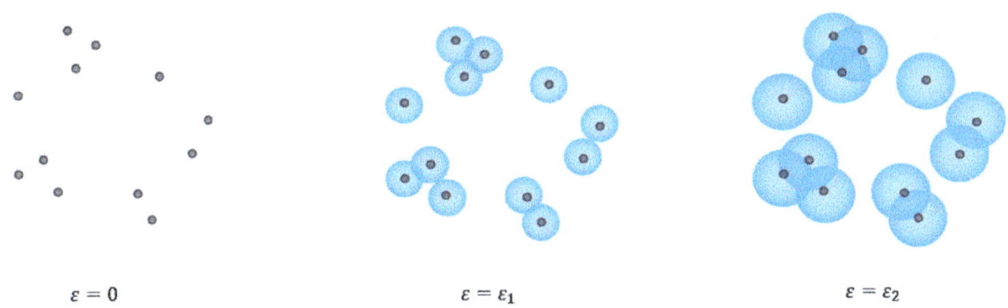


**Figure 2.** Diagram illustrating circular intersections and linked point clouds required to build a Simplicial Complex.

As $\epsilon$ gets longer, the Vietoris-Rips complex of a Point Cloud does, too. This is a filtration of simplicial complexes, i.e., a nested sequence of simplicial complexes, where $VR(X;\epsilon)$, $\epsilon \geq 0$ satisfying $VR(X;\epsilon_1) \subseteq VR(X;\epsilon_2)$ if $\epsilon_1 \leq \epsilon_2$. To represent the distance between them, balls are drawn around each point. If two balls with radius $\epsilon$ intersect with each other, the two points are at a distance at most $2\epsilon$.

The Persistence Diagram representation (PDR) is a standard way to represent PH [33,34]. K-dimensional features consist of persistence diagrams; 0-dimensional features represent components that are connected, 1-dimensional features represent holes, 2-dimensional features voids, etc. [35]. Concurrently, a PDR $W_m$ is made of $n$ features, $W_{m_i} = (b_i, d_i)$, with $i = (1, 2, \ldots, n)$. Each point corresponds to the lifespan of one topological feature, where $b_i$ and $d_i$ are its birth time and death time, respectively (birth time indicates when the geometrical structure appears, while death time indicates when the geometrical structure disappears). Points are entirely located in the half-plane above the diagonal [36] (Figure 3).

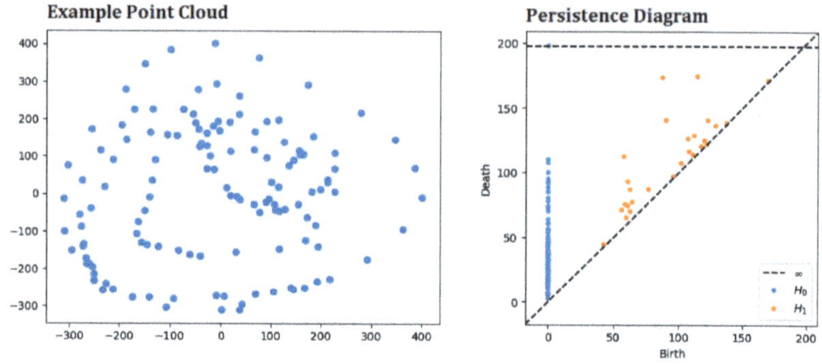


**Figure 3.** Representation of Point Cloud—RPC (**left**) on a Persistence Diagram representation—PDR (**right**).

When it comes to machine learning and statistics, a Persistence Landscapes Representation (LR) is more straightforward to work with than PDR and can be considered an alternative representation [37]. The approach takes the topological information that was previously encoded on a PDR and presents it as elements of a Hilbert space. Statistical learning methods can then be applied directly. Additionally, Persistence Silhouette representation (SR) [38] are constructed by mapping each point $z = (d, b)$ of a PDR to a piecewise linear function, namely the 'triangle' function $T_z$, which can be defined as follows:

$$T_z(y) = (y - b + d)l_{[b-d,b]}(y) + (b + d - y)l_{[b,b+d]}(y) \tag{2}$$

where $l_A(x)$ is the standard indicator function: $l_A(x) = 1$ if $x \in A$ and $l_A(x) = 0$, otherwise. A triangle function binds the points of the diagram to the diagonal, with segments parallel to the axes, and later they are rotated by 45 degrees. The triangles $T_z$ can be merged together in various manners, and if we take their $k_{Amax}$, i.e., the $k^{th}$ largest value in the set $T_z(y)$, the $k^{th}$ persistence landscape $\lambda^k = k - max_{zgD} T_z(y), k \in N^+$ results. The Persistence Landscape $\lambda_D$ is the gathering of functions $\lambda^k(y)$. Finally, the Power Weighted Silhouette representation $\Psi_p(t)$ (later named SR) is obtained by taking the weighted average of the functions $T_z(y)$, as the following equation shows.

$$SR = \Psi_p(t) = \frac{\sum_{zgD} \omega^p T_z(y)}{\sum_{zgD} \omega^p Z} \tag{3}$$

In Figure 4 we can see a representation of the PDR and the LR.

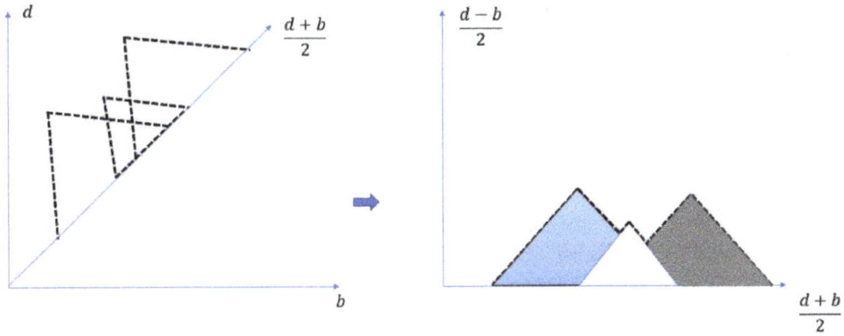


**Figure 4.** A visual example of the transformation of a persistence diagram representation (PDR) into a persistence landscape (PL). The horizontal axis represents birth time, while the vertical axis represents death time on the persistence diagram (**left**). The horizontal axis is the average of the homologies of the birth and death times, and the vertical axis is used for $(d-b)/2$ on the persistence landscape (**right**).

Another means of persistence diagram transformation is Persistence Images (PI) [23]. This allows for representations to be simply vectorized. Persistence images can be informally considered as a type of heatmap coming from a Calculate Gaussian KDE [39], which can be defined as follows:

$$\hat{f}(x) = \sum \alpha_i k(x - x_i) \tag{4}$$

where $k$ is kernel function centered at the data points $x_i$ with $i = (0, \ldots, n)$, and $\alpha_i$ are the weighting coefficients.

## 3. Materials and Methods

To use the topological data features described above, the classification procedure proposed in Figure 5 is used.

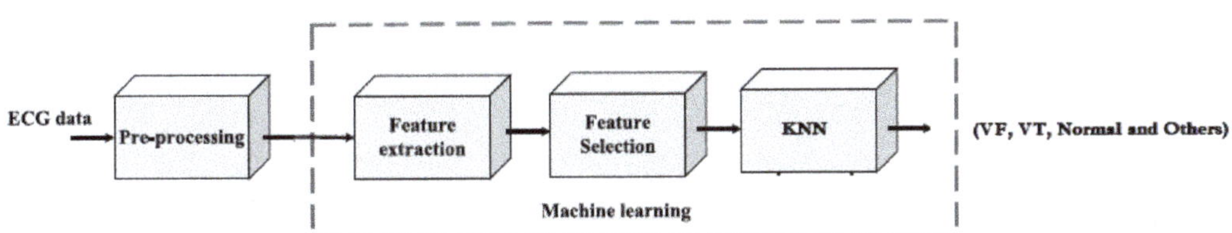


**Figure 5.** Schematic diagram illustrating the feature selection process proposed for the discrimination of ventricular arrhythmias, normal sinus, and other types of rhythms.

To provide a clear and detailed explanation of data processing, this section has been divided into different parts: Section 3.1 describes the used dataset; Section 3.2 details the noise cancellation, baseline removal, and segmentation is done as preprocessing; Section 3.3 describes the feature extraction and selection, and Section 3.4 outlines the classification procedure and parameters used to evaluate the performance of the classification.

### 3.1. Materials

Data records from two standard databases were used: MIT-BIH Malignant Arrhythmia Database [40,41] and AHA (American Heart Association) 2000 series [42]. It is important to note that no preselection of ECG episodes was done, i.e., all annotated segments from the database were used. Thus, 24 patients were analyzed (i.e., 24 records), 22 of which were from the MIT-BIH database and two additional patients from the AHA Database. Each record contained half an hour of continuous ECG recordings. According to the database annotation, each segment was assigned to a class. Four classes of rhythms were established: Ventricular Fibrillation (named *VF*), including Ventricular Fibrillation or Ventricular Flutter episodes, Ventricular Tachycardia as *VT* class, sinus rhythms were assigned to the *Normal* class, and lastly, any signal not labeled within the above classes (e.g., other non-ventricular arrhythmias, noise, etc.) was assigned to class *Others*.

### 3.2. ECG Signal Preprocessing

The performance of machine learning algorithms can be brought down due to errors that may appear due to noise interruption or other input data corruption leading to improper feature values. Thus, a signal preprocessing is required to remove unwanted data corruption of the ECG signals: breathing, skin interference, baseline wander, powerline interference, motion artifact due to electrodes, muscle artifact, white Gaussian noise, etc. [43]. Since this work proposes the full data flow analysis from acquisition to classification as in a real scenario, we add this data preprocessing step, too. The steps used in the preprocessing stage to prepare signals for later processes are:

- Reduction of the baseline wandering, aiming to provide better quality and definition of the temporal signal, which will later result in better feature extraction. This stage involves the introduction of an 8th order infinite impulse response filter (IIR) with a Butterworth bandpass type ranging from 1 Hz to 45 Hz [44,45]. Figure 6 shows the effect of applying this bandpass filter, resulting in a reduction of the baseline.

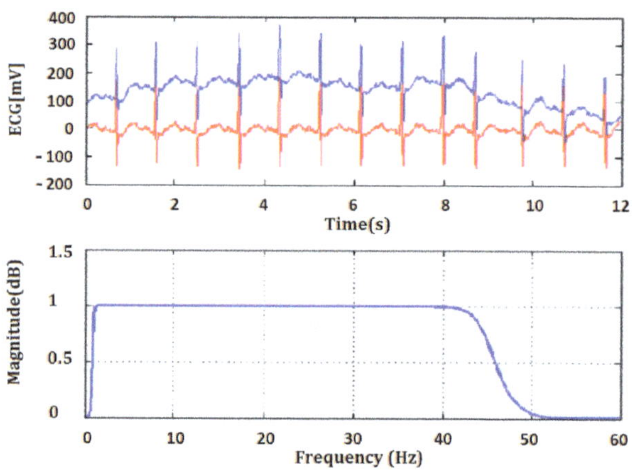


**Figure 6.** Bandpass filter application to a data segment from the *Normal* ECG class, and resulting baseline reduction.

- Later, a Window Reference Marks (WRM) and a time window ($tw$) are obtained. The mark indicates the start of each time window from which the features will be extracted. Consecutive time windows are obtained to analyze all ECG data. As the values between 50 and 120 beats per minute (bpm) can be considered a normal heart rate range [46], the minimum and maximum distance values between any two consecutive WRMs were established in 0.5 s and 1.2 s, respectively. Next, an algorithm already developed by the authors in [47] was used to obtain the calculation of WRM reference marks. A time window $tw$ of 1.2 s (150 samples) in length was obtained, starting at each WRM reference mark, as the following equation shows, with $\{j = 1, 2, \ldots, NLMC\}$ where $NLMC$ is the number of local maxima LM marks existing in the ECG signal:

$$tw_i = [WRM_j, WRM_j + 1.2s] \tag{5}$$

- For each time window, the Taken's Delay Embedding Theorem is applied to convert the ECG data (a time series) to a Representation of Point Cloud data (RPC), a Persistence Diagram Representation (PDR), a Persistence Landscape (LR) and a weighted Silhouette Representation (SR).

### 3.3. Feature Extraction and Reduction

The feature extraction stage can be regarded as the most essential stage in the detection of ventricular arrhythmias. Within the methodology proposed, several discriminatory features from TDA were extracted. The temporal signal in each window was first transformed into Point Clouds using delay embedding. Then, topological representations were extracted: Persistence Diagram (PDR), Persistence Landscape (LR), and Power Weighted Silhouettes (SR) (Tables 1 and 2).

**Table 1.** Columns $a_1$, $a_2$, $a_3$ and $a_4$ correspond to the original ECG time signal windows; columns $b_1$, $b_2$, $b_3$, $b_4$ and $c_1$, $c_2$, $c_3$ and $c_4$ show RPC and PDR, respectively. Each row, from top to bottom, corresponds to *Normal*, *Other*, *VT* and *VF* classes, respectively.

| Parameter Extraction Time Representation (TR) | Representation of Point Cloud (RPC) | Persistence Diagram Representation (PDR) |
|---|---|---|
| 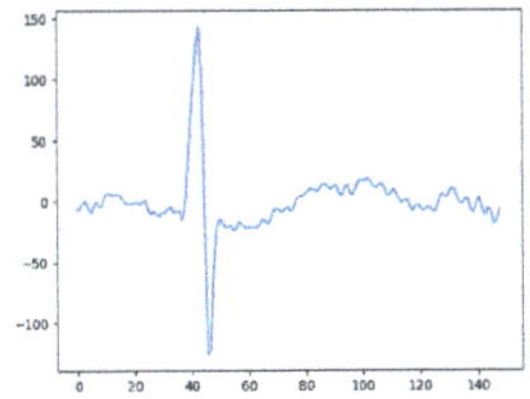 | 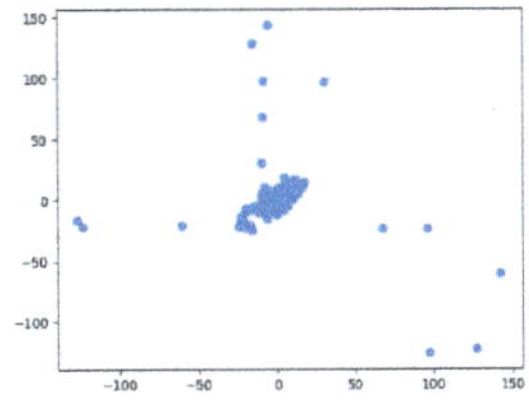 | 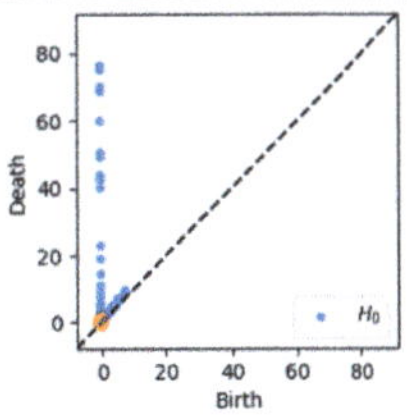 |
| $a_1$ | $b_1$ | $c_1$ |
| 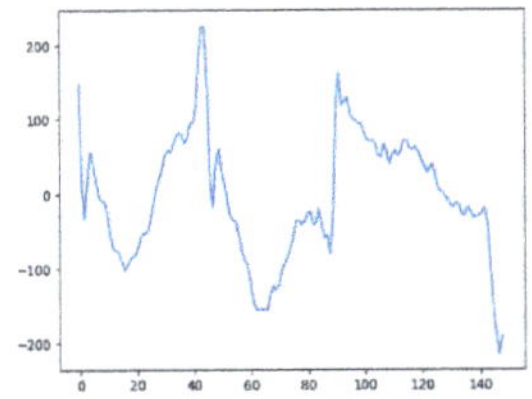 | 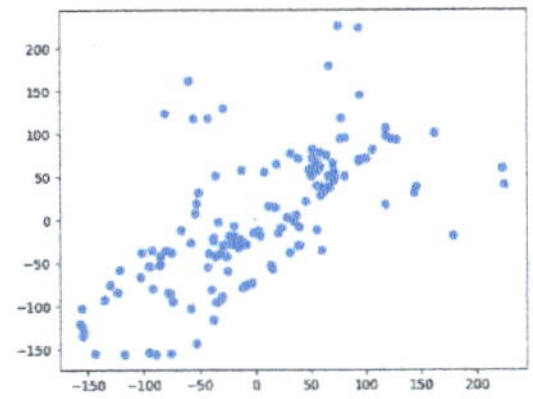 | 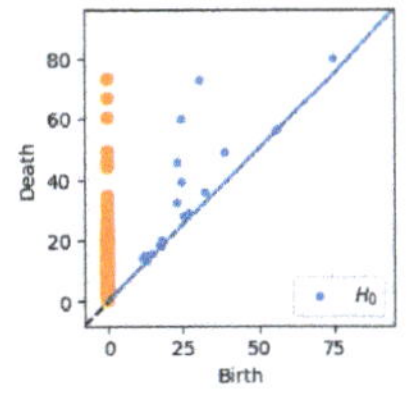 |
| $a_2$ | $b_2$ | $c_2$ |
| 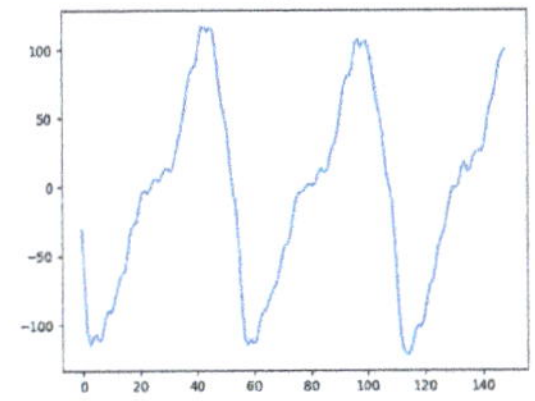 | 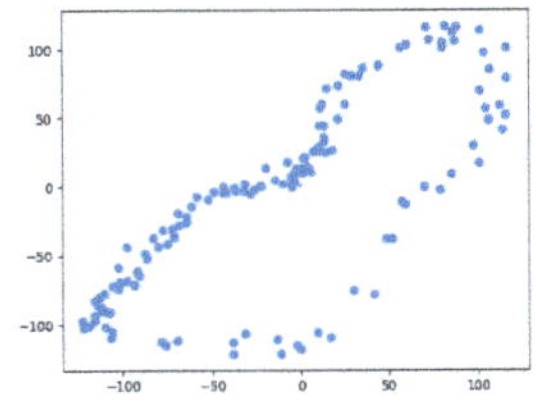 | 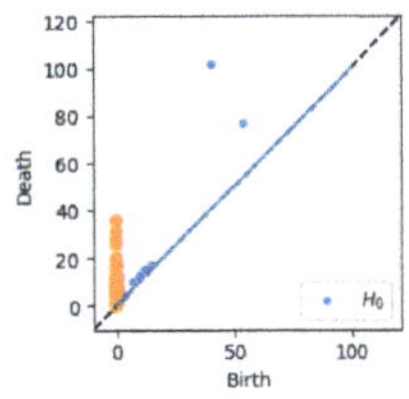 |
| $a_3$ | $b_3$ | $c_3$ |
| 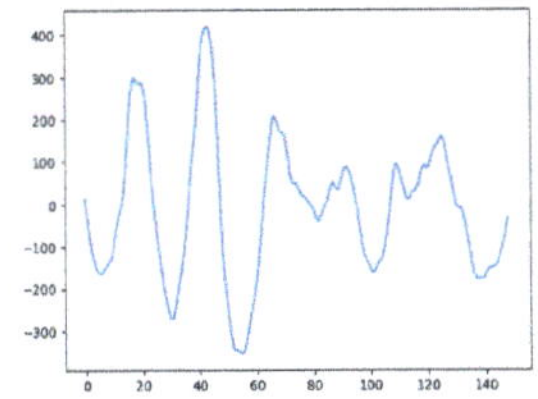 | 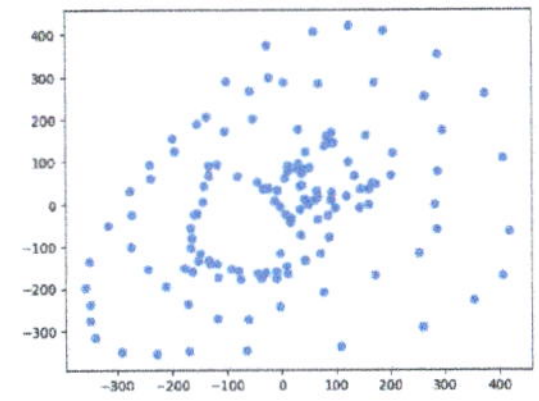 | 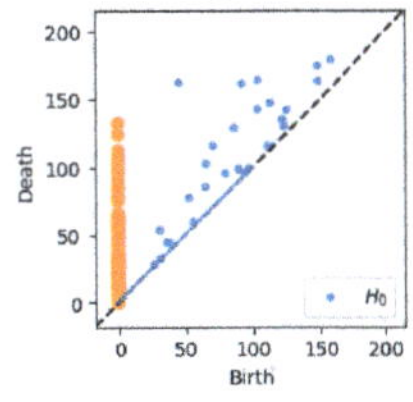 |
| $a_4$ | $b_4$ | $c_4$ |

Table 1 illustrates the data point clouds (Representation of Point Cloud - RPC) and the persistence diagram (Persistence Diagram Representation - PDR) for each class (*Normal*, *Other*, *VT*, and *VF*). As seen, those representations provide a clear difference among classes. Regarding *Normal* class, the points in RPC have a focused distribution with respect to the rest of the arrhythmias where the points are scattered. Moreover, the point distribution differs between *VT* and *VF* as a very heterogeneous cloud is observed in *VF*, in contrast with *VT*. In the case of PDR, more points are located in a high birth-death ratio for *VF*, showing a clear difference with the rest of the rhythms. In Table 2, Persistence Diagrams (PDR) are compared with Persistence Landscapes Representation (LR) and Weighted Silhouettes Representation (SR). For each class, LR and SR show different shapes.

**Table 2.** From top to bottom, left to right: $a_1$, $a_2$, $a_3$ and $a_4$ illustrate Persistence Diagrams (PDR). $b_1$, $b_2$, $b_3$ and $b_4$ show Persistence Landscapes representation (LR), while $c_1$, $c_2$, $c_3$ and $c_4$ detail Weighted Silhouette representation (SR). These all correspond to the four classes *Normal*, *Other*, *VT* and *VF*, respectively.

| Persistence Diagram (PDR) | Landscape (LR) | Silhouette (SR) |
| --- | --- | --- |
| 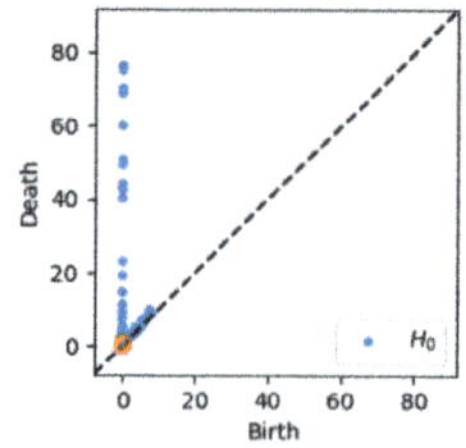 $a_1$ | 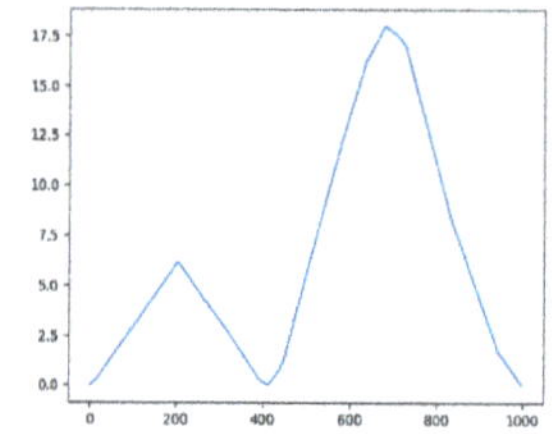 $b_1$ | 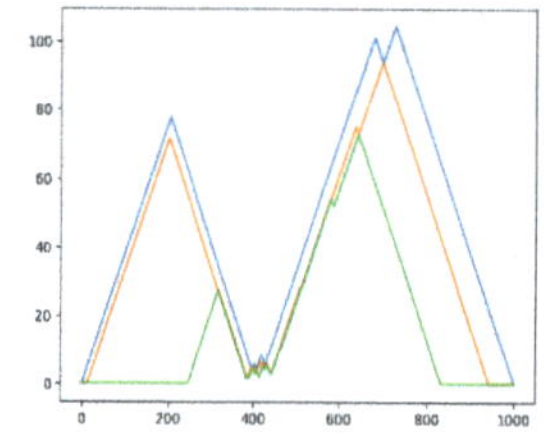 $c_1$ |
| 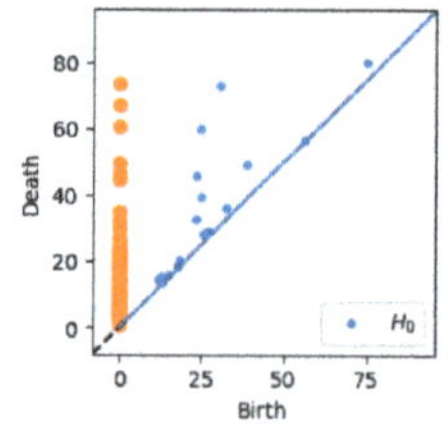 $a_2$ | 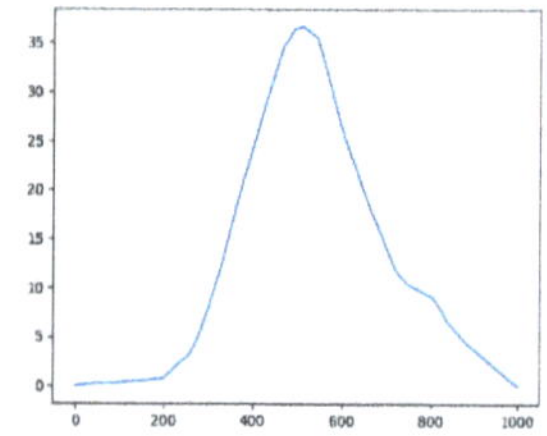 $b_2$ | 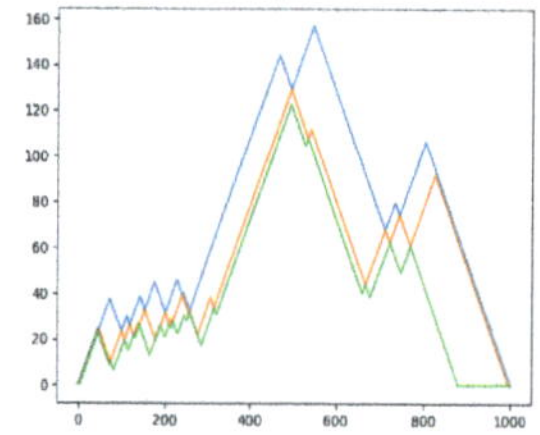 $c_2$ |
| 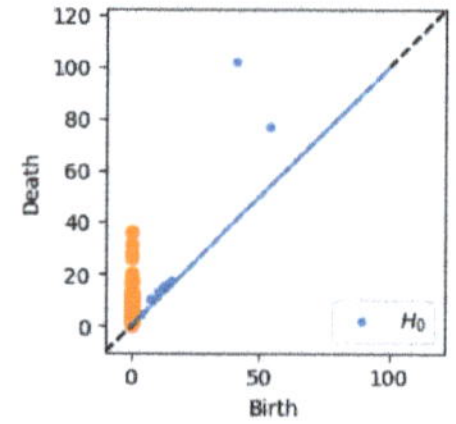 $a_3$ | 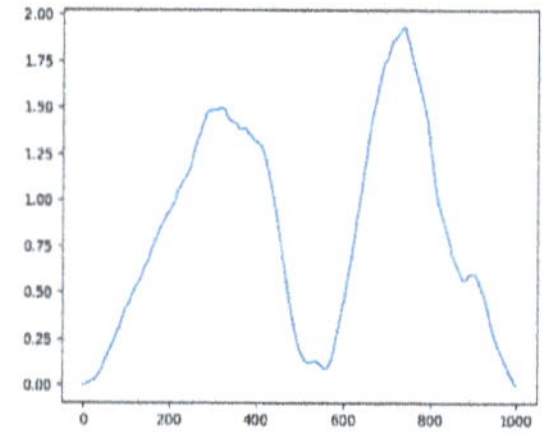 $b_3$ | 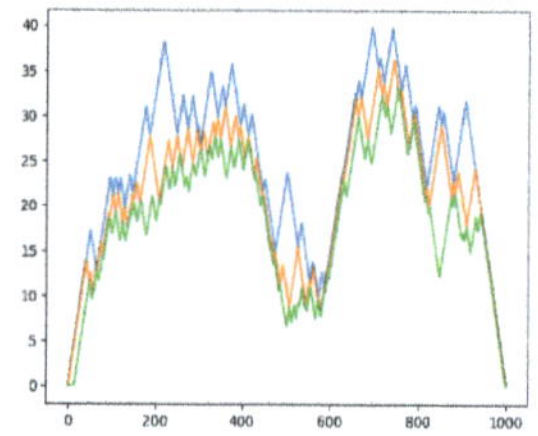 $c_3$ |
| 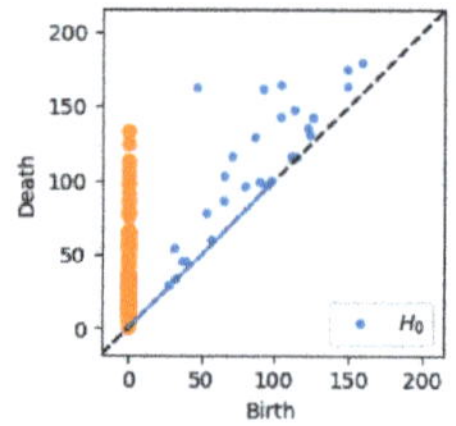 $a_4$ | 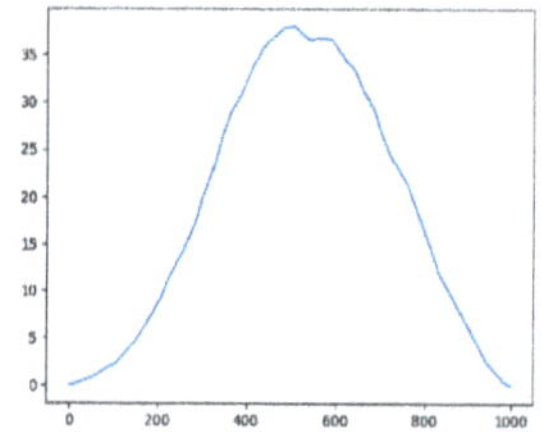 $b_4$ | 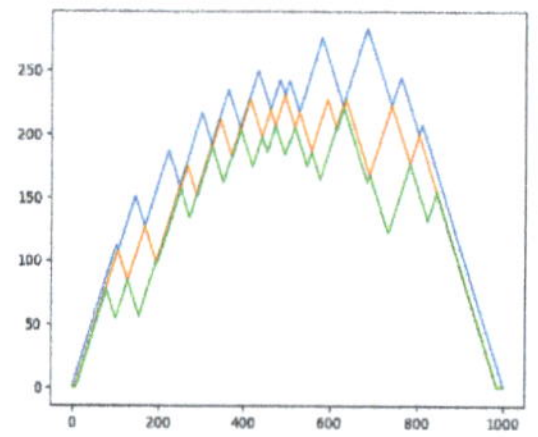 $c_4$ |

From each representation, a number of parameters are calculated, which will become the input features to the classifier. Initially, 79 parameters are evaluated, combining the most commonly used features in the bibliography with the proposed topological features. However, a feature reduction stage was performed. This stage allows the removal of any potentially redundant features existing, as well as reducing the computational complexity of the data analysis. In addition, we combined these features with other usual time-domain features representing statistical characteristics, such as variance, skewness, and kurtosis.

The feature selection was achieved using the Sequential Forward Selection (SFS) method, an iterative method that adds the best feature iteratively to the model until new additions do not improve the performance of the model. This method enables the selection of the most relevant features. Finally, a total of 27 features from all representations (time domain, RPC, PDR, LR, and SR) were selected amongst the 79 initial features. The extracted features are detailed in Table 3. This selection allows to improve the computational efficiency and reduce the generalization error of the model by removing irrelevant features or noise.

**Table 3.** List of extracted features using TDA and other time-domain parameters used as an input vector to the classifier.

| Representation or Domain | Parameter |
|---|---|
| Time Domain | Std (Standard deviation along the specified axis) [48]<br>Permutation entropy [49]<br>Spectral entropy [50]<br>Singular Value decomposition entropy [51]<br>Aproximate entropy [52]<br>Sample entropy [53]<br>Lempel-Ziv complexity [54]<br>Shannon entropy [55]<br>Petrosian fractal dimension [56]<br>Katz fractal dimension [57]<br>Higuchi fractal dimension [58]<br>Detrended fluctuation analysis [59] |
| Representation of Point Cloud (RPC) | Std (Standard deviation along the specified axis) [53]<br>Tsem (trimmed standar error of the mean) [60]<br>Nanmean [61]<br>Tvar (Tail value at risk) [62] |
| Persistence Diagram Representation (PDR) | Persistence Weighted Gaussian Kernel [63]<br>Approximate PWG kernel [63]<br>Persistence Scale SpaceKerne [64]<br>Approximate PSS kernel [64]<br>Sliced Wasserstien Distance [27]<br>Sliced Wasserstein Kernel [27] |
| Landscape Representation (LR) | Tsem (Trimmed standard error of the mean) [60]<br>Tstd (Trimmed sample standard deviation) [53]<br>Wasserstien distance [65]<br>Heat kernel distance bottleneck [53] |
| Silhouette Representation (SR) | RMS (Root Mean Square) [66] |

In addition, another input feature set was obtained. In this case, the Calculate Gaussian KDE was applied to the Persistence Diagram, and then an image-like was obtained (Table 3). The obtained image will be used as a direct input to the classifier, in a similar form as in [47]. Table 4 shows the resulting Gaussian KDE from a PDR.

**Table 4.** Persistence Diagrams (left) $a_1$, $a_2$, $a_3$, $a_4$ and representation of Calculate Gausssian KDE (right) $b_1$, $b_2$, $b_3$, $b_4$ for classes *Normal*, *Other*, *VT* and *VF*, respectively. The representation of KDE is used as input to the kNN classifier in the proposed PDI method.

| Scatterplot from Persistence Diagram | Calculate Gaussian KDE |
|---|---|
| 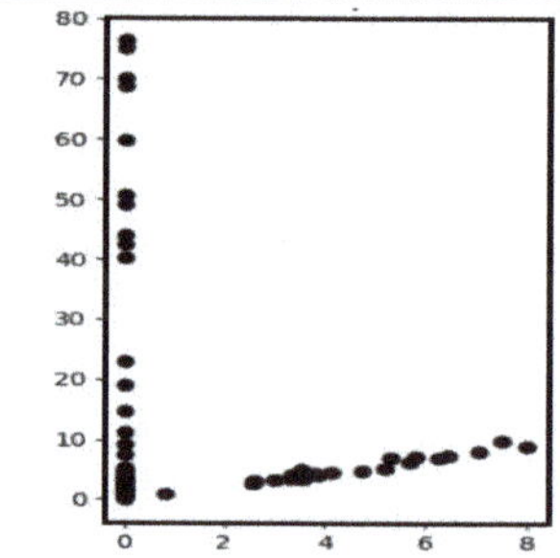 a1 | 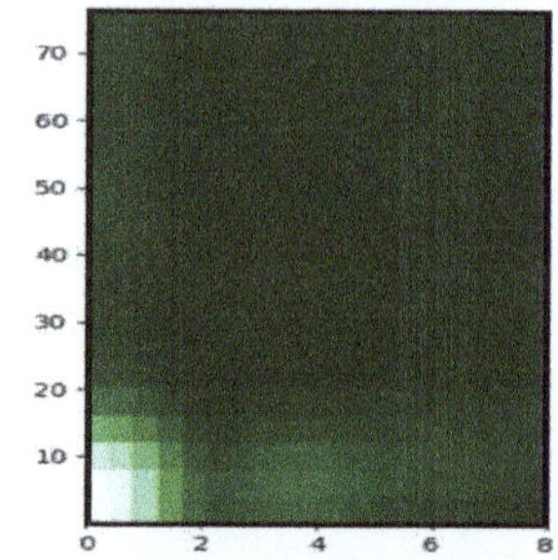 b1 |
| 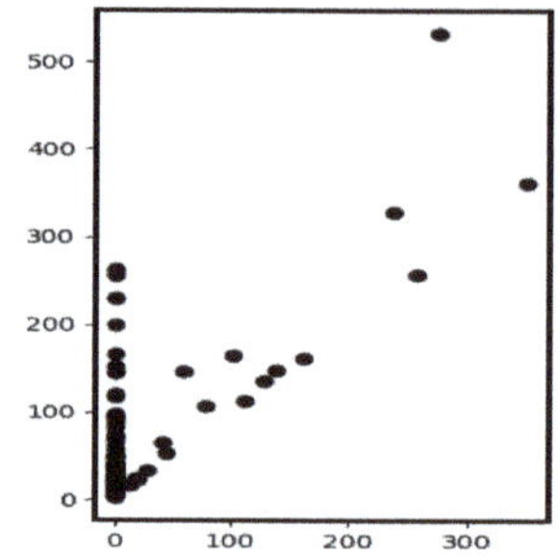 a2 | 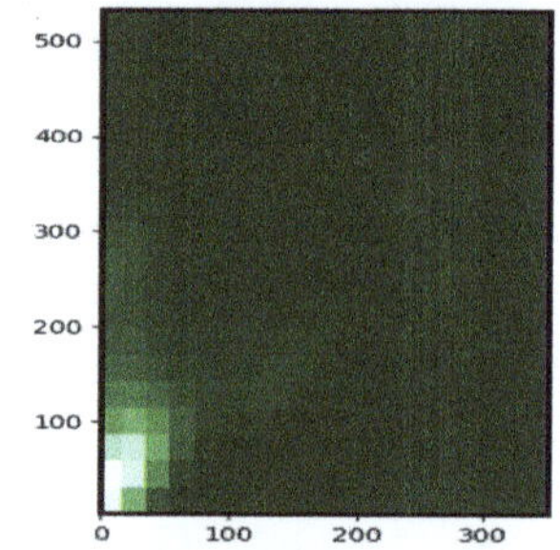 b2 |
| 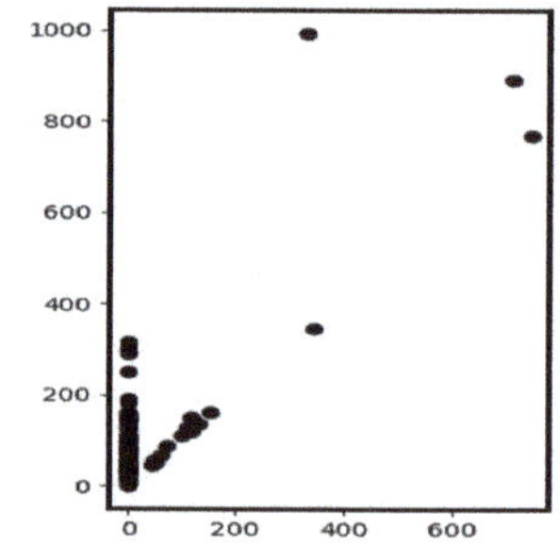 a3 | 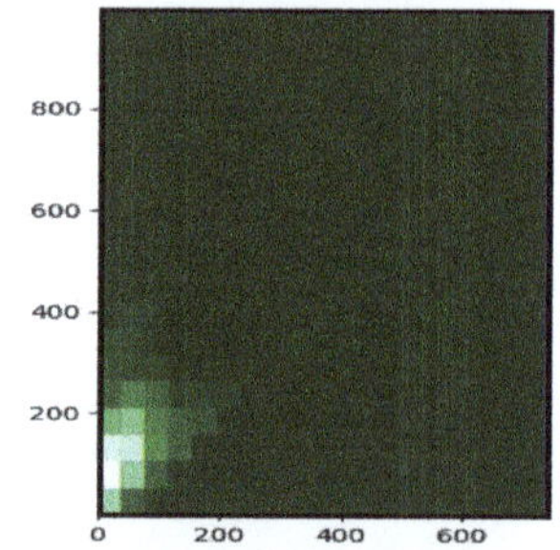 b3 |
| 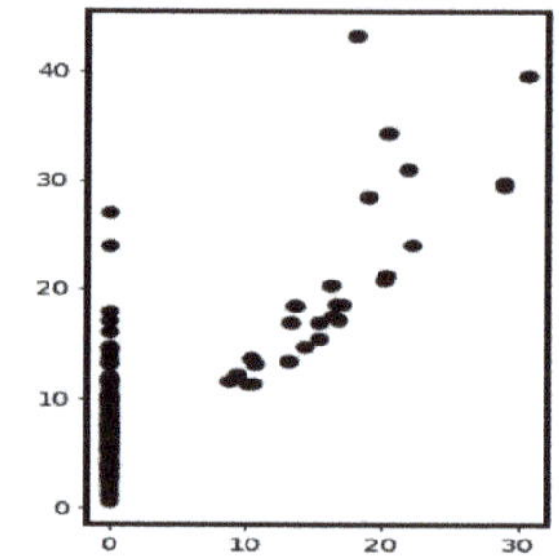 a4 | 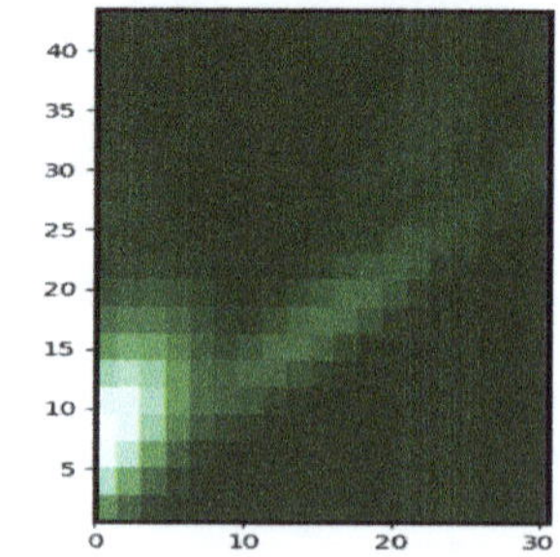 b4 |

### 3.4. Classification Procedure and Performance Evaluation

This work used supervised learning. The input to the classifier was formed by a feature vector linking together all the selected features calculated from consecutive ECG time windows. The only classifier used in this work was the k-Nearest Neighbor (kNN) algorithm, which is commonly used in the bibliography [12,67]. And the used distance is the Euclidean distance. In addition, two different input sets are evaluated. The first input set corresponds to the features described in Table 3, named the TDA method. The second input set is based on the Gaussian KDE bidimensional representation, named Persistence Diagram Image - PDI.

It has been used as a repeated random sub-sampling validation technique. Thus, for each class, 67% of data was randomly selected for training and the remaining 33% for testing. The kNN training process was done, and then the testing dataset was used to evaluate the classification performance by measuring the Sensibility (Sen), Specificity (Spe), and Accuracy (Acc). This cross-validation approach was repeated five times at random, and the performance of the classifiers was evaluated overall by taking this five iterations average. This number of iterations was chosen after some trials as it showed the lowest generalization error.

Standard statistical parameters were followed to assess the performance in accurate classification of the ECG signal into the *VF*, *VT*, *Normal*, or *Others* classes. These include the Sensitivity (Sen), Specificity (Spe), and Accuracy (Acc), which are calculated using the following equations where *TP*, *TN*, *FP*, and *FN* represent the number of true positives, true negatives, false positives, and false negatives, respectively:

$$Sensitivity = \frac{TP}{TP + FN} \times 100\% \tag{6}$$

$$Specificity = \frac{TN}{TN + FP} \times 100\% \tag{7}$$

$$Accuracy = \frac{TP + TN}{TP + FP + TN + FN} \times 100\% \tag{8}$$

## 4. Results

The experiments were carried out using signals from the MIT-BIH and AHA standard databases, Section 3.1. They were divided into four classes, namely *VF*, *VT*, *Normal* and *Others*. The preprocessing stage carries out an 8th order bandpass (1 Hz to 45 Hz) Butterworth IIR filter to denoise and reduce the baseline variation, Section 3.2, and calculates the window reference marks (WRM) of the signal, marks indicating the beginning and end of the 1.2 s time window from the temporal signals.

At the feature extraction, we have proposed two different topological techniques to extract the parameters feeding the classifier: Topological Data Analysis (TDA) and Persistence Diagrams (PDI). In the case of the TDA method, each window of temporal signals were converted first into a Point Clouds representation, using delay embedding, and then into Persistence Diagrams, Persistence Landscapes, and Power Weighted Silhouettes. Finally, some parameters were extracted from these diagrams, Section 3.3, and then combined to create the features vector feeding the input of the classifier. Concerning the PDI method, the gaussian KDE was applied to the Persistence Diagram and the whole resulting image was used as a direct input to the classifier.

The k-Nearest Neighbor (kNN) classifier was the only classifier used for both proposals.

For each class, 67% of data was randomly selected for training and the remaining 33% for testing. The kNN training process was calculated and then the testing dataset was used to evaluate the classification performance by measuring the Sensibility (Se), Specificity (Sp), and Accuracy (Acc). This approach was repeated five times at random, and the performance of the classifiers was evaluated overall by taking this five iterations average. This number of iterations was chosen after some trials because it showed the lowest generalization error.

Table 5 shows the confusion matrix for one of these iterations. It shows a great classification performance. Nevertheless, the values represented in the following tables (Tables 6–9) indicate the average performance values obtained from the repeated random validation used in this work.

**Table 5.** Confusion matrix for classification of *VF*, *VT*, *Others*, and *Normal* classes using the TDA topological method.

| Algorithms | TDA | | | |
|---|---|---|---|---|
| | **VF** | **VT** | **Normal** | **Other** |
| VF | 1701 | 36 | 12 | 3 |
| VT | 45 | 601 | 9 | 1 |
| Normal | 15 | 3 | 4957 | 28 |
| Other | 0 | 2 | 55 | 1940 |

Thus, Tables 6–9 show the testing classification results for TDA and PDI feature selection methods. As it can be seen, the TDA method shows better classification results than the PDI. On the one hand, the PDI method results in values of accuracy above 92% for all classes, having better accuracy values for *VT* and *Other* classes (97.38% and 96.19%, respectively), but curiously falling to 92.65% for the detection of *Normal* sinus rhythms. The sensitivity widely varies depending on the case, ranging from 82.25% for *VT* to 93.09% for the *Normal* classes, being more sensitive to *Normal* and *Others* classes (around 93%) than to *VT* and *VF* (around 84%). Except for the *Normal* case, the global specificity (Spe) becomes greater than sensitivity, reaching the value of 98.53% for the *VT* class.

On the other hand, the TDA method results in very high results of accuracy, around 99% for all classes, with little differences between them. The sensitivity remains above 97% except for the *VT* class, falling to 92.72% and getting the maximum sensitivity value for the *Normal* case (99.05%), with 97.07% for the *VF* class. Finally, the global specificity achieves high values: near 99% for the *Normal* class and above 99% for the rest of the classes, hitting a maximum of 99.53% for the *VT* class.

**Table 6.** Results obtained for *VF* class classification in testing.

| **Type** | **VF** | | | | | |
|---|---|---|---|---|---|---|
| **Algorithms** | **Sensitivity%** | **Specificity%** | | | | **Accuracy%** |
| | **VF** | **Global** | **VT** | **Other** | **N** | **Total** |
| TDA | 97.07 | 99.25 | 93.78 | 99.90 | 99.68 | 98.68 |
| PDI | 84.34 | 96.77 | 89.70 | 99.14 | 96.68 | 94.26 |

**Table 7.** Results obtained for *VT* class classification in testing.

| **Type** | **VT** | | | | | |
|---|---|---|---|---|---|---|
| **Algorithms** | **Sensitivity%** | **Specificity%** | | | | **Accuracy%** |
| | **VT** | **Global** | **VF** | **Other** | **Normal** | **Total** |
| TDA | 92.72 | 99.53 | 97.93 | 99.89 | 99.94 | 99.05 |
| PDI | 82.25 | 98.53 | 94.85 | 99.62 | 99.36 | 97.38 |

**Table 8.** Results obtained for *Normal* class classification in testing.

| Type | Normal | | | | | |
|---|---|---|---|---|---|---|
| Algorithms | Sensitivity% | Specificity% | | | | Accuracy% |
| | Normal | Global | VF | VT | Other | Total |
| TDA | 99.05 | 98.45 | 99.27 | 98.88 | 97.16 | 98.76 |
| PDI | 93.09 | 92.14 | 90.22 | 91.65 | 93.95 | 92.65 |

**Table 9.** Results obtained for *Other* class classification in testing.

| Type | Other | | | | | |
|---|---|---|---|---|---|---|
| Algorithms | Sensitivity% | Specificity% | | | | Accuracy% |
| | Other | Global | VT | Normal | VF | Total |
| TDA | 97.43 | 99.54 | 99.88 | 99.40 | 99.82 | 99.09 |
| PDI | 92.86 | 97.15 | 99.02 | 96.76 | 97.72 | 96.19 |

## 5. Discussion

The same as with any other classification problem, the detection of ventricular arrhythmias normally uses a feature extraction and selection stage to optimize the class separation capabilities of the classifier. This feature selection stage aims at gathering the relevant aspects of the ECG signal based on TDA. Among a wide set of features, a reduction stage is done to lower the number of features used as input to the classifier.

In this work, we hypothesized that, by using Topological Data Analysis (TDA), some geometric features containing information about the 'shape of data' could be extracted. This method condenses the relevant information about the shape of the data, resulting in very valuable for the detection and discrimination between shockable VF and VT rhythms, even in noise and complex signals cases.

The obtained results (Tables 6–9) use the kNN classifier with the input features obtained by using two topological methods (TDA and PDI). Results show that the TDA features provide better results. For this reason, the TDA method is compared with other works in the bibliography. We have used the kNN classifier, given that it is enough to prove the improvement in classification results compared to other works. Nevertheless, using other classifiers is an open topic, which may lead to reach even better classification results.

As it can be seen from Tables 6–9, the use of the proposed TDA method provides an average accuracy of 98.9% for multiclass discrimination, which differentiates *VF* and *VT* ventricular arrhythmias but also *Normal* and *Other* types of rhythms. On the other hand, Table 10 shows a two-class classification approach to show that the proposed TDA method provides an accuracy of 99.5% when used to discriminate shockable (*VT* or *VF*) and non-shockable rhythms (rest of cardiac rhythms).

Thus, it can be established that the TDA method provides a very high classification performance. Nevertheless, we show a comparison of results with other works in the bibliography. Note, however, that this comparison is difficult due to the differences in the source signals used by different works; or even in the type of discrimination, they carry out: some works discriminate between ventricular arrhythmias and non-ventricular rhythms, others between ventricular fibrillation rhythms and non-ventricular fibrillation, others between shockable rhythms and non-shockable rhythms (considering as shockable both VT and VF).

For this reason, we divide the comparison into two separate blocks: the first block focuses on the comparison with works performing rhythm discrimination, while the second focus on the comparison with those works performing shockable vs. non-shockable signal discrimination.

Table 11 shows a group of works distinguishing between VF and non-VF rhythms. In this group, Roopaei et al. [68] obtained an accuracy of 88.60% using chaotic-based reconstructed phase space features to detect VF episodes. Arafat et al. [69] achieved a high value in the specificity of detecting VF episodes (Sp = 98.51%) using an improved version of the Threshold Crossing Interval (TCI) algorithm, called TCSC, and the MIT-BIH and CUDB databases. However, this detection was carried out with a sensitivity as low as 80.97%. Later, Alonso-Atienza et al. [70] obtained high values of specificity and accuracy (Spe = 97.10% and Acc = 96.80%) for the discrimination of VF episodes, with their specific feature selection and SVM classifiers. In their case, the sensitivity got a moderate value of 91.90%. Further, Li and Rajagopalan [71] used a genetic algorithm to make the feature selection for classifying VF episodes, achieving high-performance values: Sen = 98.40%, Spe = 98.00%, and Acc = 96.30%. Next, Acharya et al. [72] obtained high-performance values of specificity (Spe = 98.19%) and accuracy (Acc = 97.88%) using a Convolutional Neural Network (CNN) for the detection of VF. However, they achieved an extremely low value of sensitivity (Sen = 56.44%). Finally, in 2019 Ibtehaz et al. [73] got the highest results in this group, using a scheme of incorporating Empirical Mode Decomposition (EMD) and SVM classifiers (Sen = 99.99%, Spe = 98.40%, Acc = 99.19%) for the classification of VF and non-VF episodes.

As it can be seen, the results of the TDA proposal in this work achieve one of the best results (Sen = 97.07%, Spe = 99.25%, and Acc = 98.68%) compared with other works of the VF-discriminating group, with the only exception of Ibtehaz [73], that obtained better results. However, to establish a fair comparison, note that Ibtehaz obtained slightly higher results (i.e., a difference of 0.51% in Accuracy) at the expense of preselecting and rejecting the noise episodes, while in this work, there was not any preselection of ECG episodes.

Furthermore, another group of works in the bibliography can be compared, distinguishing between VT and VF rhythms (Table 11). In this group, Xie et al. [74] proposed a fuzzy similarity-based approximate entropy approach, distinguishing between VT and VF and obtaining high-performance ratios (Sen = 97.98% and Spe = 97.03% to VF and Sen = 97.03% and Spe = 97.98% to VT). However, to establish a fair comparison, it must be considered that Xie was selected as input data representative and clean episodes of VF and VT, while our work was done without preselection of ECG episodes. This kind of preselection is usual in the literature, as in Kaur and Singh [75], that used a selection of VF and VT episodes from the MIT-BIH database, using Empirical Mode Decomposition (EMD) and Approximate Entropy. Kaur and Singh obtaining moderate values for classification performance (Sen = 90.47%, Spe= 91.66%, and Acc = 91.17%). Later, Xia et al. [76] obtained high performance values (Sen = 98.15% and Spe = 96.01% to VF, and Sen = 96.01% and Spe = 96.01% and Spe = 98.15% to VT) using Lempel-Ziv and Empirical Mode Decomposition (EMD). In this case, a selection of clean episodes of VT and VF was made too. Finally, the authors of the present work achieved high values of classification performance [47] feeding the complete time-frequency image as the input of different classifiers (e.g., Sen = 92.8% and Spe = 97.0% to VF and Sen = 91.8% and Spe = 98.7% to VF, using an Artificial Neural Network Classifier, ANNC).

In any case, the results of the TDA method in this work achieve the best results when compared with the rest of the works in the bibliography aiming to discriminate between VF and VT rhythms (despite the preselection of ECG episodes done by some works).

Table 10 shows a comparison focused on detecting VT/VF episodes, i.e., shockable and non-shockable. This set of works usually targets its implementation on external defibrillators (AED) and implantable cardioverter defibrillators (ICD). Thus, these works distinguish between shockable and non-shockable rhythms (considering shockable both VT and VF). In this group, Li et al. [71] achieved an Accuracy of Acc = 98.1% (Sen = 98.4% and Spe = 98.0%) using a Genetic Algorithm (GA) for feature selection and a SVM classifier. The same year, Alonso-Atienza et al. [70] also achieved high classification performance values (Acc = 98.6, Sen = 95.0%, and Spe = 99.0%) using a selection of features and a Support Vector Machine (SVM) classifier. This work obtained one of the highest accuracy and specificity values in this group. In 2016, Tripathy et al. [9] used the Variational Mode Decomposition (VMD) and the Random Forest (RF) classifier to detect and classify shockable and non-shockable ECG episodes, achieving high values of accuracy, sensitivity, and specificity of 97.23%, 96.54%, and 97.97%, respectively. Later, in 2018, Mohanty et al. [11] detected and classified ventricular arrhythmias using cubic support vector machine (SVM) and C4.5 classifiers and achieving an Accuracy of Acc = 97.02% (Sen = 90.97% and Spe = 97.86%). Acharya et al. [77] brought forward an eleven-layer Convolutional Neural Network (CNN) for the classification of shockable and non-shockable arrhythmias. They obtained a 93.18% accuracy (Sen = 91.04% and Spe = 95.32%). Finally, Mohanty et al. [11] detected and classified ventricular arrhythmia using cubic support vector machine (SVM) and C4.5 classifiers, achieving high accuracy of Acc = 97.02% (Sen = 90.97% and Spe = 97.86%).

As it can be seen, the results of the TDA proposal in this work show the highest performance values also in this group of works, achieving an accuracy of 99.51%, 99.03% sensitivity, and 99.67% specificity.

Thus, the benefits of using the geometric features extracted from Topological Data Analysis (TDA) in the classification procedure are clear. Then, we can state that TDA, and the geometric features derived from it, can be successfully used both in the detection and classification of ventricular arrhythmias and in the classification of shockable episodes. It proves that the geometric features derived from Topological Data Analysis provides a good description of the signal. Moreover, it also foresees a successful application of these features in both Automated External Defibrillation (AED) and Implantable Cardioverter Defibrillation (ICD) therapies.

**Table 10.** Performance results comparison with other works discriminating shockable and non-shockable rhythms.

| **Types** | **Shockable (VT+VF)** | | | **Data Base** |
|---|---|---|---|---|
| **Method:** | **Sens%** | **Spe%** | **Accu%** | |
| This work, TDA | 99.03 | 99.67 | 99.51 | $AHA, MITBIH$ |
| This work, PDI | 89.63 | 96.96 | 95.12 | $AHA, MITBIH$ |
| [77] Convolutional neural network (CNN) (2018) | 91.04 | 95.32 | 93.18 | $CUDB, MITBIH$ |
| [11] C4.5 classifier (2018) | 90.97 | 97.86 | 97.02 | $CUDB, MITBIH$ |
| [78] Adaptive variational and boosted CART (2018) | 97.32 | 98.95 | 98.29 | $CUDB, MITBIH$ |
| [71] SVM and bootstrap (2013) | 98.40 | 98.00 | 98.10 | $AHA, CUDB, MITBIH$ |
| [9] VMD with Random Forest (2016) | 96.54 | 97.97 | 97.23 | $CUDB, MITBIH$ |
| [70] FS and SVM (2013) | 95.00 | 99.00 | 98.60 | $CUDB, MITBIH$ |

AHA: American Heart Association ECG Database (200 series); MIBIH: MIT-BIH Malignant Ventricular Arrhythmia Database; CCU: Registers from Coronary Care Unit (CCU) of the Royal Infirmary of Edinburgh; CUDB: MIT 'cudb' (Creighton University Ventricular Tachyarrhythmia Database).

It should be taken into account that these good results occur even in the absence of preselected ECG episodes. This work performs data classification in the same form as an Automated External Defibrillator (AED) operating in an emergency situation, following the AHA recommendations for Automated External Defibrillator (AED) algorithm performance [79]. That is, data can be continuously analyzed in time windows as they are received from the electrocardiograph.

To conclude, the success of using the TDA-derived geometric features suggests that this method may overcome conventional feature selection limitations by better describing the 'shape of data' and, thus, enabling us to build better performance arrhythmia detectors.

**Table 11.** Performance results comparison with other rhythm-discriminating works.

| Types | VF | | | VT | | | Other | | | Normal | | | Data Base |
|---|---|---|---|---|---|---|---|---|---|---|---|---|---|
| Method: | Sens% | Spe% | Accu% | Sens% | Spe% | Accu% | Sens% | Spe% | Accu% | Sens% | Spe% | Accu% | |
| This work, TDA | 97.07 | 99.25 | 98.68 | 92.72 | 99.53 | 99.05 | 97.43 | 99.54 | 99.09 | 99.05 | 98.45 | 98.76 | *AHA*, *MITBIH* |
| This work, PDI | 84.34 | 96.77 | 94.26 | 82.25 | 98.53 | 97.38 | 92.86 | 97.15 | 96.19 | 93.09 | 92.14 | 92.65 | *AHA*, *MITBIH* |
| [47] time-frequency, L2-RLR (2017) | 89.60 | 96.70 | | 91.00 | 98.10 | | 92.50 | 98.10 | | 94.90 | 96.40 | | *AHA*, *MITBIH* |
| [47] time-frequency, ANNC (2017) | 92.80 | 97.00 | | 91.80 | 98.70 | | 92.90 | 99.00 | | 96.20 | 96.70 | | *AHA*, *MITBIH* |
| [47] time-frequency, SSVR (2017) | 91.00 | 97.00 | | 92.80 | 98.70 | | 92.30 | 99.20 | | 96.60 | 96.30 | | *AHA*, *MITBIH* |
| [47] time-frequency, BAGG (2017) | 95.20 | 96.40 | | 88.80 | 99.70 | | 88.60 | 99.80 | | 96.60 | 94.10 | | *AHA*, *MITBIH* |
| [75] EMD and App Entropy (2013) | 90.47 | 91.66 | 91.17 | 90.62 | 91.11 | 90.80 | | | | | | | *MITBIH* |
| [69] TCSC algorithm (2011) | 80.97 | 98.51 | 98.14 | | | | | | | | | | *CUDB*, *MITBIH* |
| [76] Lempel-Ziv and EMD (2014) | 98.15 | 96.01 | | 96.01 | 98.15 | | | | | | | | *CUDB*, *MITBIH* |
| [68] Chaotic based (2010) | | | 88.60 | | | | | | | | | | *CCU*, *MITBIH* |
| [73] EMD and SVM (2019) | 99.99 | 98.40 | 99.19 | | | | | | | | | | *CUDB*, *MITBIH* |
| [72] CNN neural network (2017) | 56.44 | 98.19 | 97.88 | | | | | | | | | | *CUDB*, *MITBIH* |
| [70] FS and SVM (2013) | 91.90 | 97.10 | 96.80 | | | | | | | | | | *CUDB*, *MITBIH* |
| [71] Genetic algorithm, SVM (2014) | 98.40 | 98.00 | 96.30 | | | | | | | | | | *AHA*, *CUDB* |
| [74] Approximated entropy (2011) | 97.98 | 97.03 | | 97.03 | 97.98 | | | | | | | | *CUDB*, *MITBIH* |

AHA: American Heart Association ECG Database (200 series); MIBIH: MIT-BIH Malignant Ventricular Arrhythmia Database; CUDB: MIT 'cudb' (Creighton University Ventricular Tachyarrhythmia Database).

## 6. Conclusions

The rapid and reliable detection of VT and VF is fundamental in patient monitoring, but also in Automated External Defibrillation (AED) or Implantable Cardioverter Defibrillation (ICD) therapies. Any incorrect interpretation of a ventricular arrhythmia, or even the confusion between VF and VT, can be dangerous for the life of the patient.

In this paper, we propose a feature extraction method based on Topological Data Analysis (TDA) that provides near 99% accuracy in the discrimination of ventricular arrhythmias, normal and other rhythms (98.68% to VF; 99.05% to VT; 99.09% to Other; and 98.76% to Normal episodes). It also provides very high accuracy of 99.5% when discriminating between shockable (VT/VF) and non-shockable rhythms.

The novelty of this work is the incorporation of geometric features proceeding from Topological Data Analysis to the detection and classification of ventricular arrhythmias. Note also that these powerful results were obtained without preselection of episodes. Taking into consideration the obtained results, we can conclude that TDA, and the geometric features derived from it, can be successfully used both in the detection and classification of ventricular arrhythmias and in the classification of shockable rhythms. Moreover, it proves that the geometric features derived from Topological Data Analysis (TDA) provide valuable features easing the task of the classifier. Finally, we can conclude that TDA features can be beneficial in other classification tasks.

**Author Contributions:** Individual contributions: conceptualization, A.M., and A.R.-M.; methodology, A.M., A.R.-M., and M.B.-M.; software, A.M.; formal analysis, A.M., A.R.-M., and J.V.F.-V.; writing—original draft preparation, A.M., and J.V.F.-V.; writing—review and editing, A.M., J.V.F.-V., A.R.-M., and M.B.-M. All authors have read and agreed to the published version of the manuscript.

**Funding:** This research received no external funding.

**Informed Consent Statement:** Patient consent was waived due to the use of publicly available Physionet databse.

**Data Availability Statement:** Data can be publicly obtained from MIT and AHA databases, described in the references section.

**Conflicts of Interest:** The authors declare no conflict of interest.

## Abbreviations

The following abbreviations are used in this manuscript:

| | |
|---|---|
| TDA | Topological Data Analysis |
| PH | Persistent Homology |
| VF | Ventricular Fibrillation |
| VT | Ventricular Tachycardia |
| AED | Automated External Defibrillator |
| ICD | Implantable Cardioverter Defibrillator |
| ECG | Electrocardiogram |
| RF | Random Forest |
| kNN | k-Nearest Neighbor |
| ICA | Independent Component Analysis |
| DT | Decision Tree |
| PC | Point Cloud |
| RPC | Representation of Point Cloud |
| PD | Persistence Diagram |
| PDI | Persistence Diagram Image |
| KDE | Kernel Density Estimation |
| PI | Persistence Images |
| WRM | Window Reference Mark |
| TR | Time Representation |
| PDR | Persistence Diagram Representation |
| SFS | Sequential Forward Selection |

| | |
|---|---|
| AHA | American Heart Association |
| CUDB | Creighton University Ventricular Tachyarrhythmia Database |
| CCU | Coronary Care Unit |
| TCI | Threshold Crossing Interval |
| CNN | Convolutional Neural Network |
| EMD | Empirical Mode Decomposition |
| VMD | Variational mode decomposition |
| SVM | Support Vector Machine |
| ANNC | Artificial Neural Network Classifier |
| GA | Genetic Algorithm |

## References

1. Yu, M.; Zhang, G.; Wu, T.; Li, C.; Wan, Z.; Li, L.; Wang, C.; Wang, Y.; Lu, H.; Chen, F. A new method without reference channels used for ventricular fibrillation detection during cardiopulmonary resuscitation. *Australas. Phys. Eng. Sci. Med.* **2016**, *39*, 391–401. [CrossRef] [PubMed]
2. Jin, D.; Dai, C.; Gong, Y.; Lu, Y.; Zhang, L.; Quan, W.; Li, Y. Does the choice of definition for defibrillation and CPR success impact the predictability of ventricular fibrillation waveform analysis? *Resuscitation* **2017**, *111*, 48–54. [CrossRef] [PubMed]
3. Amann, A.; Tratnig, R.; Unterkofler, K. Reliability of old and new ventricular fibrillation detection algorithms for automated external defibrillators. *Biomed. Eng. Online* **2005**, *4*, 1–15. [CrossRef] [PubMed]
4. Pourmand, A.; Galvis, J.; Yamane, D. The controversial role of dual sequential defibrillation in shockable cardiac arrest. *Am. J. Emerg. Med.* **2018**, *36*, 1674–1679. [CrossRef]
5. Chatterjee, S.; Thakur, R.S.; Yadav, R.N.; Gupta, L.; Raghuvanshi, D.K. Review of noise removal techniques in ECG signals. *IET Signal Process.* **2020**, *14*, 569–590. [CrossRef]
6. Hammad, M.; Kandala, R.N.; Abdelatey, A.; Abdar, M.; Zomorodi-Moghadam, M.; Tan, R.S.; Acharya, U.R.; Pławiak, J.; Tadeusiewicz, R.; Makarenkov, V.; et al. Automated detection of shockable ECG signals: A review. *Inf. Sci.* **2021**, *571*, 580–604. [CrossRef]
7. Orozco-Duque, A.; Rúa, S.; Zuluaga, S.; Redondo, A.; Restrepo, J.V.; Bustamante, J. Support Vector Machine and Artificial Neural Network Implementation in Embedded Systems for Real Time Arrhythmias Detection. In Proceedings of the International Conference on Bio-inspired Systems and Signal Processing (BIOSIGNALS-2013), Barcelona, Spain, 11–14 February 2013; pp. 310–313.
8. Pooyan, M.; Akhoondi, F. Providing an efficient algorithm for finding R peaks in ECG signals and detecting ventricular abnormalities with morphological features. *J. Med. Signals Sens.* **2016**, *6*, 218. [CrossRef]
9. Tripathy, R.; Sharma, L.; Dandapat, S. Detection of shockable ventricular arrhythmia using variational mode decomposition. *J. Med. Syst.* **2016**, *40*, 40–79. [CrossRef]
10. Jekova, I.; Krasteva, V. Real time detection of ventricular fibrillation and tachycardia. *Physiol. Meas.* **2004**, *25*, 1167. [CrossRef]
11. Mohanty, M.; Sahoo, S.; Biswal, P.; Sabut, S. Efficient classification of ventricular arrhythmias using feature selection and C4. 5 classifier. *Biomed. Signal Process. Control* **2018**, *44*, 200–208. [CrossRef]
12. Jothiramalingam, R.; Jude, A.; Patan, R.; Ramachandran, M.; Duraisamy, J.H.; Gandomi, A.H. Machine learning-based left ventricular hypertrophy detection using multi-lead ECG signal. *Neural Comput. Appl.* **2021**, 33, 4445–4455. [CrossRef]
13. Tang, J.; Li, J.; Liang, B.; Huang, X.; Li, Y.; Wang, K. Using Bayesian decision for ontology mapping. *J. Web Semant.* **2006**, *4*, 243–262. [CrossRef]
14. Kuzilek, J.; Kremen, V.; Soucek, F.; Lhotska, L. Independent component analysis and decision trees for ECG holter recording de-noising. *PLoS ONE* **2014**, *9*, e98450. [CrossRef] [PubMed]
15. Chung, M.K.; Bubenik, P.; Kim, P.T. Persistence diagrams of cortical surface data. In *International Conference on Information Processing in Medical Imaging*; Springer: Berlin/Heidelberg, Germany, 2009; pp. 386–397.
16. Hiraoka, Y.; Nakamura, T.; Hirata, A.; Escolar, E.G.; Matsue, K.; Nishiura, Y. Hierarchical structures of amorphous solids characterized by persistent homology. *Proc. Natl. Acad. Sci. USA* **2016**, *113*, 7035–7040. [CrossRef]
17. Emrani, S.; Gentimis, T.; Krim, H. Persistent homology of delay embeddings and its application to wheeze detection. *IEEE Signal Process. Lett.* **2014**, *21*, 459–463. [CrossRef]
18. Perea, J.A.; Deckard, A.; Haase, S.B.; Harer, J. SW1PerS: Sliding windows and 1-persistence scoring; discovering periodicity in gene expression time series data. *BMC Bioinform.* **2015**, *16*, 1–12. [CrossRef]
19. Kramár, M.; Levanger, R.; Tithof, J.; Suri, B.; Xu, M.; Paul, M.; Schatz, M.F.; Mischaikow, K. Analysis of Kolmogorov flow and Rayleigh–Bénard convection using persistent homology. *Phys. D Nonlinear Phenom.* **2016**, *334*, 82–98. [CrossRef]
20. Ferri, M. Persistent topology for natural data analysis—A survey. In *Towards Integrative Machine Learning and Knowledge Extraction*; Springer: Berlin/Heidelberg, Germany, 2017; pp. 117–133.
21. Nicolau, M.; Levine, A.J.; Carlsson, G. Topology based data analysis identifies a subgroup of breast cancers with a unique mutational profile and excellent survival. *Proc. Natl. Acad. Sci. USA* **2011**, *108*, 7265–7270. [CrossRef]
22. Carlsson, G.; Ishkhanov, T.; De Silva, V.; Zomorodian, A. On the local behavior of spaces of natural images. *Int. J. Comput. Vis.* **2008**, *76*, 1–12. [CrossRef]

23. Dey, T.; Mandal, S.; Varcho, W. Improved image classification using topological persistence. In Proceedings of the Conference on Vision, Modeling and Visualization, Bonn, Germany, 25–27 September 2017; pp. 161–168.
24. Berwald, J.; Gidea, M. Critical transitions in a model of a genetic regulatory system. *arXiv* **2013**, arXiv:1309.7919.
25. Seversky, L.M.; Davis, S.; Berger, M. On time-series topological data analysis: New data and opportunities. In Proceedings of the IEEE Conference on Computer Vision and Pattern Recognition Workshops, Las Vegas, NV, USA, 26 June–1 July 2016; pp. 59–67.
26. Kwitt, R.; Huber, S.; Niethammer, M.; Lin, W.; Bauer, U. Statistical topological data analysis-a kernel perspective. In Proceedings of the Advances in Neural Information Processing Systems 28 (NIPS 2015), Montreal, QC, USA, 7–12 December 2015; pp. 3070–3078.
27. Carriere, M.; Cuturi, M.; Oudot, S. Sliced Wasserstein kernel for persistence diagrams. In Proceedings of the International Conference on Machine Learning, PMLR, Sydney, Australia, 6–11 August 2017; pp. 664–673.
28. Garin, A.; Tauzin, G. A topological "reading" lesson: Classification of MNIST using TDA. In Proceedings of the 2019 18th IEEE International Conference On Machine Learning And Applications (ICMLA), Boca Raton, FL, USA, 16–19 December 2019; pp. 1551–1556. [CrossRef]
29. Ravishanker, N.; Chen, R. Topological data analysis (TDA) for time series. *arXiv* **2019**, arXiv:1909.10604.
30. Cohen-Steiner, D.; Edelsbrunner, H.; Harer, J. Stability of persistence diagrams. *Discret. Comput. Geom.* **2007**, *37*, 103–120. [CrossRef]
31. Oudot, S.Y. *Persistence Theory: From Quiver Representations to Data Analysis*; American Mathematical Society Providence: Providence, RI, USA, 2015; Volume 209.
32. Yan, Y.; Ivanov, K.; Mumini Omisore, O.; Igbe, T.; Liu, Q.; Nie, Z.; Wang, L. Gait rhythm dynamics for neuro-degenerative disease classification via persistence landscape-based topological representation. *Sensors* **2020**, *20*, 2006. [CrossRef] [PubMed]
33. Carlsson, G.; Zomorodian, A.; Collins, A.; Guibas, L.J. Persistence barcodes for shapes. *Int. J. Shape Model.* **2005**, *11*, 149–187. [CrossRef]
34. Chazal, F.; Michel, B. An introduction to topological data analysis: fundamental and practical aspects for data scientists. *Front. Artif. Intell.* **2021**, *4*, 667963. [CrossRef]
35. Zulkepli, N.F.S.; Noorani, M.S.M.; Razak, F.A.; Ismail, M.; Alias, M.A. Topological characterization of haze episodes using persistent homology. *Aerosol Air Qual. Res.* **2019**, *19*, 1614–1624. [CrossRef]
36. Karan, A.; Kaygun, A. Time series classification via topological data analysis. *Expert Syst. Appl.* **2021**, *183*, 115326. [CrossRef]
37. Bubenik, P. Statistical topological data analysis using persistence landscapes. *J. Mach. Learn. Res.* **2015**, *16*, 77–102.
38. Chazal, F.; Fasy, B.T.; Lecci, F.; Rinaldo, A.; Wasserman, L. Stochastic convergence of persistence landscapes and silhouettes. In Proceedings of the Thirtieth Annual Symposium on Computational Geometry, Kyoto, Japan, 8–11 June 2014; pp. 474–483.
39. Scott, D.W. *Multivariate Density Estimation: Theory, Practice, and Visualization*; John Wiley & Sons: Hoboken, NJ, USA, 2015.
40. Goldberger, A.L.; Amaral, L.A.; Glass, L.; Hausdorff, J.M.; Ivanov, P.C.; Mark, R.G.; Mietus, J.E.; Moody, G.B.; Peng, C.K.; Stanley, H.E. PhysioBank, PhysioToolkit, and PhysioNet: components of a new research resource for complex physiologic signals. *Circulation* **2000**, *101*, e215–e220. [CrossRef]
41. Physionet.org. The MIT-BIH Database. Available online: https://physionet.org/physiobank/database/mitdb/ (accessed on 25 October 2021).
42. Ecri.org. American Heart Association ECG Database. Available online: http://ecri.org (accessed on 25 October 2021).
43. Boda, S.; Mahadevappa, M.; Dutta, P.K. A hybrid method for removal of power line interference and baseline wander in ECG signals using EMD and EWT. *Biomed. Signal Process. Control* **2021**, *67*, 102466. [CrossRef]
44. Kaur, M.; Singh, B. Comparison of different approaches for removal of baseline wander from ECG signal. In Proceedings of the International Conference & Workshop on Emerging Trends in Technology, Maharashtra, India, 25–26 February 2011; pp. 1290–1294.
45. Narwaria, R.P.; Verma, S.; Singhal, P. Removal of baseline wander and power line interference from ECG signal-a survey approach. *Int. J. Electron. Eng.* **2011**, *3*, 107–111.
46. Viitasalo, M.; Karjalainen, J. QT intervals at heart rates from 50 to 120 beats per minute during 24-hour electrocardiographic recordings in 100 healthy men. Effects of atenolol. *Circulation* **1992**, *86*, 1439–1442. [CrossRef] [PubMed]
47. Mjahad, A.; Rosado-Muñoz, A.; Bataller-Mompean, M.; Frances-Villora, J.; Guerrero-Martinez, J. Ventricular Fibrillation and Tachycardia detection from surface ECG using time-frequency representation images as input dataset for machine learning. *Comput. Methods Programs Biomed.* **2017**, *141*, 119–127. [CrossRef]
48. Miranda, M.V.G.; Espinosa, I.P.V.; Calero, M.J.F. ECG signal features extraction. In Proceedings of the 2016 IEEE Ecuador Technical Chapters Meeting (ETCM), Guayaquil, Ecuador, 12–14 October 2016; pp. 1–6. [CrossRef]
49. Yin, J.; Xiao, P.; Li, J.; Liu, Y.; Yan, C.; Zhang, Y. Parameters analysis of sample entropy, permutation entropy and permutation ratio entropy for RR interval time series. *Inf. Process. Manag.* **2020**, *57*, 102283. [CrossRef]
50. Asgharzadeh-Bonab, A.; Amirani, M.C.; Mehri, A. Spectral entropy and deep convolutional neural network for ECG beat classification. *Biocybern. Biomed. Eng.* **2020**, *40*, 691–700. [CrossRef]
51. Li, H.; Liu, T.; Wu, X.; Chen, Q. Research on bearing fault feature extraction based on singular value decomposition and optimized frequency band entropy. *Mech. Syst. Signal Process.* **2019**, *118*, 477–502. [CrossRef]
52. Alù, F.; Miraglia, F.; Orticoni, A.; Judica, E.; Cotelli, M.; Rossini, P.M.; Vecchio, F. Approximate Entropy of Brain Network in the Study of Hemispheric Differences. *Entropy* **2020**, *22*, 1220. [CrossRef]

53. Zhan, Z.Q.; Nikus, K.; Birnbaum, Y. Different ECG patterns of left main coronary artery occlusion signifying varying degrees of ischemic severity. *J. Electrocardiol.* **2020**, *60*, 12–14. [CrossRef]
54. Zhang, Y.; Ma, Z.; Dong, W. Nonlinear quality indices based on a novel Lempel-Ziv complexity for assessing quality of multi-lead ECGs collected in real time. *J. Inf. Process. Syst.* **2020**, *16*, 508–521.
55. Espinosa, R.; Talero, J.; Weinstein, A. Effects of Tau and Sampling Frequency on the Regularity Analysis of ECG and EEG Signals Using ApEn and SampEn Entropy Estimators. *Entropy* **2020**, *22*, 1298. [CrossRef]
56. Dhanka, B.; Vijayvargiya, A.; Kumar, R.; Singh, G. A Comparative Assessment of Machine Learning Techniques for Epilepsy Detection using EEG Signal. In Proceedings of the 2020 IEEE 7th Uttar Pradesh Section International Conference on Electrical, Electronics and Computer Engineering (UPCON), Prayagraj, India, 27–29 November 2020; pp. 1–6.
57. Yazdi-Ravandi, S.; Arezooji, D.M.; Matinnia, N.; Shamsaei, F.; Ahmadpanah, M.; Ghaleiha, A.; Khosrowabadi, R. Complexity of information processing in obsessive-compulsive disorder based on fractal analysis of EEG signal. *EXCLI J.* **2021**, *20*, 462.
58. Moctezuma, L.A.; Molinas, M. Classification of low-density EEG for epileptic seizures by energy and fractal features based on EMD. *J. Biomed. Res.* **2020**, *34*, 180. [CrossRef] [PubMed]
59. Pavlov, A.; Dubrovsky, A.; Koronovskii Jr, A.; Pavlova, O.; Semyachkina-Glushkovskaya, O.; Kurths, J. Extended detrended fluctuation analysis of sound-induced changes in brain electrical activity. *Chaos Solitons Fractals* **2020**, *139*, 109989. [CrossRef]
60. Yamamoto, M.S.; Sadatnejad, K.; Tanaka, T.; Islam, R.; Tanaka, Y.; Lotte, F. Detecting EEG outliers for BCI on the riemannian manifold using spectral clustering. In Proceedings of the 2020 42nd Annual International Conference of the IEEE Engineering in Medicine & Biology Society (EMBC), Montreal, QC, Canada, 20–24 July 2020; pp. 438–441.
61. Murugan, S.; Selvaraj, J.; Sahayadhas, A. Detection and analysis: Driver state with electrocardiogram (ECG). *Phys. Eng. Sci. Med.* **2020**, *43*, 525–537. [CrossRef] [PubMed]
62. Cossette, H.; Mailhot, M.; Marceau, E.; Mesfioui, M. Vector-valued tail value-at-risk and capital allocation. *Methodol. Comput. Appl. Probab.* **2016**, *18*, 653–674. [CrossRef]
63. Kusano, G.; Hiraoka, Y.; Fukumizu, K. Persistence weighted Gaussian kernel for topological data analysis. In Proceedings of the 33rd International Conference on Machine Learning, PMLR, New York, NY, USA, 19–24 June 2016; pp. 2004–2013.
64. Reininghaus, J.; Huber, S.; Bauer, U.; Kwitt, R. A stable multi-scale kernel for topological machine learning. In Proceedings of the IEEE Conference on Computer Vision and Pattern Recognition, Boston, MA, USA, 7–12 June 2015; pp. 4741–4748.
65. Luo, T.j.; Fan, Y.; Chen, L.; Guo, G.; Zhou, C. Eeg signal reconstruction using a generative adversarial network with wasserstein distance and temporal-spatial-frequency loss. *Front. NeuroInform.* **2020**, *14*, 15. [CrossRef]
66. Frances-Villora, J.V.; Bataller-Mompean, M.; Mjahad, A.; Rosado-Muñoz, A.; Gutierrez Martin, A.; Teruel-Marti, V.; Villanueva, V.; Hampel, K.G.; Guerrero-Martinez, J.F. Real-Time Localization of Epileptogenic Foci EEG Signals: An FPGA-Based Implementation. *Appl. Sci.* **2020**, *10*, 827. [CrossRef]
67. Mjahad, A.; Rosado-Muñoz, A.; Guerrero-Martínez, J.F.; Bataller-Mompeán, M.; Francés-Villora, J.V.; Dutta, M.K. Detection of Ventricular Fibrillation Using the Image from Time-Frequency Representation and Combined Classifiers without Feature Extraction. *Appl. Sci.* **2018**, *8*, 2057. [CrossRef]
68. Roopaei, M.; Boostani, R.; Sarvestani, R.R.; Taghavi, M.A.; Azimifar, Z. Chaotic based reconstructed phase space features for detecting ventricular fibrillation. *Biomed. Signal Process. Control* **2010**, *5*, 318–327. [CrossRef]
69. Arafat, M.A.; Chowdhury, A.W.; Hasan, M.K. A simple time domain algorithm for the detection of ventricular fibrillation in electrocardiogram. *Signal Image Video Process.* **2011**, *5*, 1–10. [CrossRef]
70. Alonso-Atienza, F.; Morgado, E.; Fernandez-Martinez, L.; Garcia-Alberola, A.; Rojo-Alvarez, J.L. Detection of life-threatening arrhythmias using feature selection and support vector machines. *IEEE Trans. Biomed. Eng.* **2013**, *61*, 832–840. [CrossRef] [PubMed]
71. Li, Q.; Rajagopalan, C.; Clifford, G.D. Ventricular fibrillation and tachycardia classification using a machine learning approach. *IEEE Trans. Biomed. Eng.* **2013**, *61*, 1607–1613. [CrossRef] [PubMed]
72. Acharya, U.R.; Fujita, H.; Lih, O.S.; Hagiwara, Y.; Tan, J.H.; Adam, M. Automated detection of arrhythmias using different intervals of tachycardia ECG segments with convolutional neural network. *Inf. Sci.* **2017**, *405*, 81–90. [CrossRef]
73. Ibtehaz, N.; Rahman, M.S.; Rahman, M.S. VFPred: A fusion of signal processing and machine learning techniques in detecting ventricular fibrillation from ECG signals. *Biomed. Signal Process. Control* **2019**, *49*, 349–359. [CrossRef]
74. Xie, H.B.; Zhong-Mei, G.; Liu, H. Classification of ventricular tachycardia and fibrillation using fuzzy similarity-based approximate entropy. *Expert Syst. Appl.* **2011**, *38*, 3973–3981. [CrossRef]
75. Kaur, L.; Singh, V. Ventricular fibrillation detection using emprical mode decomposition and approximate entropy. *Int. J. Emerg. Technol. Adv. Eng.* **2013**, *3*, 260–268.
76. Xia, D.; Meng, Q.; Chen, Y.; Zhang, Z. Classification of ventricular tachycardia and fibrillation based on the lempel-ziv complexity and EMD. In *International Conference on Intelligent Computing*; Springer: Berlin/Heidelberg, Germany, 2014; pp. 322–329. [CrossRef]
77. Acharya, U.R.; Fujita, H.; Oh, S.L.; Raghavendra, U.; Tan, J.H.; Adam, M.; Gertych, A.; Hagiwara, Y. Automated identification of shockable and non-shockable life-threatening ventricular arrhythmias using convolutional neural network. *Future Gener. Comput. Syst.* **2018**, *79*, 952–959. [CrossRef]

78. Xu, Y.; Wang, D.; Zhang, W.; Ping, P.; Feng, L. Detection of ventricular tachycardia and fibrillation using adaptive variational mode decomposition and boosted-CART classifier. *Biomed. Signal Process. Control* **2018**, *39*, 219–229. [CrossRef]
79. Kerber, R.E.; Becker, L.B.; Bourland, J.D.; Cummins, R.O.; Hallstrom, A.P.; Michos, M.B.; Nichol, G.; Ornato, J.P.; Thies, W.H.; White, R.D.; et al. Automatic external defibrillators for public access defibrillation: recommendations for specifying and reporting arrhythmia analysis algorithm performance, incorporating new waveforms, and enhancing safety: A statement for health professionals from the American Heart Association Task Force on Automatic External Defibrillation, Subcommittee on AED Safety and Efficacy. *Circulation* **1997**, *95*, 1677–1682.

*applied sciences*

MDPI

*Article*

# A Machine-Learning Model Based on Morphogeometric Parameters for RETICS Disease Classification and GUI Development

**José M. Bolarín [1], F. Cavas [2,*], J.S. Velázquez [2] and J.L. Alió [3,4]**

1 Technology Centre for IT and Communications (CENTIC), Scientific Park of Murcia, 30100 Murcia, Spain; bolarin5@hotmail.com

2 Department of Structures, Construction and Graphical Expression, Technical University of Cartagena, 30202 Cartagena, Spain; jose.velazquez@upct.es

3 Keratoconus Unit of Vissum Corporation Alicante, 03016 Alicante, Spain; jlalio@vissum.com

4 Department of Ophthalmology, Miguel Hernández University of Elche, 03202 Alicante, Spain

* Correspondence: francisco.cavas@upct.es; Tel.: +34-968-328856

Received: 19 December 2019; Accepted: 2 March 2020; Published: 9 March 2020

**Featured Application: This work presents a Graphics User Interface that applies two automated learning models based on machine-procured independent variables to assist ophthalmology professionals in keratoconus disease diagnosis and classification.**

**Abstract:** This work pursues two objectives: defining a new concept of risk probability associated with suffering early-stage keratoconus, classifying disease severity according to the RETICS (Thematic Network for Co-Operative Research in Health) scale. It recruited 169 individuals, 62 healthy and 107 keratoconus diseased, grouped according to the RETICS classification: 44 grade I; 18 grade II; 15 grade III; 15 grade IV; 15 grade V. Different demographic, optical, pachymetric and eometrical parameters were measured. The collected data were used for training two machine-learning models: a multivariate logistic regression model for early keratoconus detection and an ordinal logistic regression model for RETICS grade assessments. The early keratoconus detection model showed very good sensitivity, specificity and area under ROC curve, with around 95% for training and 85% for validation. The variables that made the most significant contributions were gender, coma-like, central thickness, high-order aberrations and temporal thickness. The RETICS grade assessment also showed high-performance figures, albeit lower, with a global accuracy of 0.698 and a 95% confidence interval of 0.623–0.766. The most significant variables were CDVA, central thickness and temporal thickness. The developed web application allows the fast, objective and quantitative assessment of keratoconus in early diagnosis and RETICS grading terms.

**Keywords:** Scheimpflug; 3D cornea model; early keratoconus; Corrected Distance Visual Acuity (CDVA)

## 1. Introduction

Corneal tomography is a validated technology for evaluating the changes occurring in the corneal morphology of keratoconus (KC) disease [1], which allows for the control of the geometric decompensation driven by the asymmetry present while this disease progresses [2,3]. However, no agreement has been reached about the relative importance of the indices and technologies to be used to detect which patients may suffer post-surgical corneal iatrogenic ectasia when evaluating patients' suitability for refractive surgery [4–10].

Some studies have evaluated the objective efficiency of different indices for early stages of this disease in an individualized way. These studies are generally related to values of curvature, elevation and pachymetry [11], and to patterns based on the combination of indices that come from the same [12–16] or different technologies [7,9,17–20]. The evaluation of these patterns is based on subjective analysis [7,21], which is critical to gain a better discrimination capability according to clinical experience. Nevertheless, there are still significant discrepancies in terms of the relative value of the results obtained in their evaluation of the risk probability associated with disease development.

In KC, geometric decompensation occurs and causes disease progression [11], which means increased high-order optical aberrations and high irregular astigmatism values [22], whose main refractive sign is the impossible complete optical compensation of their ametropia by spherical-cylindrical lenses. Consequently, corrected visual acuity will diminish compared to individuals with no corneal pathology [23]. The scientific literature contains many classifications for the degree of KC severity [24–31]. However in clinical practice, it is difficult to handle the many indices on which these classifications are based for proper optical-geometrical evaluations of disease progression. From an optical point of view, patients show a deteriorated spectacle-corrected visual acuity during disease development, insofar as their visual performance worsens as the degree of KC severity progresses. Following this criterion, a classification of disease stages has been developed [23,32]: the so-called RETICS grading. This grading takes into account four geometrical parameters (Internal Astigmatism, RMS Coma-Like, $Q_{8mm}$ and Pachimetry) and a functional one (Corrected Distance Visual Acuity (CDVA)) to establish five KC degrees: I to V.

This study develops and validates a Graphics User Interface (GUI) that combines two automatic learning models based on a set of independent variables with two aims: defining a new concept of risk probability associated with the development of early-stage KC and classifying disease severity according to the RETICS (Thematic Network for Co-Operative Research in Health) scale to assist ophthalmology professionals in disease management.

## 2. Materials and Methods

### 2.1. Patients

This research work was conceived as an observational comparative study. It comprised 169 eyes of 169 subjects divided into two groups. To avoid potential biases, those cases showing any other ocular comorbidity that could affect the present study parameters, who had undergone any ocular surgical procedure, or had worn contact lenses in the 4 weeks prior to the topographical evaluation, were excluded from both groups.

The first group, called the "control" group, comprised 62 healthy eyes of 62 patients (48.4% males, 51.6% females) whose ages ranged from seven to 60. The cases included in the control group were randomly selected from the refractive surgery candidates, and the data used for this study were acquired during their pre-surgical appointments, always with the same experienced technician.

The second group was formed by 107 KC candidates (63.2% males, 36.8% females) aged from 15 to 98. They were classified into five subgroups in accordance with the RETICS grading system [32].

The procedure followed for KC group diagnosis and classification was based on state-of-the-art clinical and topographical evaluations (Figure 1), including uncorrected distance visual acuity (UDVA), CDVA, manifest refraction (sphere and cylinder), slit-lamp biomicroscopy, Goldmann tonometry, fundus evaluation and ultrasonic pachymetry [33]. In all cases, pre-surgical evidence for KC was assessed: asymmetric bowtie pattern with or without skewed axes, localized stromal thickness reduction, conical protuberance at the apex, Fleischer ring, Vogt striae or anterior stromal scar.

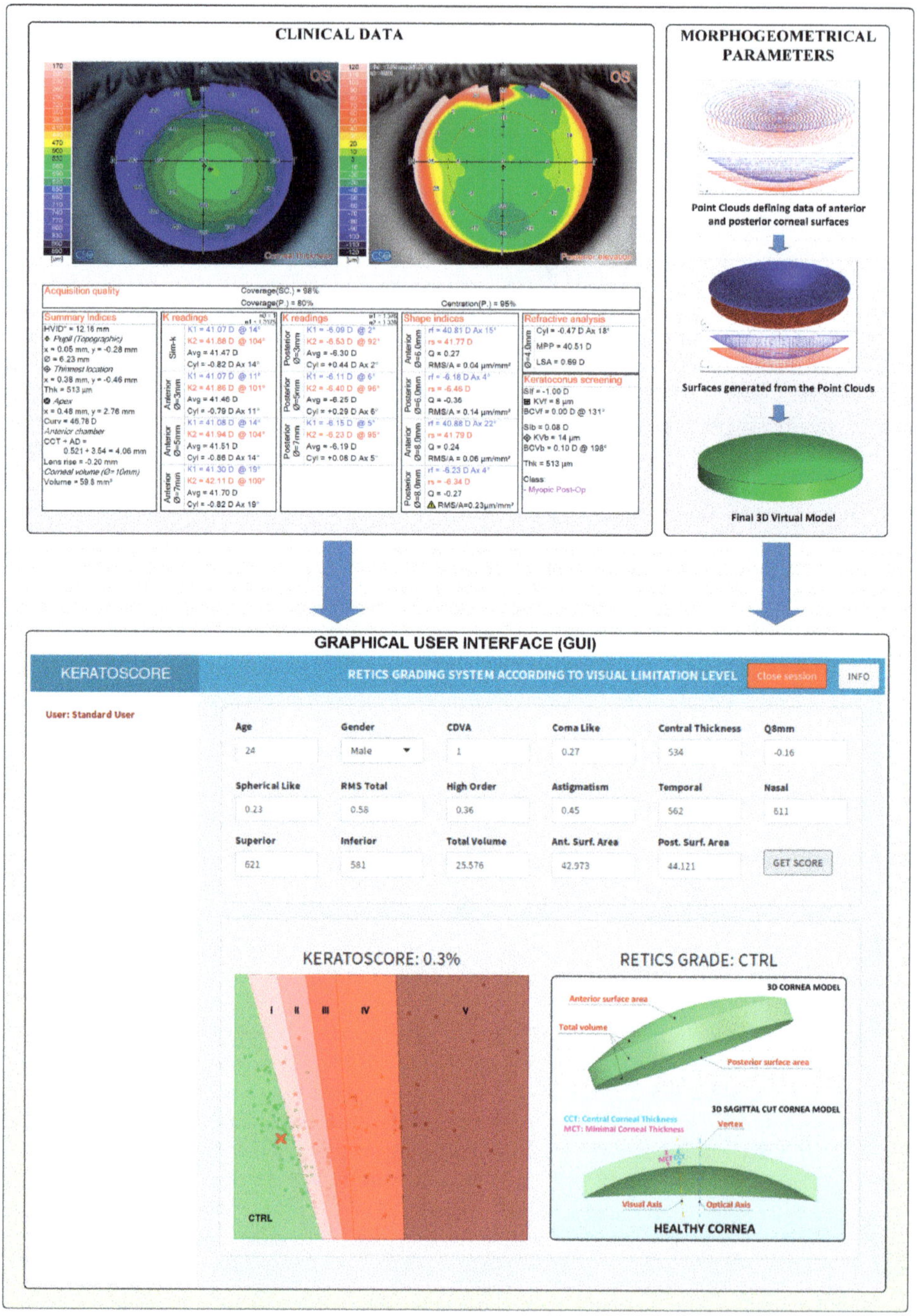


**Figure 1.** Methodology proposed for Keratoconus Score Calculator generation.

All the evaluations were made at Vissum Corporation Alicante (a centre affiliated with the Miguel Hernández University of Elche, Elche, Spain), and now form part of the official "Iberia" database of KC cases created for the National Network for Clinical Research in Ophthalmology RETICS-OFTARED.

Patients were adequately informed about the study and agreed to freely participate in it. The study was also ratified by the hospital's Ethical Committee for Clinical Research according to the ethical guidelines in the Declaration of Helsinki (7th revision, October 2013, Fortaleza, Brazil).

### 2.2. Methods

Each case selected to form part of this research was examined using a Sirius System® tomographer (Costruzione Strumenti Oftalmici, Florence, Italy), following the specifications of a validated procedure previously created by our research group, which is clearly explained in former research works [2,34]. This procedure is effective when used for both the diagnosis and characterization of KC [35,36], and comprises two phases: 3D virtual modeling, followed by a morpho-geometric analysis.

The product of this procedure is a patient-specific 3D custom corneal model, which can be studied to determine several morpho-geometric parameters that have already been described and used in a previous study [37]. Of them all, anterior corneal surface area (anterior surface area), posterior corneal surface area (posterior surface area) and total corneal volume (total volume) were selected to be used, along with demographic, pachymetry and clinical parameters (Figure 1).

### 2.3. Statistical Analysis

The quantitative variables were summarized using mean±standard deviation (SD), median and interquartile range (25th and 75th percentiles). The Shapiro–Wilk test was employed to assess if the quantitative variables followed normal distribution. The Student's *t*-test was run to compare the normally distributed variables between two groups, while an ANOVA was used when there were three groups or more. For the non-normally distributed variables, the Mann–Whitney test and the Kruskal–Wallis test were, respectively, carried out. Differences in the qualitative variables among independent groups where compared by the $\chi^2$ test. The predictive score model for early KC detection was defined using multivariate logistic regression with the control and RETICS grade I groups. Seventeen variables were included in the model: two demographic (age and gender), seven optical (CDVA, coma-like, $Q_{8mm}$, spherical-like, RMS total, high-order, astigmatism), five pachymetry (central thickness, temporal, nasal, superior, inferior) and three morpho-geometric (total volume, anterior surface area, posterior surface area). Model discriminative efficiency was evaluated by receiver operating characteristic (ROC) curves, when area under curve (AUC), sensitivity and specificity were taken as performance indicators. An internal cross-validation procedure was followed using bootstrap aggregating (bagging) [38]. This procedure works as follows: it first generates a new dataset of equal size by sampling with replacement from the original dataset. The model is then trained with these data. Finally, this model is used to make predictions on those cases not used during training. This procedure is repeated 100 times to obtain a set of quality parameters that can be averaged, and confidence intervals that can be calculated. On average, 63.2% of the original data were used in all these 100 training steps. The remaining 36.8% were used for validation. Ordinal multivariate logistic regression was utilized to determine the RETICS grade with the same predictor variables. In this case, the confusion matrix was employed to estimate model performance by means of sensitivity, specificity and balanced accuracy per group. Data were evaluated by R Statistics v3.6.1 (R Foundation for Statistical Computing, Vienna, Austria) [39].

Packages "tidyr", "dplyr", "dlookr" and "smbinning" were used for data loading, exploration and transformation. Packages "corrplot", "yarrr" and "FactoMiner" were employed for data visualization. Packages "pROC", "ROCR" and "Epi" were utilized for the ROC curve analysis and representation. Logistic regression models were trained with the "glm" function from the base package. Ordered logistic regression models were trained with the "polr" function from the MASS package. Confusion matrices were analyzed by the "caret" package. Optimal sample size calculations were made by the "rcompanion" and "pmsampsize" packages.

The statistical power analysis was conducted with simulation using the Wald test to estimate the power for each covariate according to sample size, as described in the literature [40,41].

A web application was developed using Shiny v1.3.2 (RStudio Inc., Boston, MA, USA) [42] and the ShinyAuthr v0.0.99 authentication module (Paul Campbell, Paris, France) [43]. This application was deployed in a private secure institutional network (because patients' clinical data were used, and the application was, therefore, accessible only from controlled computers to minimize the security risks linked to using patients' data).

Application landing page, shown in Figure 2, is a login form that adds a secured authentication layer. No registering possibility was included, and new users can only be directly added by the administrator. After logging in, users view a form with text boxes corresponding to all the model predictors that are filled by default with sample values from a healthy individual. Users can type in new values and, after pressing the "get score" button values, they are passed to trained models and predictions are made (Figures 3–8), including the early detection KC score (known as "keratoscore"), the RETICS grade prediction and a brief graphical description of the hypothetical cornea, with a schema indicating the representation of the different parameters.

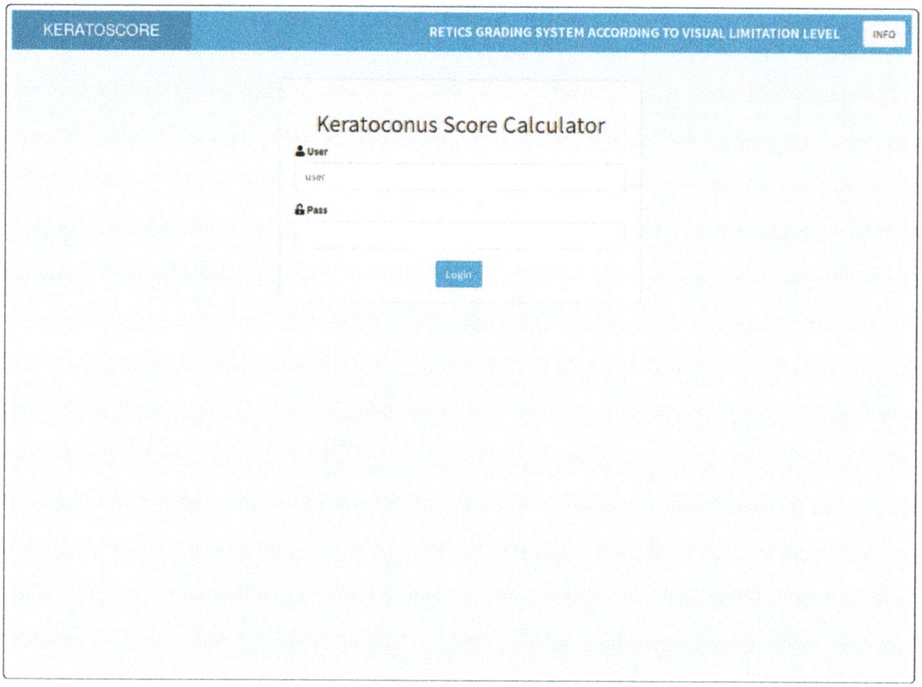


**Figure 2.** Application landing page showing the login form with a secured authentication layer.

**Figure 3.** Screenshot of healthy individuals (Ctrl) with the typical 3D virtual corneal model schematic representation.

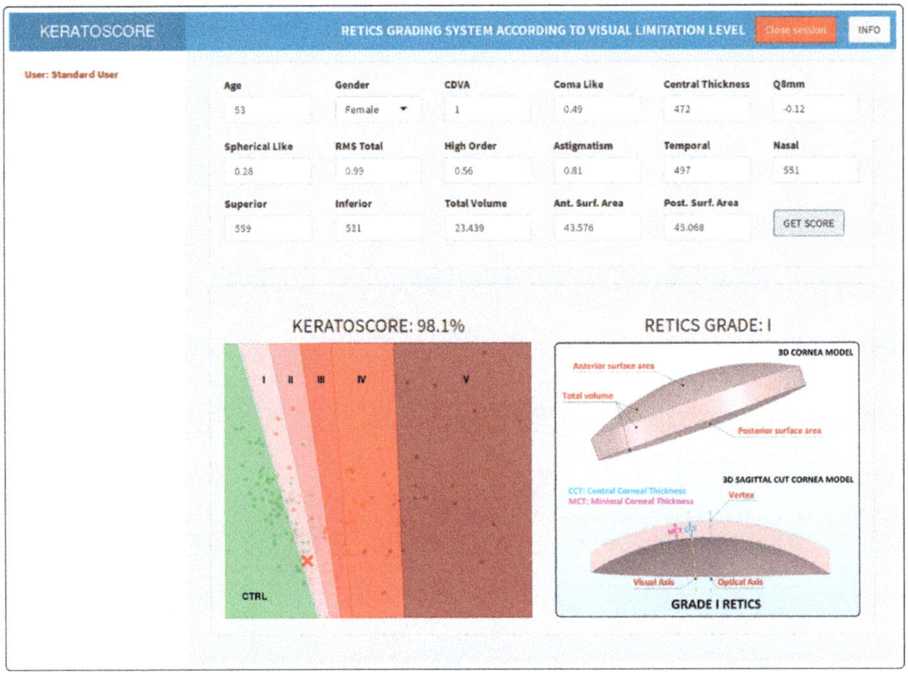


**Figure 4.** Screenshot of keratoconus grade I individuals (RETICS) with the typical 3D virtual corneal model schematic representation.

**Figure 5.** Screenshot of keratoconus grade II individuals (RETICS) with the typical 3D virtual corneal model schematic representation.

**Figure 6.** Screenshot of keratoconus grade III individuals (RETICS) with the typical 3D virtual corneal model schematic representation.

**Figure 7.** Screenshot of keratoconus grade IV individuals (RETICS) with the typical 3D virtual corneal model schematic representation.

**Figure 8.** Screenshot of keratoconus grade V-Plus individuals (RETICS) with the typical 3D virtual corneal model schematic representation.

In summary, the GUI application was developed using a responsive design, which makes it accessible from any network-connected device because it does not require any software or driver being installed, apart from an up-to-date web browser.

## 3. Results

Sixty-two healthy eyes (36.7%) (control group, Ctrl) and 107 KC eyes (63.3%) (KC group, KC) were herein considered. Several subgroups were established in the KC group depending on the disease stage according to the RETICS grading system: grade I (44 eyes, 41.1%), grade II (18 eyes, 16.9%), grade III (15 eyes, 14.0%), grade IV (15 eyes, 14.0%) and grade Plus-V (15 eyes, 14.0%). The main demographic information in the control and KC subgroups are displayed and summarized in Table 1. No significant differences were found in age and gender terms among all the groups.

**Table 1.** Demographic information for healthy individuals (Ctrl) and the RETICS-classified keratoconus patients, graded between I and V.

| | Ctrl | I | II | III | IV | V | p |
|---|---|---|---|---|---|---|---|
| Number of cases | 62 | 44 | 18 | 15 | 15 | 15 | |
| Age in years (mean ± SD) | 37.5 ± 14.4 | 41.2 ± 16.6 | 47.4 ± 22.5 | 40.3 ± 12.6 | 34.0 ± 14.3 | 36.3 ± 18.0 | 0.161 |
| Female/Male | 32/30 | 12/32 | 6/12 | 7/8 | 6/9 | 4/11 | 0.146 |

Table 2 summarizes the descriptive analysis outcomes obtained for all the quantitative variable analyses in the control vs. RETICS I and the control vs. KC groups. The descriptive analysis indicated that most variables did not follow normal distribution. Statistically significant $p$-values were found for all the variables between the control group (Ctrl) and the RETICS grade I group, except for age ($p = 0.665$). When testing for differences within all the groups (Ctrl and RETICS grade I to V), every $p$-value was significant, except for age ($p = 0.344$).

**Table 2.** Descriptive analysis for the quantitative variables and $p$-values for normality and differences between RETICS groups. CDVA: Corrected Distance Visual Acuity; Q: asphericity; RMS, root mean square

| Variables | Mean | SD | Median | Range | Normality | Ctrl vs. RETICS I | Ctrl–RETICS I–II–III–IV–V |
|---|---|---|---|---|---|---|---|
| **Demographic** | | | | | | | |
| Age | 39 | 16 | 38 | 7–98 | <0.001 | 0.665 | 0.344 |
| **Optical** | | | | | | | |
| CDVA | 0.77 | 0.32 | 0.96 | 0.05–1.20 | <0.001 | <0.001 | <0.001 |
| Coma-like | 1.85 | 2.15 | 1.01 | 0.08–13.0 | <0.001 | <0.001 | <0.001 |
| Q8mm | −0.63 | 0.70 | −0.45 | −2.80–2.82 | < 0.001 | 0.025 | <0.001 |
| Spherical-like | 0.70 | 0.93 | 0.44 | 0.15–7.20 | <0.001 | <0.001 | <0.001 |
| RMS total | 3.22 | 3.09 | 2.38 | 0.33–15.6 | <0.001 | <0.001 | <0.001 |
| High-order | 2.02 | 2.33 | 1.12 | 0.24–13.8 | <0.001 | <0.001 | <0.001 |
| Astigmatism | 2.31 | 2.25 | 1.57 | 0.04–11.22 | <0.001 | <0.001 | <0.001 |
| **Pachymetry** | | | | | | | |
| Central thickness | 499 | 62 | 508 | 285–633 | <0.001 | <0.001 | <0.001 |
| Temporal | 545 | 50 | 546 | 385–645 | 0.073 | <0.001 | <0.001 |
| Nasal | 579 | 48 | 579 | 451–692 | 0.835 | <0.001 | <0.001 |
| Superior | 591 | 50 | 590 | 408–695 | 0.062 | <0.001 | <0.001 |
| Inferior | 559 | 57 | 563 | 332–762 | <0.001 | <0.001 | <0.001 |

**Table 2.** *Cont.*

| Variables | Mean | SD | Median | Range | Normality | Ctrl vs. RETICS I | Ctrl–RETICS I–II–III–IV–V |
|---|---|---|---|---|---|---|---|
| **Morpho-Geometric** | | | | | | | |
| Total volume | 24.7 | 1.9 | 24.4 | 19.8–29.1 | 0.335 | <0.001 | <0.001 |
| Anterior surface area | 43.35 | 0.59 | 43.24 | 42.49–47.44 | <0.001 | <0.001 | <0.001 |
| Posterior surface area | 44.70 | 0.93 | 44.53 | 43.53–51.14 | <0.001 | <0.001 | <0.001 |

The multivariate logistic regression model results are summarized in Table 3 with the coefficients for each variable. The variables that made a statistically significant contribution in the model are shown below: gender, coma-like, central thickness, high-order and temporal.

**Table 3.** Summary of the multivariate logistic regression model for the Ctrl vs. RETICS I patients. CDVA: Corrected Distance Visual Acuity; Q: asphericity; RMS, root mean square

| | | Std. | Z | p | | 95% CI | |
|---|---|---|---|---|---|---|---|
| Variables | Coefficient | Error | Value | Value | OR | Lower | Upper |
| **Demographic** | | | | | | | |
| Age | 0.135 | 0.078 | 1.727 | 0.084 | 1.145 | 1.021 | 1.416 |
| Gender | 5.267 | 2.520 | 2.090 | 0.037 | 193.907 | 4.154 | $2.14 \times 10^5$ |
| **Clinical** | | | | | | | |
| CDVA | −6.345 | 4.613 | −1.376 | 0.169 | 0.002 | 0.001 | 1.821 |
| Coma-like | 4.072 | 1.917 | 2.124 | 0.034 | 58.687 | 3.615 | $1.16 \times 10^4$ |
| Q8mm | 6.027 | 4.618 | 1.305 | 0.192 | 414.459 | 0.070 | $3.02 \times 10^7$ |
| Spherical-like | 1.673 | 1.259 | 1.329 | 0.119 | 5.329 | 0.671 | $1.08 \times 10^2$ |
| RMS total | 2.482 | 1.591 | 1.560 | 0.119 | 11.969 | 1.071 | $8.65 \times 10^2$ |
| High-order | 5.534 | 2.484 | 2.227 | 0.026 | 253.143 | 5.830 | $3.12 \times 10^5$ |
| Astigmatism | −1.054 | 1.218 | −0.866 | 0.387 | 0.348 | 0.014 | 2.963 |
| **Pachymetry** | | | | | | | |
| Central thickness | -0.242 | 0.116 | −2.083 | 0.037 | 0.785 | 0.577 | 0.941 |
| Temporal | 0.028 | 0.012 | 2.248 | 0.025 | 1.028 | 1.009 | 1.064 |
| Nasal | 0.007 | 0.011 | 0.641 | 0.522 | 1.007 | 0.987 | 1.034 |
| Superior | 0.092 | 0.058 | 1.587 | 0.113 | 1.096 | 0.995 | 1.263 |
| Inferior | −0.051 | 0.034 | −1.479 | 0.139 | 0.951 | 0.870 | 0.995 |
| **Morpho-Geometric** | | | | | | | |
| Total volume | 0.067 | 0.547 | 0.124 | 0.901 | 1.070 | 0.349 | 3.908 |
| Anterior surface area | −0.326 | 0.532 | −0.614 | 0.539 | 0.722 | 0.218 | 2.105 |
| Posterior surface area | −1.189 | 0.779 | −1.525 | 0.127 | 0.305 | 0.044 | 1.199 |
| **Constant** | 135.027 | 61.595 | 2.192 | 0.028 | - | - | - |

The ROC curve in Figure 9 shows an optimal cut-off point of 0.475 with a training AUC of 0.990, a sensitivity of 0.977 and a specificity of 0.919, corresponding to 59 true negative cases, 41 true positive cases, three false-positive cases and three false-negative cases. The bootstrapped validation values corresponding to the 95% CI are 0.843 ± 0.058 for AUC, 0.844 ± 0.095 for sensitivity and 0.838 ± 0.081 for specificity.

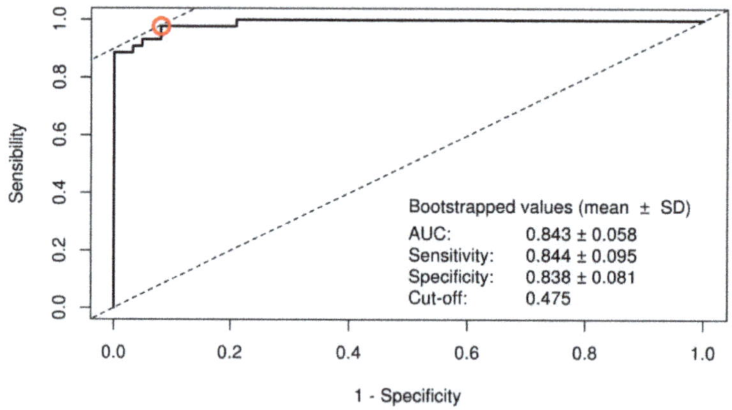


**Figure 9.** Logistic regression model ROC curve indicating the optimal cut-off point and the bootstrapped validation parameters.

Figure 10 shows the distribution of the calculated scores in both groups, where the vast majority of the control cases obtained a score between 0 and 0.1, and between 0.9 and 1 for RETICS I.

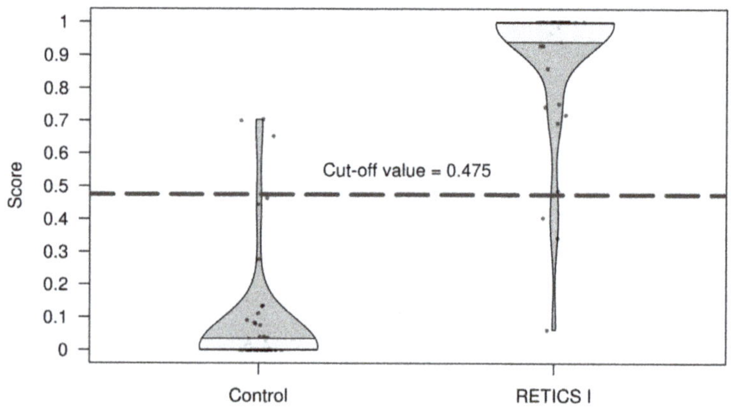


**Figure 10.** Distribution of the logistic regression scores between the Control and RETICS I patients, and the optimal cut-off.

The parameters for the multivariate ordinal logistic regression model are summarized in Table 4. In this case, the variables that made a statistically significant contribution to the model were: CDVA, central thickness and temporal.

The confusion matrix for the training dataset is shown in Table 5, with an overall accuracy of 0.698 at a 95% CI between 0.623 and 0.766. Sensitivity, specificity and balanced accuracy are also shown for each group.

Figure 11 depicts power according to the simulated sample size for both models, including only the variables with maximum power over 0.9 at some point within the range, plus Age and Gender. The remaining variables and their respective maxima for both the binary logistic regression and ordinal logistic regression models were Coma-Like Deviation (0.61 and 0.62), $Q_{8mm}$ (0.55 and 0.57), Spherical-Like Deviation (0.49 and 0.55), Nasal Thickness (0.81 and 0.83), Superior Thickness (0.88 and 0.89), Inferior Thickness (0.72 and 0.77), Volume (0.69 and 0.71), Anterior Surface (0.83 and 0.89) and Posterior Surface (0.58 and 0.60). For both models, the variables of High-Order Aberration, CDVA, Central Thickness, Total RMS and Temporal Thickness have power values over 0.80 for the sample sizes exceeding 150, and the powers for the ordinal logistic model at small sample sizes (below 100) are somewhat lower than for the binary logistic regression models.

**Table 4.** Summary of the ordinal logistic regression model for the Ctrl and RETICS grade I-V patients. CDVA: Corrected Distance Visual Acuity; Q: asphericity; RMS, root mean square

| | | Std. | *t* | *p* | | 95% CI | |
|---|---|---|---|---|---|---|---|
| **Variables** | **Coefficient** | **Error** | **Value** | **Value** | **OR** | **Lower** | **Upper** |
| **Demographic** | | | | | | | |
| Age | 0.016 | 0.011 | 1.475 | 0.140 | 1.016 | 0.995 | 1.037 |
| Gender | −0.073 | 0.361 | −0.203 | 0.839 | 0.929 | 0.458 | 1.885 |
| **Clinical** | | | | | | | |
| CDVA | −5.495 | 0.865 | −6.355 | 0.001 | 0.004 | 0.001 | 0.022 |
| Coma-like | 0.106 | 0.348 | 0.304 | 0.761 | 1.111 | 0.562 | 2.197 |
| Q8mm | 0.184 | 0.359 | 0.512 | 0.609 | 1.201 | 0.595 | 2.423 |
| Spherical-like | −0.063 | 0.250 | −0.250 | 0.803 | 0.940 | 0.575 | 1.535 |
| RMS total | −0.422 | 0.298 | 1.416 | 0.157 | 1.526 | 0.850 | 2.737 |
| High-order | 0.621 | 0.345 | 1.797 | 0.072 | 1.860 | 0.945 | 3.660 |
| Astigmatism | −0.250 | 0.239 | −1.046 | 0.296 | 0.779 | 0.488 | 1.244 |
| **Pachymetry** | | | | | | | |
| Central thickness | −0.024 | 0.007 | −3.298 | 0.001 | 0.977 | 0.963 | 0.991 |
| Temporal | 0.006 | 0.002 | 2.662 | 0.008 | 1.006 | 1.002 | 1.010 |
| Nasal | −0.002 | 0.003 | −0.771 | 0.441 | 0.998 | 0.992 | 1.004 |
| Superior | 0.006 | 0.007 | 0.933 | 0.351 | 1.006 | 0.993 | 1.019 |
| Inferior | −0.002 | 0.005 | −0.522 | 0.601 | 0.998 | 0.989 | 1.007 |
| **Morpho-Geometric** | | | | | | | |
| Volume | 0.071 | 0.141 | 0.500 | 0.617 | 1.073 | 0.814 | 1.414 |
| Anterior area | −0.204 | 0.133 | −1.533 | 0.125 | 0.816 | 0.623 | 1.058 |
| Posterior area | 0.046 | 0.129 | 0.361 | 0.718 | 1.048 | 0.814 | 1.348 |
| **Intercepts** | | | | | | | |
| Ctrl vs. RETICS I | −16.769 | 0.017 | −98.428 | <0.001 | - | - | - |
| RETICS I vs. II | −13.926 | 0.396 | −35.205 | <0.001 | - | - | - |
| RETICS II vs. III | −12.270 | 0.522 | −23.530 | <0.001 | - | - | - |
| RETICS III vs. IV | −9.998 | 0.765 | −13.072 | <0.001 | - | - | - |
| RETICS IV vs. V | −6.691 | 1.057 | −6.330 | <0.001 | - | - | - |

**Table 5.** The ordinal logistic regression confusion matrix showing sensitivity, specificity and balanced accuracy for each group. The total sum of cells by rows shows the total number of true cases present, while each column represents how many cases the model classified in that category.

| | Predicted Value | | | | | |
|---|---|---|---|---|---|---|
| **True Value** | **Ctrl** | **I** | **II** | **III** | **IV** | **V** |
| Ctrl | 55 | 7 | 0 | 0 | 0 | 0 |
| I | 12 | 26 | 4 | 2 | 0 | 0 |
| II | 2 | 9 | 6 | 3 | 1 | 0 |
| III | 0 | 3 | 3 | 6 | 3 | 0 |
| IV | 0 | 1 | 0 | 1 | 12 | 1 |
| V | 0 | 0 | 0 | 1 | 1 | 13 |
| Sensitivity | 0.887 | 0.591 | 0.333 | 0.400 | 0.800 | 0.867 |
| Specificity | 0.869 | 0.840 | 0.954 | 0.955 | 0.968 | 0.994 |
| Balanced accuracy | 0.878 | 0.715 | 0.643 | 0.677 | 0.884 | 0.930 |

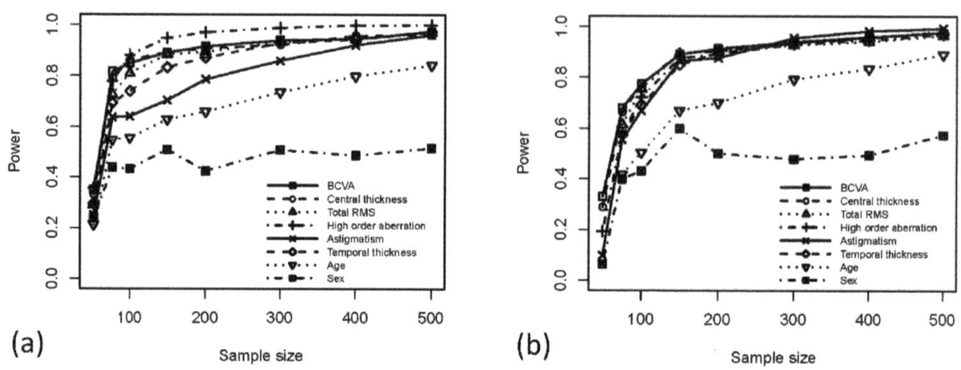


**Figure 11.** Power analysis results for (**a**) the binary logistic regression model; (**b**) the ordinal logistic regression model.

*Graphics User Interface*

Figure 3 is a screenshot corresponding to one healthy individual (Ctrl). It includes a 3D image of a characteristic cornea showing a schematic representation of how different predictors are calculated based on physical measurements.

Figures 4–8 are five screenshots corresponding to five representatives of the five RETICS grades I–V. The score rapidly rises when passing from Ctrl to grade I and remains at 100% for the other grades, which is consistent with the high sensitivity and specificity in the model. Each prediction includes a 3D image of a characteristic cornea from the corresponding RETICS group, along with a schematic representation of how different predictors are calculated based on physical measurements.

The four examples in Figure 12 indicate the difficulty of detecting early-stage KC. Indeed, some cases are wrongly classified. We must bear in mind that this early KC detection model was trained using diagnostics made by ophthalmological professionals as a "gold standard", and some inevitable undetermined amount of subjective information was taken for granted. During the fitting process, the model attempted to find some generalization to bind predictors and prediction with the best possible performance, but some samples might not match any kind of generalization given by training data's subjective nature. Therefore, it is reasonable to expect some lack of accuracy, which does not necessarily mean failure in the model's fitting capability. Our model quantitatively confirmed the difficulty of discerning both control and grade I groups as 17 of 106 cases (16%) and obtained a score between 0.1 and 0.9 which is, therefore, in the aforementioned "halfway" situation, and thus confirmed this tool's utility.

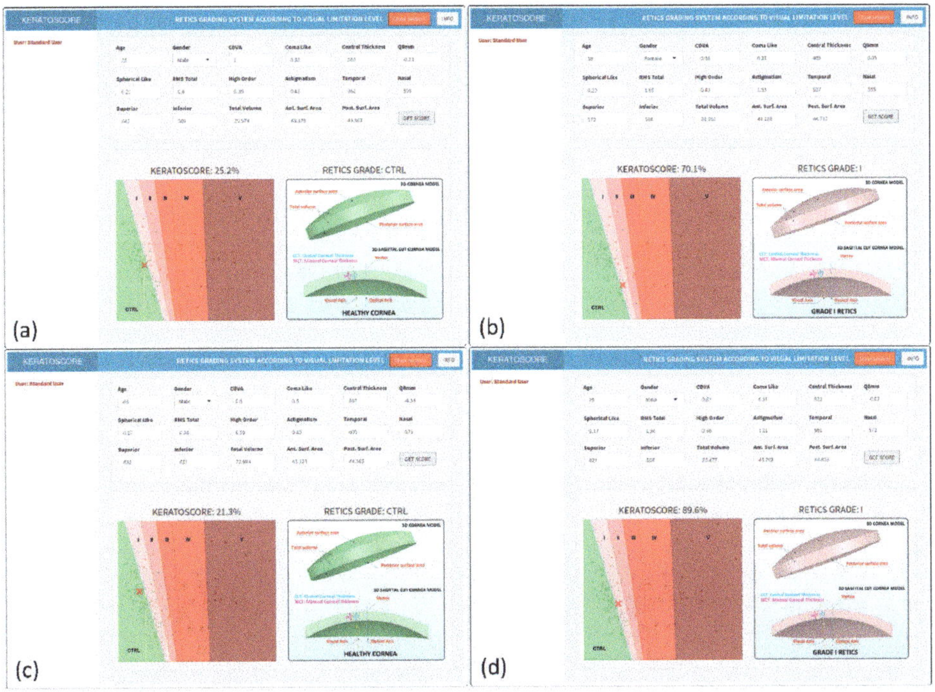


**Figure 12.** Screenshots showing: (**a**) a correctly classified healthy individual (true negative); (**b**) an incorrectly classified healthy individual (false-positive); (**c**) a correctly classified grade I KC individual (true positive); (**d**) an incorrectly classified grade I KC individual (false-negative).

## 4. Discussion

The range of techniques that allow the characterization and evaluation of the degree of KC severity vary, but the ultimate trends use machine-learning [44] and neural networks [45].

This study defines two automated learning models based on a set of independent variables (demographic, optical, pachymetric, morpho-geometric) to characterize the optical-geometrical cornea structure in different KC phases. More specifically, it considered a probability model of the risk associated with suffering early-stage KC, as well as another model to classify the degree of KC severity depending on patients' visual limitation levels.

This approach offers some very useful advantages. First, it summarizes information that derives from many parameters of different natures (qualitative or quantitative, measured in distinct units) in a single number that can be easily read and understood, which minimizes the risk of some key pieces of information going unnoticed. Otherwise, this risk is fairly high, as common analytical reports usually contain many printed pages of different parameters, which often include no associated normality range and must be read in a matter of seconds. Second, it evaluates the combined action of these different parameters, which could imply a high value in a particular key parameter, clearly indicative of disease being present, but might be less evident when the increase in many different key parameters is slight. In this situation, a score can assist health professionals in their decision-making process as it provides an objective and quantitative scale that takes into account the joint action of a set of diverse parameters [3].

A multivariate logistic regression model was fitted using healthy (Ctrl) and RETICS grade I individuals, which always made predictions ranging between 0 and 1 when applied to new data. This prediction came closer to 0 when input data were similar to those that characterized healthy controls, and close to 1 if they were similar to grade I KC patients. All those individuals with an intermediate score were considered to be more or less halfway between both groups, i.e., they showed some kind of abnormality vs. healthy controls, which indicates KC-developing patients, but had not yet reached the cut-off point to be diagnosed as diseased. Therefore, this multivariate logistic regression output could be considered to be proportionally associated with the probability of suffering KC. A higher model output indicated a stronger probability. Even though this relationship is not linear, such scores are frequently used as predicting tools to assess patient status and prognosis in different fields [3,46–48].

For initial KC development, the analyzed model-based studies generally showed a good discrimination between normal and early KC eyes. However, one of the main challenges faced by ophthalmologists today is that no consensus has been reached by experts about the characterization patterns of early corneal ectasia [7,49]. This is mainly the result of the diversity of designations used to refer to subtler KC manifestations. It is also due to the fact that the wide variety of indices used for detection are technology-specific, which does not make them easily interchangeable [19,49]. As it is difficult to directly compare these studies, we resorted to using authors' own writings to compare them to our work.

Several studies have reported similar results to ours about models based on Scheimpflug metrics [5–7,20]. Hwang et al. [7] proposed a first model based on the combination of five metrics (AUC: 0.86, Sensitivity: 83%, Specificity: 83%). Similar results have been obtained by other authors [5,6,20] in the model development phase and based on the exclusive metrics of the same Scheimpflug technology. In contrast, our study obtained higher performance values for model development (AUC: 0.95) and similar ones in the validation phase (AUC: 0.85), and was composed of metrics that are not exclusive to the same Scheimpflug technology.

Smadja et al. [50] used a machine-learning algorithm based on decision trees to analyze 55 parameters deriving from anterior and posterior corneal measurements. They found that the most discriminant variables related to posterior surface asymmetry and thickness spatial distribution achieved 93.6% sensitivity and 97.2% specificity when discriminating between normal and forme fruste KC. However, some authors [51] considered that including eyes from patients who had already been diagnosed with KC in one eye was inappropriate because it biases the sample if we contemplate that the genetic determinants for KC appearing are already present in them.

In our study, the performance measurements of the early detection model in AUC, specificity and sensitivity terms indicated very high performance with the training dataset, with all three values reaching around 0.95, and very few false positives and false negatives. These figures are significantly lower for the validation procedure, which dropped to about 0.85, but were well over 80% in all cases. These findings indicate good model-validated performance, but also suggest the presence of some overfitting. This is otherwise reasonable if we take into account the relatively few training cases (62 healthy individuals and 44 RETICS grade I patients), which were significantly lower than the training values.

Other authors have proposed using a multivariate system based on combining two different technologies. Saad et al. [9] combined two technologies to propose a model based on 54 variables and six discriminant functions, and reported 93% sensitivity and 92% specificity in the model development phase. It was validated in a later study with 92% sensitivity and 96% specificity [52]. Other studies have suggested combining several different technologies [53–55], but the authors defined a more advanced form of KC when they included patients with manifest inferior steepening [49,56].

The only reference found in the scientific literature with a classification system that uses visual acuity as a parameter is that by Wisse et al. [3], who established a scoring system that relied on five parameters (age, quality of vision, uncorrected distance visual acuity, refraction difference, maximum keratometry difference). However, this score did not classify disease grades, but disease progression. Consequently, its aim was to determine if crosslinking treatment would be necessary or not.

Our work presents and validates a probability model of the risk associated with suffering KC. Our research group is unaware of any previous study that combined demographic, optical, pachymetric and morpho-geometric variables successfully and in real-time to detect early KC.

Finally regarding the clinical KC development phase, experts' criteria have converged to diagnose this disease as the degree of severity of progressing KC [24]. Although there are several classifications that characterize clinical KC, to the authors' knowledge, there is no disease severity classifier based on patients' visual limitation. Our classifier is based on an ordinary logistic regression model that combines 17 variables and presents an overall accuracy of 0.698; that is, our model correctly classifies almost 70% of patients. The performance indicators for ordinal logistic regression are fairly lower, particularly for the RETICS grade I, II and III patients, with balanced accuracies ranging between 0.623 and 0.766. Grade II patients present the worst accuracy, with nine in every 18 patients being wrongly classified as grade I. Once again, a small sample size is the most probable explanation for such behavior, along with its non-homogeneity (controls and grade I patients are much more abundant that grade II–V patients). It could also indicate that these three groups are not clearly differentiated. Therefore, all the conclusions drawn from the ordinal logistic regression model must be considered very carefully. Even when the results are reasonably good, the training process should be repeated with a bigger and more homogeneous sample. It would also be desirable to validate our results with other ethnicities and populations.

Another limitation of our study is the proven dependence that clinical metrics has on the technology employed to measure it [19], which means that our results are only valid for those eyes tested with a Sirius tomographer (CSO, Florence, Italy).

## 5. Conclusions

A web application was developed and deployed that combines two machine-learning models to support ophthalmologic professionals: a multivariate logistic regression model for early KC predictions and an ordinal logistic regression model to assign diagnosis grades on the RETICS scale. This application has a responsive design, and it allows any sort of device to be used (computer, tablet or smartphone). It also incorporates security measurements (authentication layer and accession from intranet only). The early KC prediction model shows high-performance indicators, even though some overfitting appears, while the RETICS grading prediction model's accuracy is remarkably lower, particularly for grade I, II and III patients. In both cases, repeating the training process with a bigger sample should be considered. This falls in line with recently published recommendations for sample size calculations by multivariate prediction models [57] which, for this case, proposes an optimal sample size of 374 individuals, with a minimum of 125. An optimal figure could not be reached, given our biobank's limited database size. Moreover, even though multivariate and ordinal logistic regressions are state-of-the-art and widely used techniques for modeling biomedical research data, many other powerful artificial intelligence techniques are available (particularly deep learning techniques), and their use is strongly advised for improving the quality of results.

**Author Contributions:** Conceptualization, J.M.B. and F.C.; methodology, J.M.B., F.C. and J.L.A.; validation, J.S.V. and J.M.B.; analysis, F.C. and J.S.V.; research, J.M.B., F.C. and J.L.A.; resources, J.M.B. and F.C.; data curation, J.S.V. and J.L.A.; writing and preparing the original draft, J.M.B. and F.C.; writing—reviewing and editing F.C., J.S.V. and J.L.A.; supervision, J.L.A.; project administration, F.C. and J.L.A.; funding acquisition J.L.A. All authors have read and agree to the published version of the manuscript.

**Funding:** This publication has been carried out as part of the Thematic Network for Co-Operative Research in Health (RETICS), reference number RD16/0008/0012, financed by the Carlos III Health Institute-General Subdirection of Networks and Cooperative Investigation Centers (R&D&I National Plan 2013-2016), European Regional Development Funds (FEDER), and the Results Valorization Program financed by the Technical University of Cartagena (PROVALOR-UPCT).

**Conflicts of Interest:** The authors declare no conflict of interest.

## References

1. Pinero, D.P. *Technologies for Anatomical and Geometric Characterization of the Corneal Structure and Anterior Segment: A Review*; Informa Healthcare: London, UK, 2015; pp. 161–170.

2. Cavas-Martínez, F.; Bataille, L.; Fernández-Pacheco, D.G.; Cañavate, F.J.F.; Alio, J.L. Keratoconus detection based on a new corneal volumetric analysis. *Sci. Rep.* **2017**, *7*, 15837. [CrossRef] [PubMed]
3. Wisse, R.P.L.; Simons, R.W.P.; van der Vossen, M.J.B.; Muijzer, M.B.; Soeters, N.; Nuijts, R.M.M.A.; Godefrooij, D.A. Clinical Evaluation and Validation of the Dutch Crosslinking for Keratoconus Score. *JAMA Ophthalmol.* **2019**, *137*, 610–616. [CrossRef] [PubMed]
4. Binder, P.S. Risk factors for ectasia after LASIK. *J. Cataract Refract. Surg.* **2008**, *34*, 2010–2011. [CrossRef]
5. Binder, P.S.; Trattler, W.B. Evaluation of a risk factor scoring system for corneal ectasia after LASIK in eyes with normal topography. *J. Refract. Surg.* **2010**, *26*, 241–250. [CrossRef]
6. Chan, C.; Ang, M.; Saad, A.; Chua, D.; Mejia, M.; Lim, L.; Gatinel, D. Validation of an Objective Scoring System for Forme Fruste Keratoconus Detection and Post-LASIK Ectasia Risk Assessment in Asian Eyes. *Cornea* **2015**, *34*, 996–1004. [CrossRef]
7. Hwang, E.S.; Perez-Straziota, C.E.; Kim, S.W.; Santhiago, M.R.; Randleman, J.B. Distinguishing Highly Asymmetric Keratoconus Eyes Using Combined Scheimpflug and Spectral-Domain OCT Analysis. *Ophthalmology* **2018**, *125*, 1862–1871. [CrossRef]
8. Randleman, J.B.; Woodward, M.; Lynn, M.J.; Stulting, R.D. Risk assessment for ectasia after corneal refractive surgery. *Ophthalmology* **2008**, *115*, 37–50. [CrossRef]
9. Saad, A.; Gatinel, D. Topographic and tomographic properties of forme fruste keratoconus corneas. *Investig. Ophthalmol. Vis. Sci.* **2010**, *51*, 5546–5555. [CrossRef]
10. Seiler, T.; Quurke, A.W. Iatrogenic keratectasia after LASIK in a case of forme fruste keratoconus. *J. Cataract Refract. Surg.* **1998**, *24*, 1007–1009. [CrossRef]
11. Cavas-Martinez, F.; De la Cruz Sanchez, E.; Nieto Martinez, J.; Fernandez Canavate, F.J.; Fernandez-Pacheco, D.G. Corneal topography in keratoconus: State of the art. *Eye Vis.* **2016**, *3*, 5. [CrossRef] [PubMed]
12. Awad, E.A.; Abou Samra, W.A.; Torky, M.A.; El-Kannishy, A.M. Objective and subjective diagnostic parameters in the fellow eye of unilateral keratoconus. *BMC Ophthalmol.* **2017**, *17*, 186. [CrossRef] [PubMed]
13. Bae, G.H.; Kim, J.R.; Kim, C.H.; Lim, D.H.; Chung, E.S.; Chung, T.Y. Corneal topographic and tomographic analysis of fellow eyes in unilateral keratoconus patients using Pentacam. *Am. J. Ophthalmol.* **2014**, *157*, 103–109. [CrossRef] [PubMed]
14. De Sanctis, U.; Loiacono, C.; Richiardi, L.; Turco, D.; Mutani, B.; Grignolo, F.M. Sensitivity and specificity of posterior corneal elevation measured by Pentacam in discriminating keratoconus/subclinical keratoconus. *Ophthalmology* **2008**, *115*, 1534–1539. [CrossRef]
15. Li, Y.; Chamberlain, W.; Tan, O.; Brass, R.; Weiss, J.L.; Huang, D. Subclinical keratoconus detection by pattern analysis of corneal and epithelial thickness maps with optical coherence tomography. *J. Cataract Refract. Surg.* **2016**, *42*, 284–295. [CrossRef] [PubMed]
16. Reddy, J.C.; Rapuano, C.J.; Cater, J.R.; Suri, K.; Nagra, P.K.; Hammersmith, K.M. Comparative evaluation of dual Scheimpflug imaging parameters in keratoconus, early keratoconus, and normal eyes. *J. Cataract Refract. Surg.* **2014**, *40*, 582–592. [CrossRef]
17. Randleman, J.B.; Lynn, M.J.; Perez-Straziota, C.E.; Weissman, H.M.; Kim, S.W. Comparison of central and peripheral corneal thickness measurements with scanning-slit, Scheimpflug and Fourier-domain ocular coherence tomography. *Br. J. Ophthalmol.* **2015**, *99*, 1176–1181. [CrossRef]
18. Reinstein, D.Z.; Archer, T.J.; Urs, R.; Gobbe, M.; RoyChoudhury, A.; Silverman, R.H. Detection of Keratoconus in Clinically and Algorithmically Topographically Normal Fellow Eyes Using Epithelial Thickness Analysis. *J. Refract. Surg.* **2015**, *31*, 736–744. [CrossRef]
19. Savini, G.; Carbonelli, M.; Sbreglia, A.; Barboni, P.; Deluigi, G.; Hoffer, K.J. Comparison of anterior segment measurements by 3 Scheimpflug tomographers and 1 Placido corneal topographer. *J. Cataract Refract. Surg.* **2011**, *37*, 1679–1685. [CrossRef]
20. Shajari, M.; Jaffary, I.; Herrmann, K.; Grunwald, C.; Steinwender, G.; Mayer, W.J.; Kohnen, T. Early tomographic changes in the eyes of patients with keratoconus. *J. Refract. Surg.* **2018**, *34*, 254–259. [CrossRef]
21. Li, X.; Rabinowitz, Y.S.; Rasheed, K.; Yang, H. Longitudinal study of the normal eyes in unilateral keratoconus patients. *Ophthalmology* **2004**, *111*, 440–446. [CrossRef]
22. Ferdi, A.C.; Nguyen, V.; Gore, D.M.; Allan, B.D.; Rozema, J.J.; Watson, S.L. Keratoconus Natural Progression: A Systematic Review and Meta-analysis of 11 529 Eyes. *Ophthalmology* **2019**, *126*, 935–945. [CrossRef] [PubMed]

23. Alio, J.L.; Pinero, D.P.; Aleson, A.; Teus, M.A.; Barraquer, R.I.; Murta, J.; Maldonado, M.J.; Castro de Luna, G.; Gutierrez, R.; Villa, C.; et al. Keratoconus-integrated characterization considering anterior corneal aberrations, internal astigmatism, and corneal biomechanics. *J. Cataract Refract. Surg.* **2011**, *37*, 552–568. [CrossRef]
24. Martinez-Abad, A.; Pinero, D.P. New perspectives on the detection and progression of keratoconus. *J. Cataract Refract. Surg.* **2017**, *43*, 1213–1227. [CrossRef]
25. Romero-Jimenez, M.; Santodomingo-Rubido, J.; Wolffsohn, J.S. Keratoconus: A review. *Cont. Lens Anterior Eye* **2010**, *33*, 157–166. [CrossRef]
26. Belin, M.W.; Duncan, J.K. Keratoconus: The ABCD Grading System. *Klin. Mon. Augenheilkd.* **2016**, *233*, 701–707. [CrossRef] [PubMed]
27. Alio, J.L.; Shabayek, M.H. Corneal higher order aberrations: A method to grade keratoconus. *J. Refract. Surg.* **2006**, *22*, 539–545. [CrossRef] [PubMed]
28. Smadja, D. Topographic and tomographic indices for detecting keratoconus and subclinical keratoconus: A systematic review. *Int. J. Keratoconus Ect. Corneal Dis.* **2013**, *2*, 60. [CrossRef]
29. McMahon, T.T.; Szczotka-Flynn, L.; Barr, J.T.; Anderson, R.J.; Slaughter, M.E.; Lass, J.H.; Iyengar, S.K. A new method for grading the severity of keratoconus: The Keratoconus Severity Score (KSS). *Cornea* **2006**, *25*, 794–800. [CrossRef] [PubMed]
30. Kanellopoulos, A.J.; Asimellis, G. Revisiting keratoconus diagnosis and progression classification based on evaluation of corneal asymmetry indices, derived from scheimpflug imaging in keratoconic and suspect cases. *Clin. Ophthalmol.* **2013**, *7*, 1539–1548. [CrossRef]
31. Krumeich, J.H.; Daniel, J.; Knülle, A. Live-epikeratophakia for keratoconus. *J. Cataract Refract. Surg.* **1998**, *24*, 456–463. [CrossRef]
32. Vega-Estrada, A.; Alio, J.L.; Brenner, L.F.; Javaloy, J.; Plaza Puche, A.B.; Barraquer, R.I.; Teus, M.A.; Murta, J.; Henriques, J.; Uceda-Montanes, A. Outcome analysis of intracorneal ring segments for the treatment of keratoconus based on visual, refractive, and aberrometric impairment. *Am. J. Ophthalmol.* **2013**, *155*, 575–584.e571. [CrossRef] [PubMed]
33. Huseynli, S.; Salgado-Borges, J.; Alio, J.L. Comparative evaluation of Scheimpflug tomography parameters between thin non-keratoconic, subclinical keratoconic, and mild keratoconic corneas. *Eur. J. Ophthalmol.* **2018**, *28*, 521–534. [CrossRef] [PubMed]
34. Cavas-Martinez, F.; Fernandez-Pacheco, D.G.; De la Cruz-Sanchez, E.; Nieto Martinez, J.; Fernandez Canavate, F.J.; Vega-Estrada, A.; Plaza-Puche, A.B.; Alio, J.L. Geometrical custom modeling of human cornea in vivo and its use for the diagnosis of corneal ectasia. *PLoS ONE* **2014**, *9*, e110249. [CrossRef] [PubMed]
35. Cavas-Martínez, F.; Fernández-Pacheco, D.; Cañavate, F.; Velázquez-Blázquez, J.; Bolarín, J.; Alió, J. Study of Morpho-Geometric Variables to Improve the Diagnosis in Keratoconus with Mild Visual Limitation. *Symmetry* **2018**, *10*, 306. [CrossRef]
36. Cavas-Martinez, F.; Fernandez-Pacheco, D.G.; Cañavate, F.J.F.; Velázquez-Blázquez, J.S.; Bolarin, J.M.; Tiveron, M.; Alio, J.L. Detección del queratocono temprano mediante modelado 3D personalizado y análisis de sus parámetros geométricos. *DYNA Ing. Ind.* **2019**, *2*, 175–181.
37. Cavas-Martínez, F.; Fernández-Pacheco, D.G.; Parras, D.; Cañavate, F.J.F.; Bataille, L.; Alió, J. Study and characterization of morphogeometric parameters to assist diagnosis of keratoconus. *Biomed. Eng. Online* **2018**, *17*, 161. [CrossRef]
38. Efron, B.; Tibshirani, R. Improvements on cross-validation: The 632+ bootstrap method. *J. Am. Stat. Assoc.* **1997**, *92*, 548–560.
39. R Core Team. R: A Language and Environment for Statistical Computing. Available online: https://www.R-project.org/ (accessed on 6 January 2020).
40. Aberson, C.L. *Applied Power Analysis for the Behavioral Sciences*, 2nd ed.; Taylor & Francis: Abingdon, UK, 2019.
41. Demidenko, E. Sample size determination for logistic regression revisited. *Stat. Med.* **2007**, *26*, 3385–3397. [CrossRef]
42. Chang, W.; Cheng, J.; Allaire, J.; Xie, Y.; Jonathan, M. Shiny: Web Application Framework for R. R Package Version 1.3.2. Available online: https://CRAN.R-project.org/package=shiny (accessed on 6 January 2020).
43. Campbell, P. Shinyauthr: Shiny Authentication Modules. R Package Version 0.0.99. Available online: https://rdrr.io/github/PaulC91/shinyauthr/ (accessed on 6 January 2020).

44. Yousefi, S.; Yousefi, E.; Takahashi, H.; Hayashi, T.; Tampo, H.; Inoda, S.; Arai, Y.; Asbell, P. Keratoconus severity identification using unsupervised machine learning. *PLoS ONE* **2018**, *13*, e0205998. [CrossRef]
45. Lavric, A.; Valentin, P. KeratoDetect: Keratoconus Detection Algorithm Using Convolutional Neural Networks. *Comput. Intell. Neurosci.* **2019**, *2019*, 9. [CrossRef]
46. Larrosa, J.M.; Moreno-Montañés, J.; Martinez-de-la-Casa, J.M.; Polo, V.; Velázquez-Villoria, Á.; Berrozpe, C.; García-Granero, M. A Diagnostic Calculator for Detecting Glaucoma on the Basis of Retinal Nerve Fiber Layer, Optic Disc, and Retinal Ganglion Cell Analysis by Optical Coherence TomographyDiagnostic Calculator of OCT for Detecting Glaucoma. *Investig. Ophthalmol. Vis. Sci.* **2015**, *56*, 6788–6795. [CrossRef]
47. Moreno-Montañés, J.; García-Nieva, A.; Osio, I.A.; Guarnieri, A.; Morilla-Grasa, A.; García-Granero, M.; Antón, A. Evaluation of RETICs Glaucoma Diagnostic Calculators in Preperimetric Glaucoma. *Transl. Vis. Sci. Technol.* **2018**, *7*, 13. [CrossRef] [PubMed]
48. Schmidl, D.; Garhöfer, G.; Schmetterer, L. A New Scoring System for Progressive Keratoconus. *JAMA Ophthalmol.* **2019**, *137*, 617. [CrossRef] [PubMed]
49. Lin, S.R.; Ladas, J.G.; Bahadur, G.G.; Al-Hashimi, S.; Pineda, R. A Review of Machine Learning Techniques for Keratoconus Detection and Refractive Surgery Screening. *Semin. Ophthalmol.* **2019**, *34*, 317–326. [CrossRef] [PubMed]
50. Smadja, D.; Touboul, D.; Cohen, A.; Doveh, E.; Santhiago, M.R.; Mello, G.R.; Krueger, R.R.; Colin, J. Detection of subclinical keratoconus using an automated decision tree classification. *Am. J. Ophthalmol.* **2013**, *156*, 237–246. [CrossRef] [PubMed]
51. Klyce, S.D. Chasing the suspect: Keratoconus. *Br. J. Ophthalmol.* **2009**, *93*, 845–847. [CrossRef] [PubMed]
52. Saad, A.; Gatinel, D. Validation of a new scoring system for the detection of early forme of keratoconus. *Age* **2012**, *37*, 37–38. [CrossRef]
53. Qin, B.; Chen, S.; Brass, R.; Li, Y.; Tang, M.; Zhang, X.; Wang, X.; Wang, Q.; Huang, D. Keratoconus diagnosis with optical coherence tomography-based pachymetric scoring system. *J. Cataract Refract. Surg.* **2013**, *39*, 1864–1871. [CrossRef]
54. Rabinowitz, Y.S.; Li, X.; Canedo, A.L.; Ambrosio, R., Jr.; Bykhovskaya, Y. Optical coherence tomography combined with videokeratography to differentiate mild keratoconus subtypes. *J. Refract. Surg.* **2014**, *30*, 80–87. [CrossRef]
55. Silverman, R.H.; Urs, R.; RoyChoudhury, A.; Archer, T.J.; Gobbe, M.; Reinstein, D.Z. Combined tomography and epithelial thickness mapping for diagnosis of keratoconus. *Eur. J. Ophthalmol.* **2017**, *27*, 129–134. [CrossRef]
56. Arbelaez, M.C.; Versaci, F.; Vestri, G.; Barboni, P.; Savini, G. Use of a support vector machine for keratoconus and subclinical keratoconus detection by topographic and tomographic data. *Ophthalmology* **2012**, *119*, 2231–2238. [CrossRef] [PubMed]
57. Riley, R.D.; Snell, K.I.; Ensor, J.; Burke, D.L.; Harrell, F.E., Jr.; Moons, K.G.; Collins, G.S. Minimum sample size for developing a multivariable prediction model: PART II - binary and time-to-event outcomes. *Stat. Med.* **2019**, *38*, 1276–1296. [CrossRef] [PubMed]

applied sciences

MDPI

Article

# Wavelia Breast Imaging: The Optical Breast Contour Detection Subsystem

**Julio Daniel Gil Cano [1,*], Angie Fasoula [2], Luc Duchesne [2] and Jean-Gael Bernard [2]**

1 MVG Industries, 29280 Plouzané, France
2 MVG Industries, 91140 Villejust, France; angie.fasoula@mvg-world.com (A.F.); luc.duchesne@mvg-world.com (L.D.); jean-gael.bernard@mvg-world.com (J.-G.B.)
* Correspondence: julio_daniel.gil_cano@mvg-world.com

Received: 21 November 2019; Accepted: 10 February 2020; Published: 12 February 2020

**Abstract:** Wavelia is a low-power electromagnetic wave breast imaging device for breast cancer diagnosis, which consists of two subsystems, both performing non-invasive examinations: the Microwave Breast Imaging (MBI) subsystem and the Optical Breast Contour Detection (OBCD) subsystem. The Wavelia OBCD subsystem is a 3D scanning device using an infrared 3D stereoscopic camera, which performs an azimuthal scan to acquire 3D point clouds of the external surface of the breast. The OBCD subsystem aims at reconstructing fully the external envelope of the breast, with high precision, to provide the total volume of the breast and morphological data as a priori information to the MBI subsystem. This paper presents a new shape-based calibration procedure for turntable-based 3D scanning devices, a new 3D breast surface reconstruction method based on a linear stretching function, as well as the breast volume computation method that have been developed and integrated with the Wavelia OBCD subsystem, before its installation at the Clinical Research Facility of Galway (CRFG), in Ireland, for first-in-human clinical testing. Indicative results of the Wavelia OBCD subsystem both from scans of experimental breast phantoms and from patient scans are thoroughly presented and discussed in the paper.

**Keywords:** breast surface reconstruction; 3D scanning system; turntable calibration; breast cancer diagnosis; microwave imaging; medical radar; breast phantoms

## 1. Introduction

Wavelia is a prototype medical device which has recently been developed by MVG Industries. The purpose of Wavelia is to offer an alternative and/or complementary imaging modality to X-ray mammography for breast cancer diagnosis. As it involves microwave technology, so non-ionizing radiation, it may allow for diagnosis and follow-up of rapidly evolving pathologies by making it possible to safely schedule regular 3D scans of the breast as often as required [1,2]. Given the incorporated radar detection technology [3], this novel imaging modality is also meant to provide the clinicians with a useful tool for automated Computer-Aided Diagnosis (CAD). The hardware modules integrated in Wavelia are partly inherited from the technical background and expertise of the company in antenna measurement systems [4,5]. A detailed presentation of the actual R&D status of the system has been reported in [3].

Wavelia is now installed at Galway University Hospital, Ireland, for a first-in-human clinical investigation which started running in the last quarter of 2018. Prior to the initiation of the clinical investigation, on-site validation [3] has been performed using breast phantoms [6]. The pilot study holds full approval by the Local Ethics Committee, and the Health Products Regulatory Authority (HPRA), in Dublin, Ireland. Details on the clinical protocol of the pilot study can be found in the ClinicalTrials.gov repository to which the trial has been registered with identifier NCT03475992 [7].

The Wavelia prototype medical device consists of two subsystems, both performing a non-invasive examination: the Microwave Breast Imaging (MBI) subsystem, which is the main part of the system and the Optical Breast Contour Detection (OBCD) subsystem, which plays an auxiliary role. The MBI subsystem illuminates the breast with low-power electromagnetic waves in the microwave frequency range, which penetrate the breast under examination. The subsystem collects the scattered electromagnetic waves and recovers useful information on the breast tissue consistency, given the dielectric contrast of these tissues; this information can be represented as 3D volumetric.

The OBCD subsystem performs an azimuthal scan of the breast using a 3D infrared stereoscopic camera. The objective of the OBCD subsystem is triple [3]:

- compute the volume of the patient's breast, thus indirectly deriving the required volume of transition liquid such that the container of the MBI subsystem is optimally filled after immersion of the breast;
- compute the distance (vertical extent) from the lowest point of the pendulous breast to the lower plane of the examination table, in order to optimally define the vertical scan positions for the MBI scan of each breast; the number of vertical scan positions and thus the time duration of the MBI scan will vary accordingly to the size of each breast;
- reconstruct the external envelope of the breast, with millimetric precision; such information will further serve to quantify the level of deformation of the breast due to immersion in the transition liquid during the microwave imaging scan, as introduced in [8]; in the case of breasts minimally deformed when immersed in the transition liquid (younger patients with breasts not very large), the OBCD-reconstructed breast surface, after registration with the MBI scan, is also meant to improve the localization of the abnormalities detected with MBI in the breast, by means of revealing the exact orientation of the pendulous breast under scan, as well as details such as the nipple location, which cannot be defined using the MBI-derived reconstruction of the breast surface alone, due to its centimetric level of spatial resolution. With no access to such level of detail, the annotation of the breast quadrant and clock-position at which an abnormality has been detected with MBI cannot be very accurate, when using the MBI system stand-alone, thus inhibiting the comparison and ultimate validation of the microwave breast imaging modality against conventional reference breast imaging data (X-Ray mammography and Ultrasound breast scan).

For the third purpose to be achieved at the maximal possible level with the actual prototype #1 implementation of the Wavelia system, which involves operation of the OBCD and MBI scans on two separate identical examination tables, it is important that the patient lies in the same position for both scans. A standardized position for the arms and the overall body posture of the patient has been established and respected on both scanners during the ongoing first-in-human clinical investigation, such that rigid registration of the two breast surface reconstructions (OBCD and MBI) of each patient's breast is feasible. Two indicative results of OBCD reconstructions of patient breast surfaces, after rigid registration with the MBI scan, are presented in Section 3.4 of the paper.

The importance of retrieving the external breast envelope with high accuracy as a priori information for efficient microwave breast imaging has been extensively highlighted by the microwave breast imaging research community so far [8–12]. The significant added-value of the use of an auxiliary (laser) system to further enhance the maximal achievable accuracy of the estimated breast envelope has been analyzed in [13,14]. Such a laser system has been incorporated in the TSAR microwave breast imaging system and has been clinically trialed on a small group of patients in Calgary, Canada [13,14]. For the Wavelia system, an infrared 3D stereoscopic camera has been integrated, after appropriate selection and configuration, to perform optimally in the interior of the Wavelia examination table and for the case of the human skin texture. To the authors' knowledge, this is the first time that such an optical system is incorporated in the setting of clinical breast imaging.

The relevant state-of-the-art methods from which the Wavelia OBCD design could have been inspired are the existing 3D Surface Imaging methods, mainly designed for post-surgical and/or aesthetic breast reconstruction. An overview of these methods, the set of commercial devices in which they are used, as well the achievable accuracy of the volume computation can be found in [15–18].

Due to their application to aesthetic breast reconstruction, most of those systems and methods are designed to image the patient's torso, while at regular standing or sitting position. To the authors' knowledge, other than the MRI and CT scanners, there are no well-known systems and methods for surface reconstruction and volume computation of the pendulous breast (patient lying at prone position).

Other particularities of the Wavelia OBCD system, which justify the need for a new, system-tailored development, are:

- the requirement for integration of the OBCD system in a specific examination table, the layout of which is driven by the architecture of the MBI subsystem; this is the main reason why system-specific algorithms for denoising and extraction of the useful portion of the recorded 3D point clouds are required to be designed;
- the need to operate in an obscure, closed environment, which justifies the specific choice of a stereoscopic 3D camera to scan the breast. The procedure that led to the selection of the specific 3D camera is presented in Appendix A.

This paper will focus on the presentation of the Wavelia OBCD subsystem. An overview of the Wavelia system architecture is presented in Section 2.1. Given that the full 3D reconstruction of the breast is performed using multiple camera measurements taken along an azimuth scan of the breast, the design of an accurate calibration algorithm, resulting in a high-quality alignment of the multiple point clouds on a unique reference coordinate system is critically important for the overall performance of the imaging system. The new calibration algorithm which has been designed and integrated in the Wavelia breast imaging prototype device, in view of its first-in-human clinical test, is presented in the Section 2.2 of this paper. The new method developed for 3D breast surface reconstruction from 3D point clouds, as well as the breast volume computation method, are presented in Section 2.3. The method used to validate the Wavelia OBCD subsystem in terms of achievable accuracy of breast surface reconstruction using static breast phantoms is presented in Section 2.4. Indicative performance results of the calibration algorithm, breast surface reconstruction results from both phantom and patient scans, as well as some information concerning the envisioned evolutions of the Wavelia OBCD subsystem towards a real-time support tool for the Wavelia MBI examination and breast cancer diagnosis, are presented in Section 3 of the paper. The main conclusions that can be drawn from this study can be found in Section 4. Some supplementary technical material has been included in Appendix A.

## 2. Materials and Methods

### 2.1. Wavelia Breast Examination: OBCD Subsystem Architecture

As already mentioned in the introduction, the MBI and OBCD subsystems of the Wavelia investigational medical device are housed in two separate examination tables. This is a temporary technical limitation of the Wavelia prototype #1 system, which is planned to be resolved in the future with integration of the two subsystems in the same examination table. Such an integration is not trivial though, mainly due to two well-identified technical challenges:

- The circular array of microwave sensors moves along an opaque cylindrical container filled with a creamy opaque transition liquid. These opaque materials have dielectric properties appropriately selected to favor the penetration of the electromagnetic waves in the breast, while being biocompatible and acceptable in the clinical setting. Their replacement by semi-transparent counterparts with the desired properties is feasible, but not evident or straightforward;

- An appropriate motorization system needs to be designed, such that the MBI and OBCD scans can be performed sequentially, while the sensors (optical and microwave) are not in close vicinity to each other during the respective scans.

Given the temporary technical constraint of housing the MBI and OBCD subsystems in two separate examination tables and in order for the 3D data generated by the OBCD subsystem to be useful a priori information for the MBI subsystem, the two examination tables have been manufactured to be identical. In addition, they both use the same 3D reference system (Wavelia reference system), located on each examination table plane and centered on the circular opening, as seen in Figure 1.

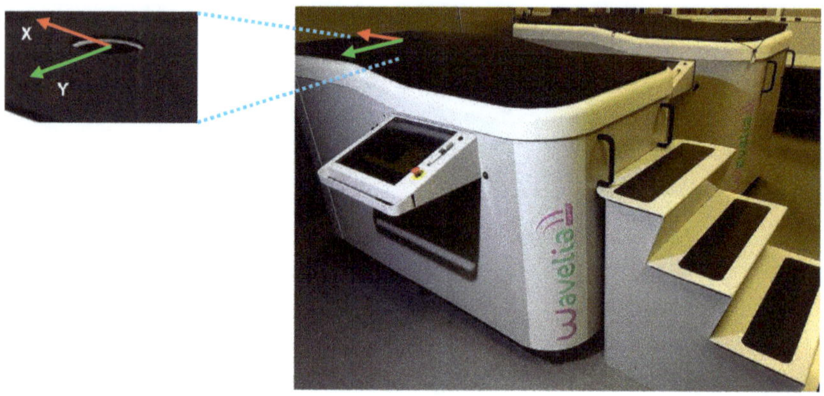


**Figure 1.** The Wavelia prototype medical device: the OBCD subsystem (on the Front) and the MBI subsystem (on the Rear). The Wavelia reference coordinate system is also shown (red and green axes).

The examination procedure is divided into two steps. In the first step the patient is lying in prone position on the examination table of the OBCD subsystem. The dedicated circular opening on the examination table permits a 3D optical scan of the breast and then a priori information for the MBI subsystem is computed. The opening is fitted with a plastic ring to protect the patient's breast.

In the second step the patient is moved to the MBI subsystem and asked to lie in prone position on the examination table. The breast is introduced in the dedicated circular opening and immersed in a tube containing a specific liquid, which will serve as coupling (transition) medium between the imaging system and the breast. A horizontal circular probe array located outside the tube, slides vertically along the tube to reach a set of predefined vertical scan positions.

It is important that the patient is lying in the same prone position during both examinations. The objective is to facilitate the registration between the 3D data generated by the OBCD subsystem and the 3D data generated by the MBI subsystem.

For the examination on the OBCD subsystem, the patient is lying on the examination table, with her breast under examination inserted in the circular opening of the examination table. For this examination, the breast is hanging below the examination table; no coupling liquid is used. An infrared 3D stereoscopic camera is placed below the examination table, at a distance of several tens of centimeters below the breast. A motorization system enables azimuthal motion of the camera in one single horizontal plane with an angular step of 30° (see Figure 2). The azimuthal scan of the 3D camera permits to acquire twelve 3D point clouds and then to reconstruct the external envelope of the breast with sub-millimetric precision. The working environment inside the OBCD subsystem where the 3D breast scan is performed is small, closed and with no light. The small closed environment is essentially a multi-reflection environment, for which stereoscopic cameras are meant to be efficient. In addition, the obscurity of the environment justifies the selection of an infrared camera. The procedure that has been followed to select a specific 3D infrared stereoscopic camera, among other candidates, for the Wavealia OBCD subsystem is reported in Appendix A.

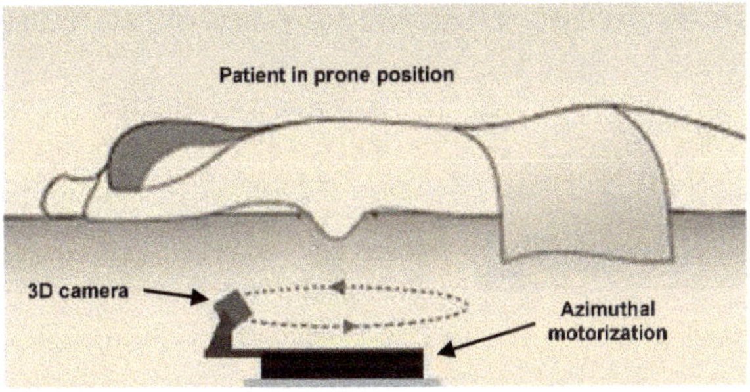


**Figure 2.** Internal schematic view of the OBCD subsystem.

### 2.2. Optical Breast Contour Detection (OBCD) Subsystem: The Calibration Procedure

The OBCD subsystem can be associated to a turntable-based 3D scanning system [19], however, the only difference in our case is that the camera is the moving element and not the patient's breast. After the 3D data acquisitions are performed by the OBCD subsystem, several point clouds are obtained; these point clouds need to be registered into the Wavelia reference coordinate system and ultimately merged. In the case of turntable-based 3D scanning systems, two main approaches can be used to perform the registration operation. The first one consists in using iterative methods like Iterative Closest Point (ICP) to perform this registration on the acquired point clouds [20]. The second approach consists in performing a preliminary calibration step of the turntable, by using a known reference object [21]. The aim of the calibration operation is to compute the transformation matrix, in order to pass from the camera coordinate system to the turntable reference coordinate system, for each angular position of the turntable. Due to the closed and obscure environment of the OBCD subsystem and the expected significant variability of shapes and sizes of the breasts to be scanned, we decided to use the second approach. In this context, we developed a new calibration procedure with no iterative process, based on the use of a cross-shaped calibration tool. The cross-shaped object was chosen because it is a simple shape that can be easily defined as the intersection of two linear segments perpendicular to each other, thus it is easy to associate a 3D reference system to this shape.

The first objective of this procedure is to compute the $T_c$ transformation which allows the passage of the point cloud acquired at 30° (azimuthal angular acquisition step) from the camera reference system at 30° to the camera reference system at 0°. $T_c$ is further used to compute the cumulative transformations to transform the point clouds acquired at the other azimuthal angular positions (60° to 330°) to the camera reference system at 0°, as follows:

$$\begin{gathered}
T_{c60} = T_c \times T_c \\
T_{c90} = T_c \times T_{c60} \\
T_{c120} = T_c \times T_{c90} \\
T_{c150} = T_c \times T_{c120} \\
T_{c180} = T_c \times T_{c150} \\
T_{c210} = T_c \times T_{c180} \\
T_{c240} = T_c \times T_{c210} \\
T_{c270} = T_c \times T_{c240} \\
T_{c300} = T_c \times T_{c270} \\
T_{c330} = T_c \times T_{c300}
\end{gathered}$$

The second objective is to compute the Wavelia reference system, as defined in Figure 1.

The calibration procedure is divided into two steps: at the first step, a 3D scan of the lower plane of the examination table is performed; at the second step a 3D scan of a cross-shaped object is performed. The point clouds acquired during both steps are used to compute $T_c$ and the Wavelia reference coordinate system, as defined in Sections 2.2.1 and 2.2.2.

### 2.2.1. The Calibration Procedure: Step A

At this step, the circular opening on the examination table is covered with a lid. Then, four 3D acquisitions are performed at 0°, 90°, 180° and 270°. A bounding box referenced to the camera reference system is used to define the volume of interest to extract the points corresponding to the lower plane of the examination table from each acquired point cloud (see points inside the bounding box in Figure 3). The dimensions and positions of the bounding box are defined manually by using the software module that was developed to perform the full calibration procedure. The four extracted point clouds are further named $Pc_{A0}$, $Pc_{A90}$, $Pc_{A180}$ and $Pc_{A270}$.

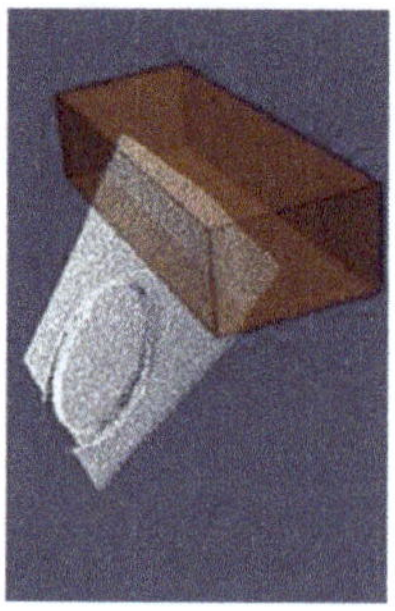

**Figure 3.** Selection of points corresponding to the lower plane of the examination table.

### 2.2.2. The Calibration Procedure: Step B

At this step a cross-shaped reference object is used. Its dimensions are slightly longer than 100 mm vertically and horizontally (see Figure 4a). The cross is fixed to a square bar. The inclination angle of the cross is the same as the inclination of the camera. The cross is centered and fixed to a cylinder that will fit into the circular opening of the examination table (see Figure 4b), making it possible to move the bar vertically or to make it rotate. The aim is to position the cross optimally, such that it can be seen by the camera at 0°, 30° and 330° (see Figure 4c).

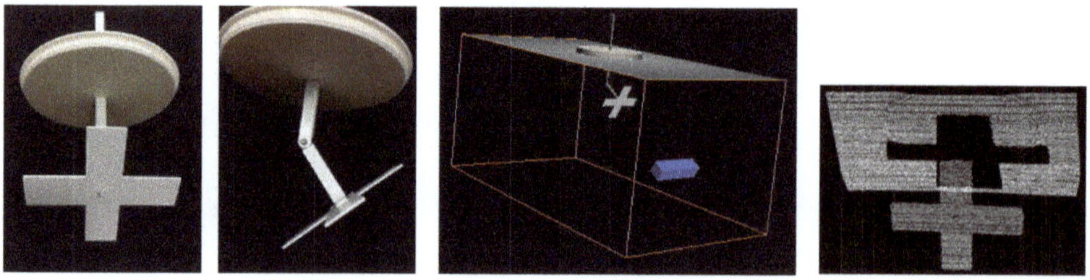

(**a**) (**b**) (**c**)

**Figure 4.** (**a**) The cross-shaped reference object; (**b**) Schematic visualization of the cross-shaped reference object, as installed in the Wavelia examination table, for camera calibration; (**c**) The point cloud corresponding to the cross-shaped reference object as seen by the camera at 0°.

After the cross-shaped reference object is mounted, four 3D acquisitions are performed at 0°, 30°, 180° and 330°. Then, the points corresponding to the cross-shaped object at 0°, 30° and 330° and the points corresponding to the vertical square bar at 180° are extracted. To ensure that the extracted points at 0°, 30° and 330° correspond to the cross-shaped object, the best fitting plane to each extracted point cloud is computed, by using a Least-Squares (LS) method and the points in the close vicinity to each plane are selected. The final extracted point clouds are further named $Pc_{B0}$, $Pc_{B30}$, $Pc_{B180}$ and $Pc_{B330}$. Figure 5 shows these point clouds in relation to the xy plane of the camera reference system and their optimal orientations in relation to the x and y camera axes. These optimal orientations are the key of our method to compute the axes of each cross-shaped object point cloud, as explained below.

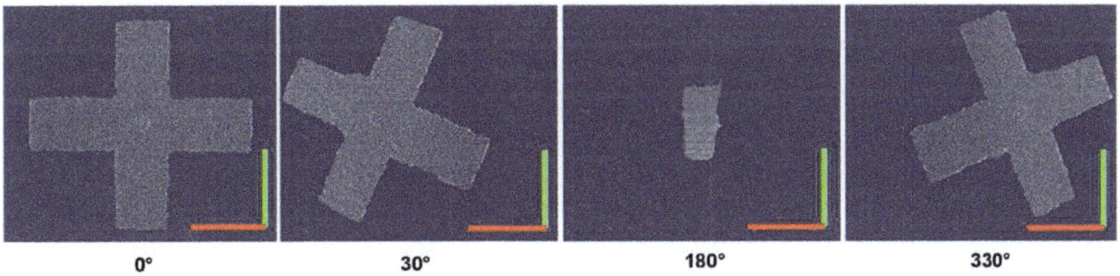


**Figure 5.** Extracted point clouds at the four angular positions of the camera, during the calibration step B. The red axis corresponds to the x camera axis and the green axis corresponds to the y camera axis.

To compute the transformation $T_c$, the 3D reference systems associated with the cross-shaped object at 0°, 30° and 330° need to be determined. To do this, each of the point clouds $Pc_{B0}$, $Pc_{B30}$, and $Pc_{B330}$ is fitted to a plane (cross-shaped object plane) by using a LS method. The unitary cross-shaped plane vector at each angular position is used as z unitary vector and is named $\hat{v}_{z0}$ (0°), $\hat{v}_{z30}$ (30°) and $\hat{v}_{z330}$ (330°). Then, four point-cloud sections corresponding to the vertical and horizontal axes are extracted from $Pc_{B0}$, $Pc_{B30}$, and $Pc_{B330}$, according to the following conditions:

- upper section extraction conditions:

$$(pm_x - size_x \cdot UpperXMinFactor) < px_i < (pm_x + size_x \cdot UpperXMaxFactor)$$

$$\left(pm_y + size_y \cdot UpperLowerYMinFactor\right) < py_i < \left(pm_y + size_y \cdot UpperLowerYMaxFactor\right)$$

- lower section extraction conditions:

$$(pm_x - size_x \cdot LowerXMinFactor) < px_i < (pm_x + size_x \cdot LowerXMaxFactor)$$

$$\left(pm_y - size_y \cdot UpperLowerYMaxFactor\right) < py_i < \left(pm_y - size_y \cdot UpperLowerYMinFactor\right)$$

- left section extraction conditions:

$$(pm_x + size_x \cdot LeftRightXMinFactor) < px_i < (pm_x + size_x \cdot LeftRightXMaxFactor)$$

$$\left(pm_y - size_y \cdot LeftYMinFactor\right) < py_i < \left(pm_y + size_y \cdot LeftYMaxFactor\right)$$

- right section extraction conditions:

$$(pm_x - size_x \cdot LeftRightXMaxFactor) < px_i < (pm_x - size_x \cdot LeftRightXMinFactor)$$

$$\left(pm_y - size_y \cdot RightYMinFactor\right) < py_i < \left(pm_y + size_y \cdot RightYMaxFactor\right)$$

where $pm_x$ and $pm_y$ correspond to the x and y coordinates of the centroid of the point cloud, $px_i$ and $py_i$ correspond to the x and y coordinates of each point on the point cloud and $size_x$, $size_y$ correspond to the size of the bounding box of the point cloud along the x and y camera axes respectively. The factor values (e.g., *UpperLowerYMaxFactor*) are positives and lower than 0.5, they have been defined empirically for each angular position of the camera, considering several 3D scans of the cross-shaped object at 0°, 30° and 330° degrees. Figure 6 shows the four extracted sections, at each angular position.

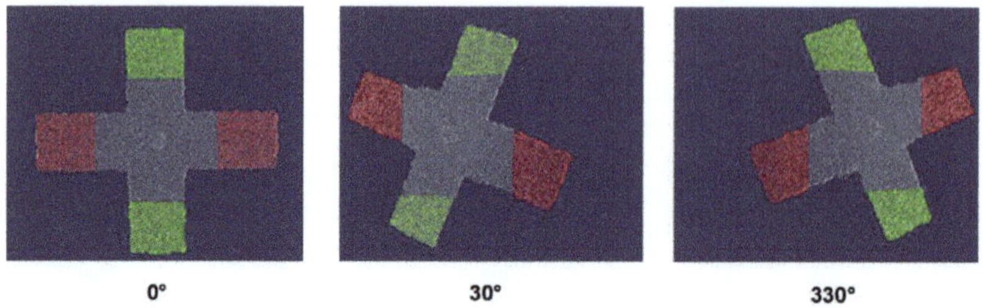


**Figure 6.** The cross-shaped object point clouds at three angular positions of the camera: the upper and lower sections are highlighted with green color, the left and right sections are highlighted with red color.

The extracted point cloud sections are used to compute the x and y axes for each angular position of the camera, as depicted in Figure 7, by using a LS method. The unitary orientation vectors, $\hat{\vartheta}_{x0}$ and $\hat{\vartheta}_{y0}$ for 0°, $\hat{\vartheta}_{x30}$ and $\hat{\vartheta}_{y30}$ for 30° and $\hat{\vartheta}_{x330}$ and $\hat{\vartheta}_{y330}$ for 330°, are used as x and y unitary vectors at each camera position. The closest point between the x and y axes at each camera position (see Figure 7) is further determined. This point is projected on the cross-shaped object plane. The resulting points on the cross-shaped object planes, at each angular position, are named $p_0$, $p_{30}$ and $p_{330}$.

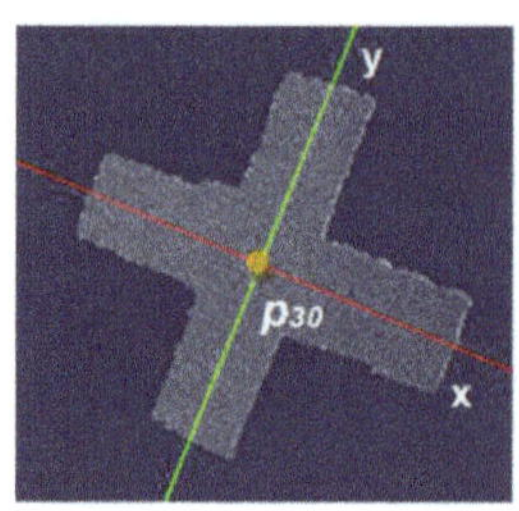


**Figure 7.** The x and y axes of the cross-shaped object point cloud at 30° and the closest point between these axes, highlighted with yellow color.

The 3D reference systems $Ref_0$, $Ref_{30}$ and $Ref_{330}$ for each camera position respectively and the transforms $T_0$, $T_{30}$ and $T_{330}$ are computed as defined in Equations (1)–(6).

$$\hat{e}_{z0} = \hat{\vartheta}_{z0}\hat{e}_{y0} = \hat{e}_{z0} \times \hat{\vartheta}_{x0}\hat{e}_{x0} = \hat{e}_{y0} \times \hat{e}_{z0} \tag{1}$$

$$Ref_0 = \left(\hat{e}_{x0},\ \hat{e}_{y0},\ \hat{e}_{z0}, p_0\right)$$

$$\hat{e}_{z30} = \hat{\vartheta}_{z30}\hat{e}_{y30} = \hat{e}_{z30} \times \hat{\vartheta}_{x30}\hat{e}_{x30} = \hat{e}_{y30} \times \hat{e}_{z30} \tag{2}$$

$$Ref_{30} = \left(\hat{e}_{x30},\ \hat{e}_{y30},\ \hat{e}_{z30}, p_{30}\right)$$

$$\hat{e}_{z330} = \hat{\vartheta}_{z330}\hat{e}_{y330} = \hat{e}_{z330} \times \hat{\vartheta}_{x330}\hat{e}_{x330} = \hat{e}_{y330} \times \hat{e}_{z330} \tag{3}$$

$$Ref_{330} = \left(\hat{e}_{x330},\ \hat{e}_{y330},\ \hat{e}_{z330}, p_{330}\right)$$

$$T_0 = \begin{bmatrix} \hat{e}_{x0} & \hat{e}_{y0} & \hat{e}_{z0} & p_0 \\ 0 & 0 & 0 & 1 \end{bmatrix} \tag{4}$$

$$T_{30} = \begin{bmatrix} \hat{e}_{x30} & \hat{e}_{y30} & \hat{e}_{z30} & p_{30} \\ 0 & 0 & 0 & 1 \end{bmatrix} \tag{5}$$

$$T_{330} = \begin{bmatrix} \hat{e}_{x330} & \hat{e}_{y330} & \hat{e}_{z330} & p_{330} \\ 0 & 0 & 0 & 1 \end{bmatrix} \tag{6}$$

The transforms $T_{30to0}$, $T_{0to330}$ and $T_{mean}$ are computed as defined in Equations (7)–(9).

$$T_{30to0} = T_0 \times T_{30}^{-1} \tag{7}$$

$$T_{0to330} = T_{330} \times T_0^{-1} \tag{8}$$

$$T_{mean} = \frac{T_{30to0} + T_{0to330}}{2} \tag{9}$$

$T_{30to0}$, $T_{0to330}$ and $T_{mean}$ allow to switch the 3D reference coordinate system of the camera from 30° to 0°. $T_c$ is further defined as the most accurate among these three estimated transforms. To evaluate accuracy, a control point $p_{ctrl}$ is used (see green sphere in Figure 8). It is referenced to the 3D coordinate system at the camera angular position 0° (in theory the same coordinates at 360°). Its position has been manually defined at about 130 mm from the azimuth rotation axis and under the lower plane of the examination table, then a cumulative transform to switch the 3D camera reference coordinate system from 360° to 0° is computed as $T_{c360} = T_c \times T_{c330}$. This cumulative transform is applied to $p_{ctrl}$ to obtain $p'_{ctrl}$. The Eucledian distance between $p_{ctrl}$ and $p'_{ctrl}$ is used as a criterion to select the most accurate transform among $T_{30to0}$, $T_{0to330}$ and $T_{mean}$ and to define the calibration transform error.

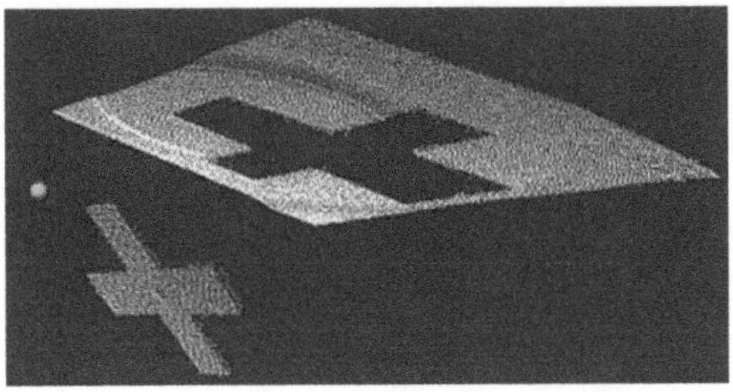

**Figure 8.** Control point in green for the calibration transform error computation.

After the $T_c$ transformation has been determined, the Wavelia reference system can be computed at 0°. To do this, the cumulative transformations $T_{c90}$, $T_{c180}$ and $T_{c270}$ (see Section 2.2) are respectively applied to the point clouds $Pc_{A90}$, $Pc_{A180}$ and $Pc_{A270}$. The result is merged to the point cloud $Pc_{A0}$ (see Figure 9).

**Figure 9.** Cumulative point cloud of the lower plane of the examination table.

The cumulative point cloud is fitted to a plane and the unitary plane vector is used as z unitary vector of the lower plane of the examination table. The unitary plane vector is defined to point upwards, and it is named $\hat{w}_z$. Then, the x and y unitary vectors are computed, according to the following condition: the y unitary vector is defined to point towards the camera.

The x, y and z unitary vectors of the Wavelia reference coordinate system are defined as: $\hat{w}_x$, $\hat{w}_y$ and $\hat{w}_z$, as shown in Equation (10).

$$\begin{aligned}\hat{e}_y &= (0,\ 0, -1)\\ \hat{w}_x &= \frac{\hat{e}_y \times \hat{w}_z}{|\hat{e}_y \times \hat{w}_z|}\\ \hat{w}_y &= \frac{\hat{w}_z \times \hat{w}_x}{|\hat{w}_z \times \hat{w}_x|}\end{aligned} \quad (10)$$

To compute the center of the Wavelia reference coordinate system, $T_{c180}$ is applied to the point cloud $Pc_{B180}$ (see Figure 5). The new point cloud is then fitted to a plane. The unitary plane vector is defined to point towards the camera. It is named $\hat{v}$. The mean point $p_{mean}$ of the new point cloud is also computed. Then, an axis is generated by using $\hat{w}_z$ and $p_{mean}$. We determine the intersection point $p_{itr}$ between this axis and the lower plane of the examination table, as computed previously. Then, the origin $p_w$ of the Wavelia reference coordinate system is computed by using half of the side length of the square bar $L_{\text{bar}}$, as follows: $p_w = p_{itr} + 0.5 \cdot L_{\text{bar}} \cdot \hat{v}$. The Wavelia reference system is defined by $Ref_{wavelia} = \left(\hat{w}_x,\ \hat{w}_y,\ \hat{w}_z,\ p_w\right)$.

### 2.3. Optical Breast Contour Detection (OBCD) Subsystem: The 3D Breast Surface Reconstruction Procedure

When the 3D scan of the patient's breast is finished, the calibration parameters are applied to the acquired point clouds. The resulting cumulative point cloud is named $Pc_{Acc}$ (see Figure 10).

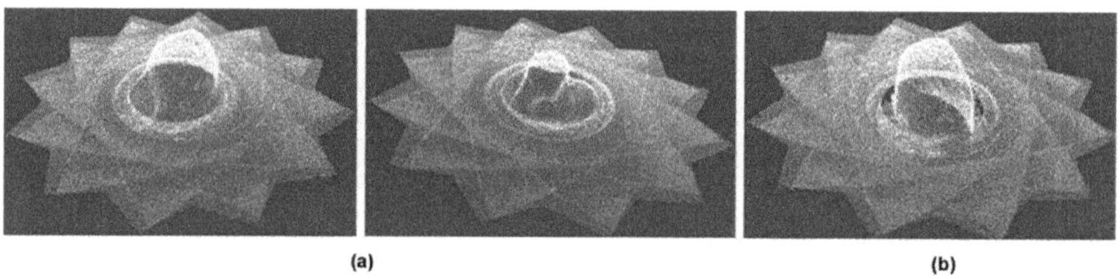


**Figure 10.** Cumulative point clouds corresponding to three 3D breast scans with the Wavelia OBCD subsystem: (**a**) cumulative point clouds corresponding to two human breasts: Patient 002 and Patient 004; (**b**) cumulative point cloud corresponding to a test breast phantom.

Figure 10 shows the point cloud of the breast when it is introduced into the circular opening, the lower plane of the examination table and the lower section of the plastic ring used to protect the patient's breast. To perform the 3D reconstruction of the breast, we need to remove the points corresponding to the ring and to extract a point cloud corresponding to the breast surface. The methods for 3D surface reconstruction from point clouds generally comprise three main processing steps: denoising, smoothing and mesh generation by triangulation, as presented in [19]. In most of the state-of-the-art methods, the input point cloud is preserved during the processing, while in other methods a deformation step is added before the standard processing [22]. Considering the shape of the pendulous breast during examination at prone patient position (see Figure 10), a new method for 3D breast surface reconstruction has been developed. This new method includes a linear stretching deformation step to transform the input point cloud into an elevation point cloud. Starting from an elevation point cloud, it is easier to transform it into a rectangular grid, thus enabling the use of 2D image processing methods, instead of 3D point cloud processing, to perform the denoising and smoothing steps. Moreover, this rectangular grid allows to define easier the topological connection relationships between points, in the meshing step.

As said before, the points related to the ring shall be removed and a point cloud corresponding to the breast surface has to be extracted from $Pc_{Acc}$ to obtain the input point cloud $Pc_B$ for the 3D breast surface reconstruction method. The procedure to generate $Pc_B$ is detailed in Appendix B.

The 3D breast surface reconstruction method for the OBCD subsystem consists in the following processing steps:

- linear stretching procedure;
- denoising procedure;
- smoothing procedure;
- meshing procedure.

Each of the above processing steps as well as the breast volume computation are specified in a separate sub-section below.

2.3.1. Linear Stretching Procedure

The aim of this processing step is to transform the point cloud $Pc_B$, obtained after removal of the ring as explained in the previous sub-section, into an elevation point cloud to ensure the generation of a surface with no overlaps (see illustration in the right-hand images in Figure 11a–c). A surface with no overlaps is defined as a surface where there is one single z coordinate for each coordinates pair (x, y).

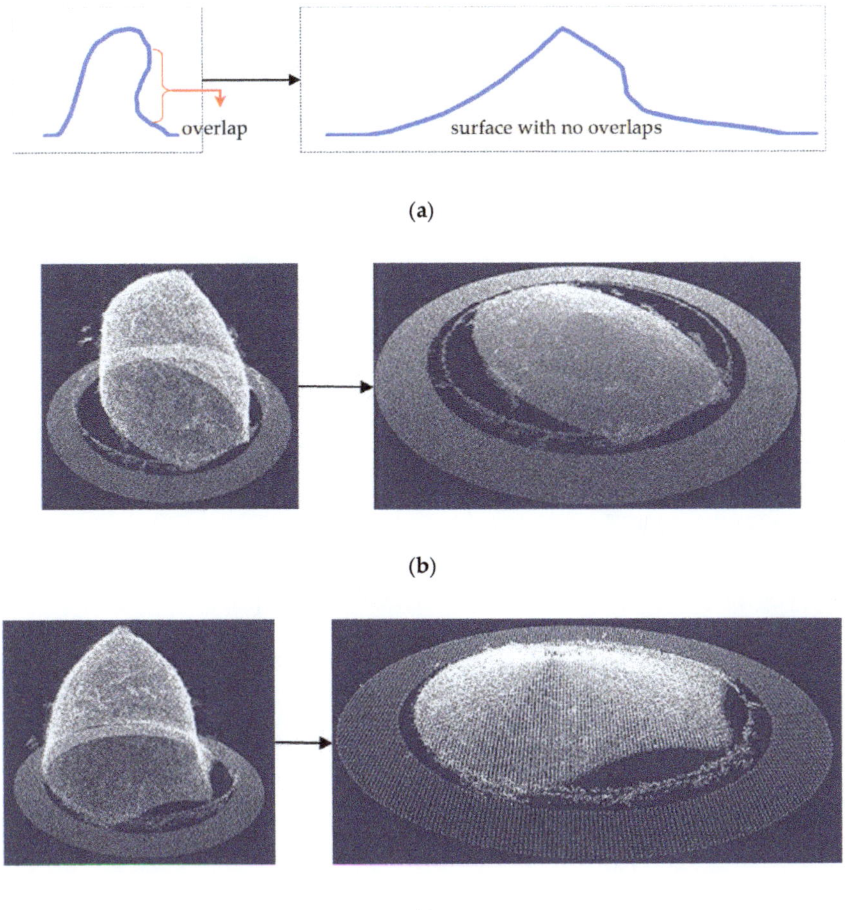


**Figure 11.** Linear stretching transformation: (**a**) example of stretching procedure applied to a surface with overlaps, the result is a surface without overlaps; (**b**) stretching procedure applied to a breast phantom point cloud; (**c**) stretching procedure applied to Patient 002 breast point cloud.

Overlaps, as defined below, are naturally very common in the geometry of the pendulous breast, especially in the case of elder patients. The stretching of the breast surface, such that it has no overlap, is a processing step which enables the use of 2D image processing methods for de-noising of the point cloud, before meshing. Obviously, the stretching operation is compensated before meshing, such that the original breast shape is ultimately reconstructed.

To perform this transformation, a new origin point $p_{str}$ needs to be defined. This point must be located at the bottom of the breast (close to the chest wall) and the z axis starting from it must intersect the non-stretched surface only once, to successfully stretch the x and y coordinates of the point cloud, as illustrated graphically in Figure 12.

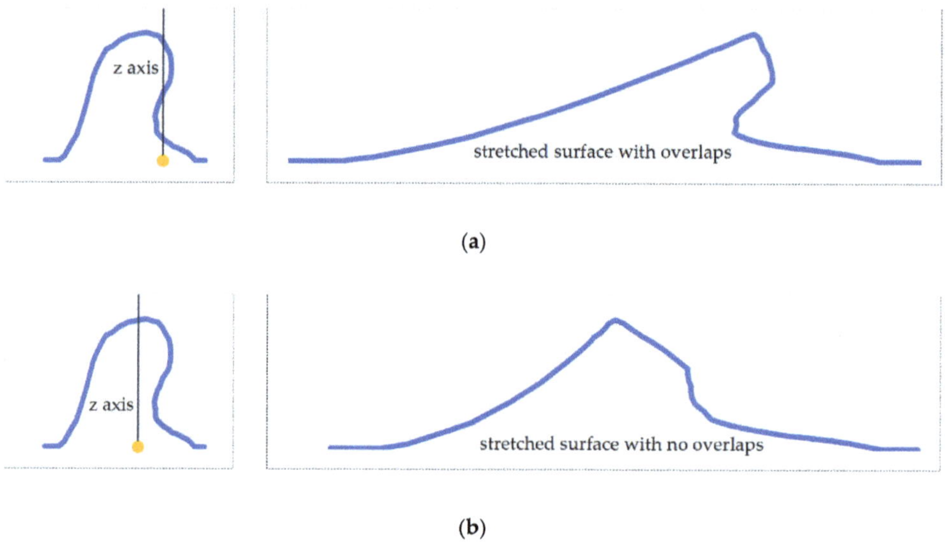


**Figure 12.** (**a**) Example of a bad choice of origin point (illustrated with yellow color): the resulting surface after the stretching process was applied is shown on the right; (**b**) example of a good choice of origin point (illustrated with yellow color): the resulting surface after the stretching process was applied is shown on the right.

In the following, the procedure put in place to define a linear stretching function $Sf(z)$ and compute the stretching factor for the x and y coordinates of each point as a function of the z coordinate (see Figure 13b), is detailed. The function is defined in the Wavelia reference coordinate system (see Figure 13a) and it is specified in Equation (11), where $s_{max}$ is the maximal stretching factor, $z_{min}$ and $z_{max}$ are the minimum and maximum z coordinates of the bounding box of $Pc_B$, and z is the z coordinate of the point to be stretched. $z_{max}$ is equal to zero, as shown in Figure 13b.

$$Sf(z) = \frac{(s_{max} - 1)}{(z_{max} - z_{min})} \cdot z + s_{max} \quad (11)$$

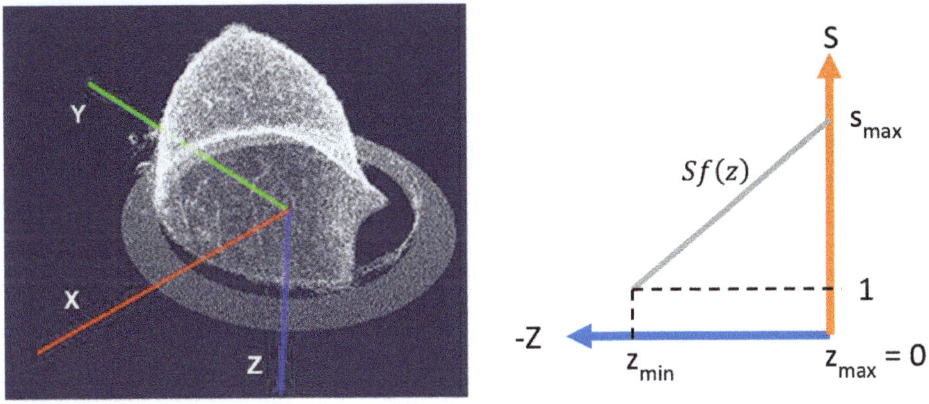


(a) (b)

**Figure 13.** Linear stretching function $Sf(z)$: (**a**) the Wavelia reference coordinate system; (**b**) graphical representation of the $Sf(z)$ function.

The following processing steps are applied to compute $s_{max}$:

1. the number of points of the point cloud is reduced by voxelization (Figure 14a).
2. the points associated to the chest wall are removed from the voxelized point cloud, by removing the points where the z-coordinate is in range from 0 to $z_{chest}$ (white points in Figure 14b); $z_{chest}$ has been empirically determined, considering a series of tests on both human and phantom breasts. The aim of this step is to preserve the points corresponding to the pendulous breast.
3. the point cloud obtained in the previous step is divided in five vertical sections (red, blue, cyan, magenta and yellow in Figure 14b).
4. the orientation axis of the breast is computed, by applying a LS method on the centroids of the five vertical sections (see Figure 14c).
5. the $\theta$ angle between this axis and the z axis of the Wavelia reference coordinate system is computed (see Figure 14d).
6. Finally, $p_{str}$ is determined, by projecting the centroid of the middle section on the orientation axis of the breast (cyan sphere in Figure 14e) and setting to 0 the z coordinate of the projected point. The result corresponds to $p_{str}$ (orange sphere in Figure 14e).

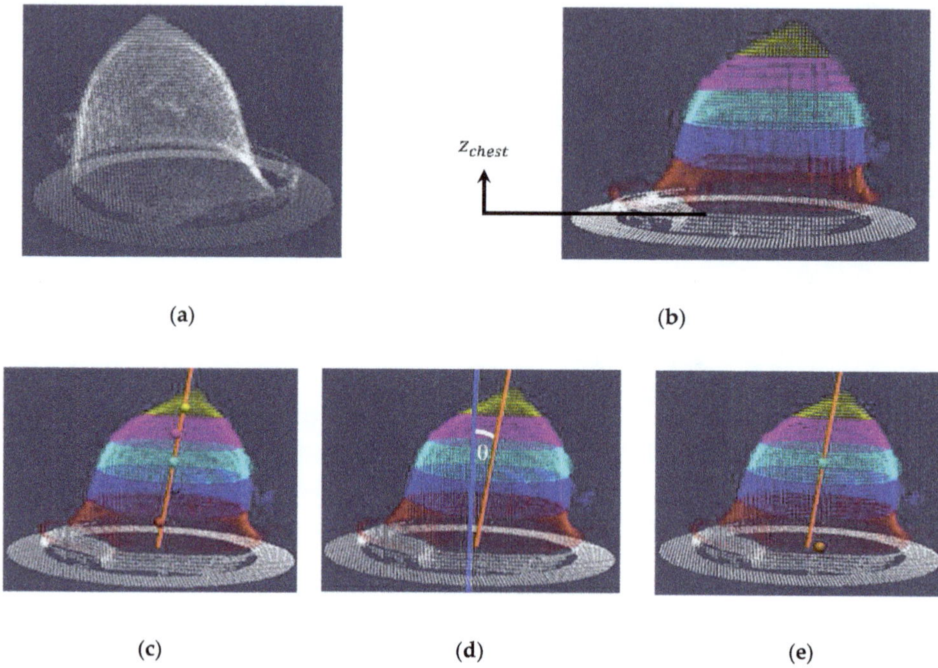


**Figure 14.** (**a**) Voxelized point cloud; (**b**) The five vertical sections of the breast, highlighted with different colors; (**c**) Orientation axis of the breast and centroids (colored spheres) of the pre-defined vertical sections; (**d**) The angle θ between the orientation axis of the breast (orange color) and the z axis of the Wavelia reference coordinate system (blue color); (**e**) New origin point $p_{str}$ (orange sphere).

Based on several tests performed both on human and phantom breasts, the minimum stretching value $s_{ref} = 8.0$ that can be applied to generate non-overlapping point clouds in all performed tests was determined empirically. The maximum value of the angle θ for which the minimum stretching value $s_{ref}$ is applied by default has been empirically defined from the same tests. The notation $\theta_{ref}$ is used in the sequel of the paper to refer to this value. $\theta_{ref}$ is further used to compute the reference value for the tilt of the breast $b\_tilt_{ref}$ as defined in Equation (12).

$$b\_tilt_{ref} = 1 + \tan\theta_{ref} \quad (12)$$

The specific $s_{ref}$ and $b\_tilt_{ref}$ values can be used to compute the maximal stretching factor $s_{max}$ as defined in Figure 15 and Equation (13), where $b\_tilt$ is the tilt of the breast which is computed by using Equation (12) while $\theta_{ref}$ is replaced by θ.

$$s_{max} = \frac{b\_tilt \cdot s_{ref}}{b\_tilt_{ref}} \quad (13)$$

The full procedure to compute $s_{max}$ is presented in Figure 15 above. This procedure ensures that $s_{max}$ is always superior than $s_{ref}$.

Finally, the point cloud $Pc_B$ is centered at the position $p_{str}$ and the stretching function $Sf(z)$ is applied (see Equation (11)). The stretched point cloud (right-hand images in Figure 11b,c) is named $Pc_C$.

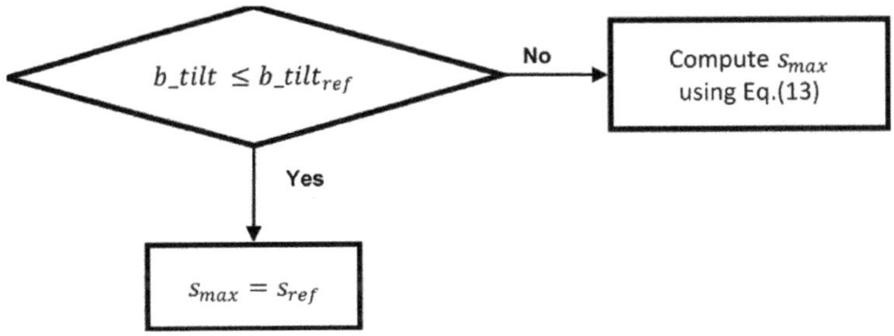


**Figure 15.** Procedure to compute $s_{max}$.

### 2.3.2. Denoising Procedure

Until the previous step, the pendulous breast point cloud has been transformed into an elevation point cloud $Pc_C$ on the xy plane, while preserving the z coordinates. The aim of the denoising procedure is to remove the artifacts and the ring remaining points and to fill the zones with no points. To do this, the denoising procedure combines morphological information of the pendulous breast, and 2D imaging processing.

At the beginning of this procedure $Pc_C$ is used to generate a *reference rectangular grid for denoising* which is then used to perform a first artifacts removal process on $Pc_C$. To perform this process, the following parameters are used:

- $x_{min}$, $x_{max}$, $y_{min}$ and $y_{max}$: these values correspond to the minimum and maximum x and y coordinates of the bounding box of $Pc_C$;
- *mesh resolution (Dxy)*: it corresponds to the desired resolution on the xy plane for the 3D breast reconstructed mesh, before the linear stretching transformation (see definition also in Appendix B.2);
- $s_{max}$: this is the maximal stretching factor, as computed in the previous step (see Section 2.3.1);
- $x_{size}$ and $y_{size}$: these are integer values. They correspond to the size of the rectangular grid along the x and y axes after the linear stretching transformation. These values are computed by using Equations (14) and (15), where $Dxy{\cdot}s_{max}$ corresponds to the stretched mesh resolution, after the stretching transformation. To preserve small parts like the nipple, after the stretching transformation, it was decided to increase the stretched mesh resolution by a factor of 2, as indicated in Equations (14) and (15).

$$x_{size} = \left\lceil \frac{x_{max} - x_{min}}{Dxy{\cdot}s_{max}{\cdot}0.5} \right\rceil \quad (14)$$

$$y_{size} = \left\lceil \frac{y_{max} - y_{min}}{Dxy{\cdot}s_{max}{\cdot}0.5} \right\rceil \quad (15)$$

By using $x_{size}$ and $y_{size}$ a rectangular grid on the xy plane is generated (see Figure 16b). Each of the z values in the grid (green points in Figure 16a) is computed as the mean value of all the z coordinates of the points located inside the associated small rectangular section (small blue rectangles in Figure 16a). If there are no points inside the small section, the associated z value is set to 0. This ensures a single z coordinate for each coordinates pair (x, y) on the grid.

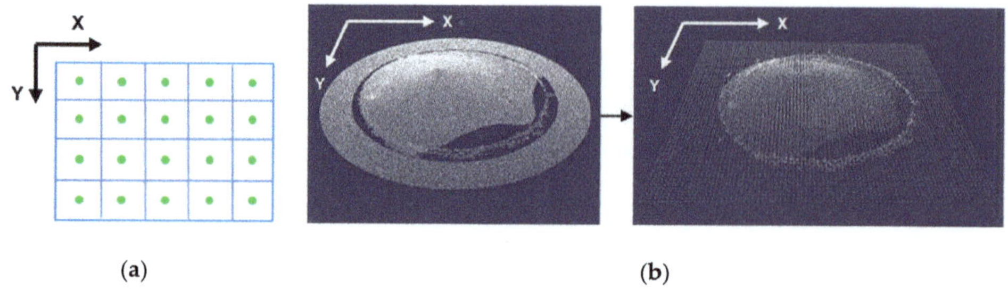


**Figure 16.** Procedure to generate a rectangular grid from a point cloud: (**a**) the rectangular grid configuration; (**b**) example of rectangular grid generation from the stretched point cloud corresponding to Patient 002.

The obtained rectangular grid can be considered as a 2D image, where the intensity at each point (x,y) corresponds to the z-value (in mm) of the point (x,y,z) in a new stretched point cloud (see Figure 17a). To reduce the noise, remove the larger artifacts and obtain a very smooth grid without completely removing the nipple, a median filter is applied with a large kernel size (11 × 11) to the rectangular grid. The result is the *reference rectangular grid for denoising* (see Figure 17b). The kernel size (11 × 11) has been selected among various kernel sizes, after testing on a series of both human and phantom breasts (test range from 9 × 9 to 17 × 17).

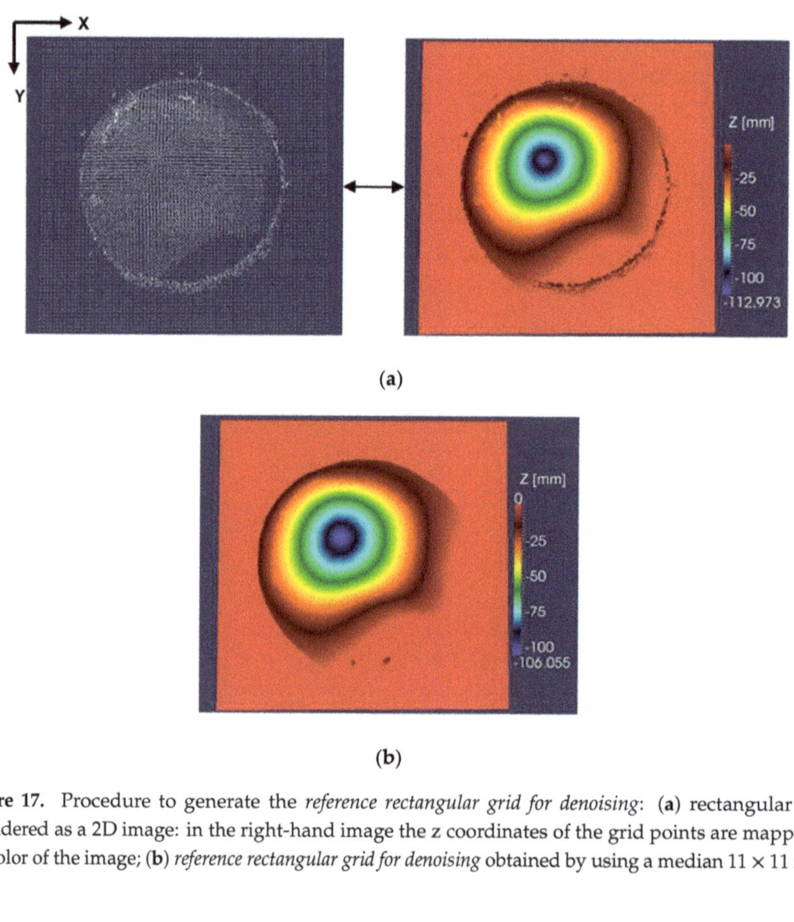


**Figure 17.** Procedure to generate the *reference rectangular grid for denoising*: (**a**) rectangular grid considered as a 2D image: in the right-hand image the z coordinates of the grid points are mapped to the color of the image; (**b**) *reference rectangular grid for denoising* obtained by using a median 11 × 11 filter.

The signed vertical distance (along the z axis) to the *reference rectangular grid for denoising* is used as a criterion for the removal of points (positive distance for points above the grid and negative distance for points below the grid). The denoising is performed after splitting the breast into two vertical sections. These are defined in $Pc_C$ by using the *z limit* of the nipple section (see Figure 18).

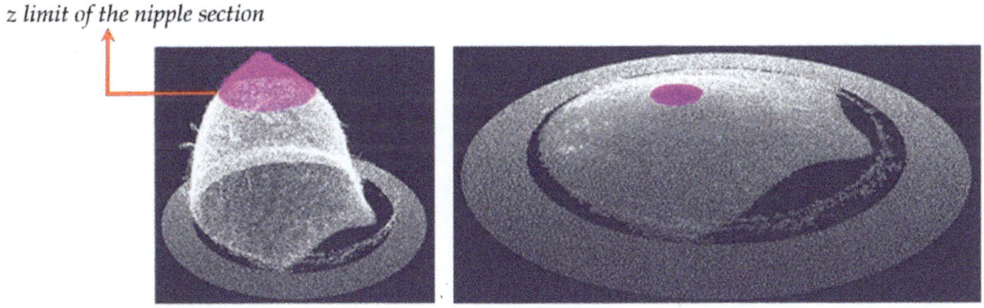


(**a**)

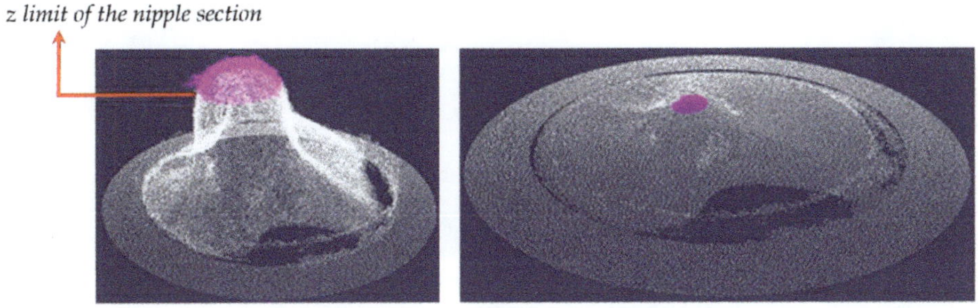


(**b**)

**Figure 18.** Nipple section (magenta points) in the not stretched point cloud (left-hand images) and in the stretched point cloud $Pc_C$ (right-hand images): (**a**) point cloud corresponding to Patient 002; (**b**) point cloud corresponding to Patient 004.

To apply the point removing criterion, the following parameters are defined:

- $z_{min}$ and $z_{max}$: these values correspond to the minimum and maximum z coordinates of the bounding box of $Pc_C$.
- $z_{nipple}$: this parameter concerns the z limit of the nipple section (see Figure 18). Its value is computed by using Equation (16), where $Dz_{nipple}$ is an empirically defined value within the range 0 to 1 based on tests performed on human and phantom breasts.

$$z_{nipple} = z_{min} + (z_{max} - z_{min}) \cdot Dz_{nipple} \tag{16}$$

- *valid distance:* this parameter is applied to the points outside the nipple section. It corresponds to the maximal signed vertical distance to the reference rectangular grid for the denoising operation to preserve points corresponding to the breast surface.
- *valid distance for nipple section:* this parameter is applied to the points inside the nipple section, where the nipple points are above the breast surface. It corresponds to the maximal signed vertical distance to the reference rectangular grid for the denoising operation to preserve points corresponding to the nipple and areola vicinity on the breast surface.

The removing criterion is applied to $Pc_C$ by using $z_{nipple}$, *valid distance for nipple section* and *valid distance*. The resulting point cloud, after denoising, is named $Pc_D$. Figure 19a,b illustrates the denoising process applied to the data from two patient scans. In Figure 19, two sample zones (yellow delimited zones) are highlighted before the denoising procedure ($Pc_C$) and after the denoising procedure ($Pc_D$). Observing the impact of the denoising procedure on these zones permits to clearly illustrate the efficacy of the denoising procedure to significantly reduce the noise.

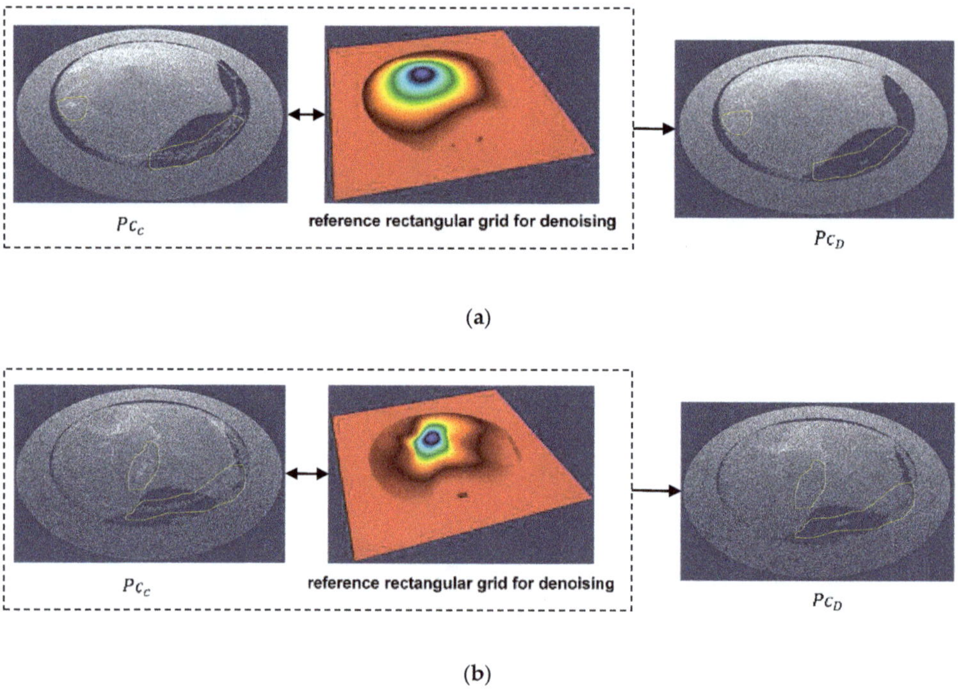


**Figure 19.** Denoising step by using the *reference rectangular grid for denoising*. Two yellow delimited zones are shown in $Pc_C$ (before denoising) and $Pc_D$ (after denoising): (**a**) point cloud corresponding to Patient 002; (**b**) point cloud corresponding to Patient 004.

When the first artifacts removal process is finalized, $Pc_D$ is transformed into a full filled rectangular grid $Pcg_A$. This point could $Pcg_A$ can be considered as a 2D image which allows using 2D imaging processing.

To generate the full filled rectangular grid $Pcg_A$ from $Pc_D$, the size of the rectangular grid is defined in the x and y axes by using Equations (14) and (15). Then, a rectangular grid is generated on the xy plane. To improve the robustness against the noise along the z axis, each of the z values of the grid (green points in Figure 20a) is computed as the median value of all the z coordinates of the points located inside the associated small rectangular section (small blue rectangles in Figure 20a). Some points of the rectangular grid will remain with no defined z value. These points are named "empty points", and the sections of empty points are named "holes" (see red points in Figure 20b). The points with a defined z value are named "valid points" (see green points in Figure 20b).

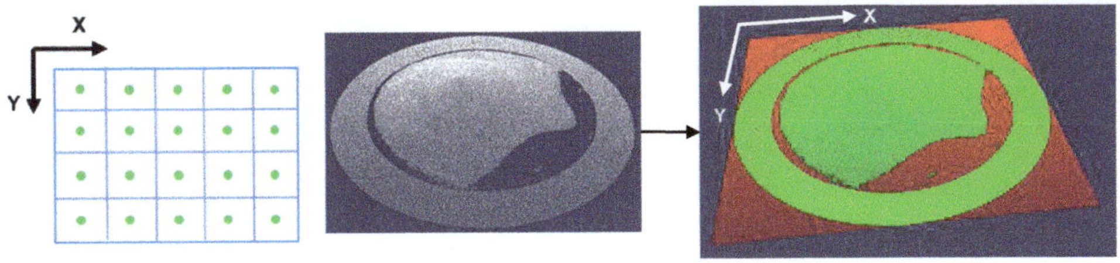


(a) (b)

**Figure 20.** Procedure to generate a rectangular grid from a point cloud by preserving the holes: (**a**) the rectangular grid configuration; (**b**) example of rectangular grid generation by preserving the holes (red point sections) from the stretched point cloud corresponding to Patient 002.

To define the z coordinate of the empty points in the rectangular grid (the process to fill the holes), the parameter *ring height* defined in Appendix B.2 is required. By using this parameter, we divide the valid points of the grid (green points in Figure 21) in two vertical sections: the first one concerns the points the z coordinates of which are in range of 0 to *ring height* (bright green points in Figure 21), and the second one concerns the points the z coordinates of which are outside the first interval (dark green points in Figure 21). The left-hand image in Figure 21 shows the two sections on the equivalent non-stretched point cloud for a better understanding.

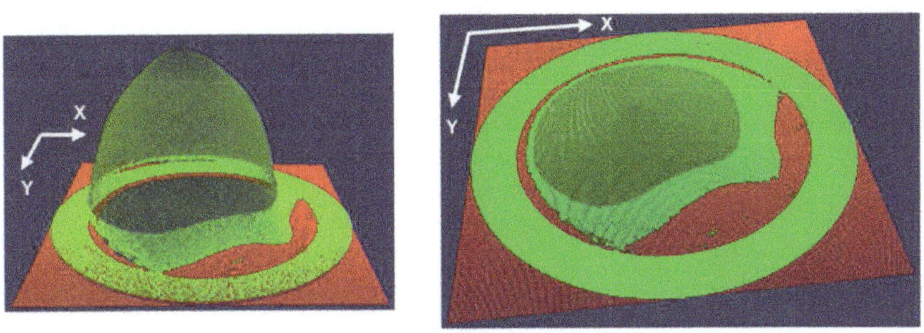


**Figure 21.** Valid points (green points) of the rectangular grid divided into two sections (bright green and dark green) by using the *ring height* parameter, in the non-stretched point cloud (left-hand image) and in the stretched point cloud $Pc_D$ (right-hand images). Illustration on Patient 002.

Three types of holes are further defined:

- Type-A hole: the holes bordered by valid points the z coordinates of which are located inside the first interval (bright green points in Figure 22). This type of hole is composed of at least two consecutive empty points along the x axis or along the y axis (see yellow delimited zones in Figure 22). These holes correspond to the zones located between the internal vertical wall of the ring and the border of the breast surface. The z coordinate value of these empty points is set to 0.
- Type-B hole: all the holes not completely bordered by valid points (see Figure 23). These holes correspond to the zones located beyond the ring, where no section of the breast surface is present. The z coordinate value of these empty points is set to 0.
- Type-C hole: the holes bordered by valid points the z coordinates of which are outside the first interval (holes located inside the dark green section).

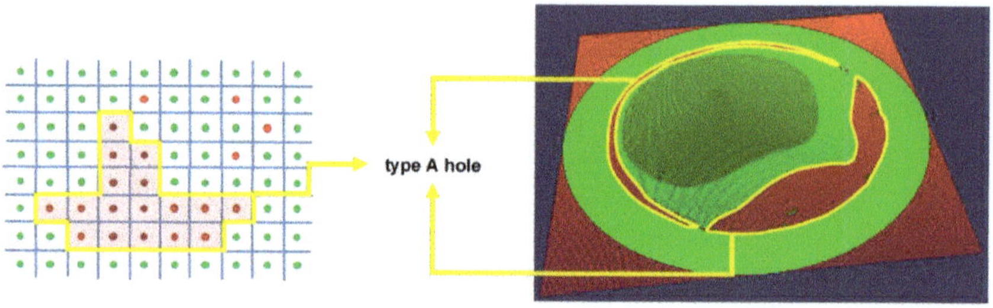


**Figure 22.** Type-A hole. Illustration on Patient002.

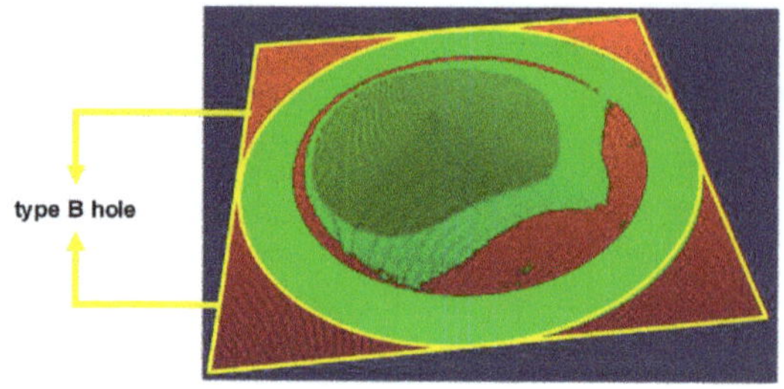


**Figure 23.** Type-B hole. Illustration on Patient 002.

To fill the type-C holes and the other remaining empty points on the rectangular grid, a bi-linear interpolation is performed by using the closest points for which the z coordinate is defined. Finally, a smoothing procedure is performed by using two b-spline approximations [23]. The first approximation is carried out along the x axis, and the second one is carried out along the y axis. The result is the mean of the two approximations. The resulting point cloud is named $Pcg_A$ and it can be considered as a full filled rectangular grid.

At this stage it is possible to use 2D image processing to finalize the denoising procedure. Two 2D processing are performed. The first one is based on erosion and dilation morphological operations by using $Pcg_A$ as input. The result is a binary mask, which is applied to $Pcg_A$ by a multiplication operation. Then the second one consists in the use of a median filter of several kernel sizes to remove the last remaining artefacts. The final resulting point cloud is named $Pcg_C$. (see Appendix C).

2.3.3. Smoothing Step

The aim of this step is to smooth $Pcg_C$ while preserving the breast surface details around the nipple. To do this, $Pcg_C$ is considered as a 2D image. $Pcg_C$ is divided into four vertical sections by using the procedure presented in Appendix C.2, and then a bi-lateral filter [24] is applied with variable kernel size, as defined below:

- a (3 × 3) kernel size for the nipple section (this kernel size allows to smooth this section while sufficiently preserving the breast surface details around the nipple);
- a (5 × 5) kernel size for the second breast section;
- a (7 × 7) kernel size for the third breast section;
- a (9 × 9) kernel size for the fourth breast section.

This, empirically defined, combination of kernel sizes allowed to preserve a good continuity along the four breast sections in all the breast scans we have performed so far. The result is a point cloud we name $Pcg_D$.

### 2.3.4. Meshing Procedure

$Pcg_D$ is a rectangular grid that can be easily presented as a 2D matrix of 3D points. A triangular mesh is further generated, where the points of each triangle are connected in clockwise order, as illustrated in Figure 24.

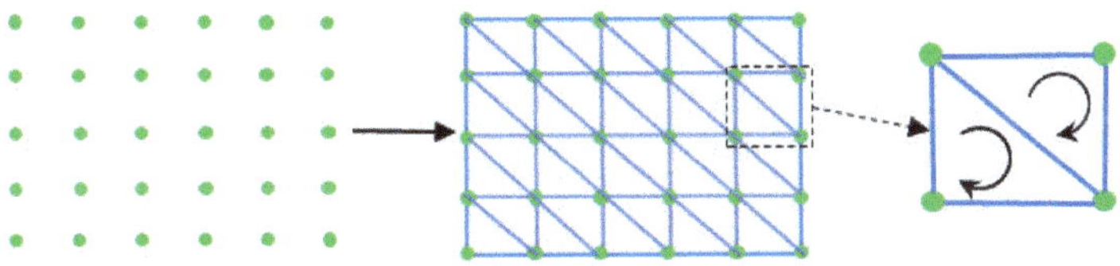

**Figure 24.** Generation of triangular mesh from a rectangular grid.

To perform the final meshing, we compute the inverse of the stretching function earlier presented in Section 2.3.1, as defined in Equation (17). Then, we apply Equation (17) to the points of the grid, and finally the point $p_{str}$ is added, such that the original position of the breast is retrieved.

$$Sf(z)^{-1} = \frac{1}{Sf(z)} = \frac{1}{\frac{(s_{max}-1)}{(z_{max}-z_{min})} \cdot z + s_{max}} \tag{17}$$

Indicative results of the meshing procedure are presented in Figure 25.

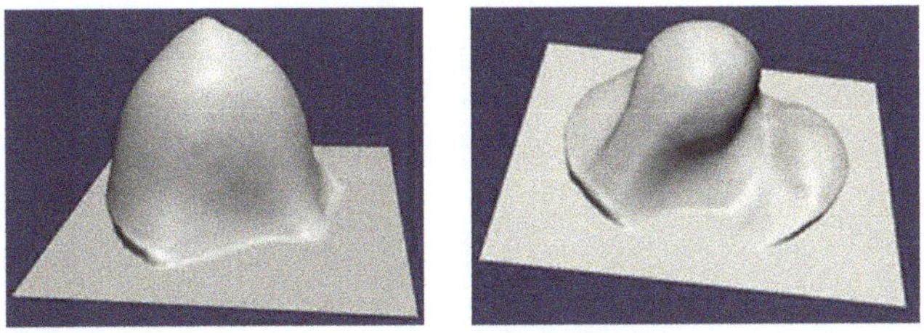

(a)

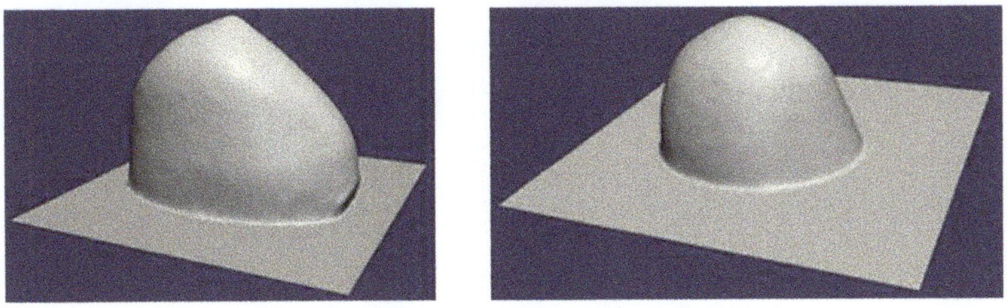

(b)

**Figure 25.** Meshing procedure results: (**a**) 3D surface reconstructions corresponding to two human breasts: Patient 002 and Patient 004; (**b**) 3D surface reconstructions corresponding to two test breast phantoms: Test Breast phantom B1 and Test Breast phantom B2.

### 2.3.5. Breast Volume Computation

To compute the volume of the 3D triangular mesh of the patient's breast (*Final Mesh*) generated in the previous step we use as reference the cumulative method presented by Zhang and Chen [25] but instead of using a tetrahedron as elementary calculation object, we use a truncated triangular prism [26], which is perpendicular to the xy plane (see Figure 26). This condition permits to reduce the number of arithmetic operations.

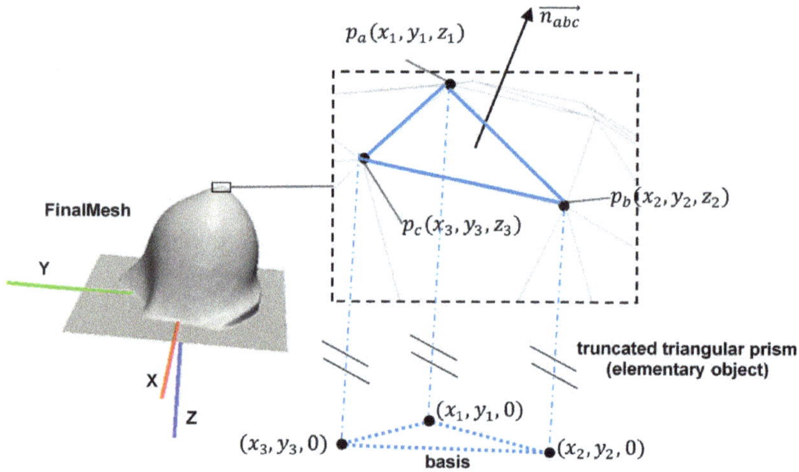


**Figure 26.** 3D mesh volume computation by using a truncated triangular prism as elementary object.

The signed volume $V$ of the elementary truncated triangular prism is computed for each triangle ( $p_a$, $p_b$, $p_c$ in Figure 26) by using Equation (18), where $A$ is the area of the basis of the truncated triangular prism and $sign$ corresponds to the sign (1 or −1) of V. The value of $sign$ is determined by the inner product $\overrightarrow{n_{abc}}.\hat{e}_z$, where $\hat{e}_z$ is the unitary vector (0,0,1) defining the orientation of the z axis in the Wavelia reference coordinate system (see Figure 26).

$$V = \left(A \cdot \frac{(|\,z_1| + |\,z_2| + |\,z_3|)}{3}\right) \cdot sign \tag{18}$$

where:

$$A = \left|\frac{x_1 \cdot (y_2 - y_3) + x_2 \cdot (y_3 - y_1) + x_3 \cdot (y_1 - y_2)}{2}\right|$$

To compute the total volume of *Final Mesh* we use Equation (19).

$$V_{total} = \left|\sum V_i\right| \tag{19}$$

where: $V_i = \left(A_i \cdot \frac{(|\,z_{i1}| + |\,z_{i2}| + |\,z_{i3}|)}{3}\right) \cdot sign_i$, $A_i = \left|\frac{x_{i1} \cdot (y_{i2} - y_{i3}) + x_{i2} \cdot (y_{i3} - y_{i1}) + x_{i3} \cdot (y_{i1} - y_{i2})}{2}\right|$ and i stands for the index of triangles of elementary truncated triangular prisms. $(x_{i1}, y_{i1}, z_{i1})$, $(x_{i2}, y_{i2}, z_{i2})$ and $(x_{i3}, y_{i3}, z_{i3})$ are the 3D coordinates of the vertices associated with the triangle i.

### 2.4. *Description of the Method to Quantify the 3D Breast Surface Reconstruction Accuracy*

#### 2.4.1. Generation of 3D Breast Reference Surfaces

To assess the accuracy of the 3D breast surface reconstruction and the breast volume computation, two MRI-based 3D printed breast phantoms named B1 and B2 have been used. The two anthropomorphic breast phantoms, which have been also used all along the Wavelia prototype medical device design, experimental testing and validation, are presented in Figure 27. As detailed in [6], the geometry of the phantoms which have been manufactured in-house by MVG Industries for this purpose, originated from real MRI images of patients, publicly made available by the University of Wisconsin (UWCEM database) [27].

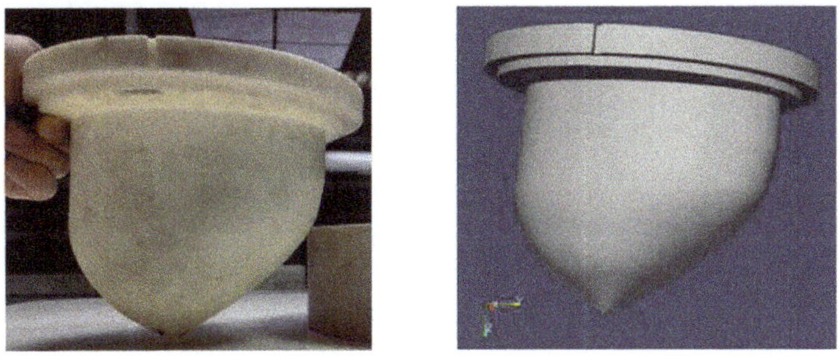

(**a**)

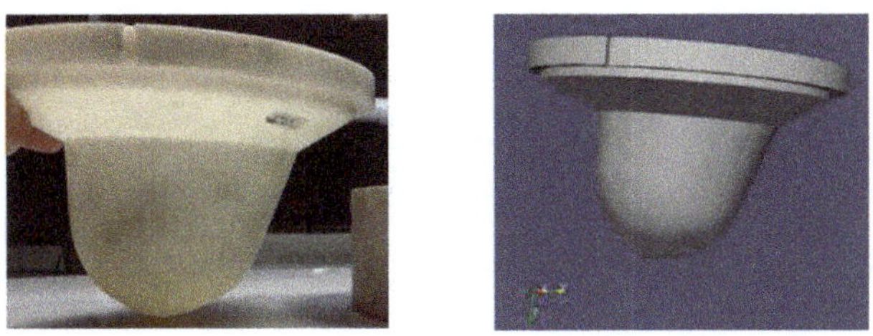

(**b**)

**Figure 27.** MRI-based 3D printed breast phantoms and their 3D representations: (**a**) Test breast phantom B1; (**b**) Test breast phantom B2.

Before starting the comparison procedure, a pre-processing step was performed on the 3D breast phantom representations in order to extract the external surface of each breast phantom (see Figure 28). To perform the pre-processing step, the MeshLab open source software has been used [28]. The resulting reference breast phantom representations have been named Model_B1_Ref and Model_B2_Ref.

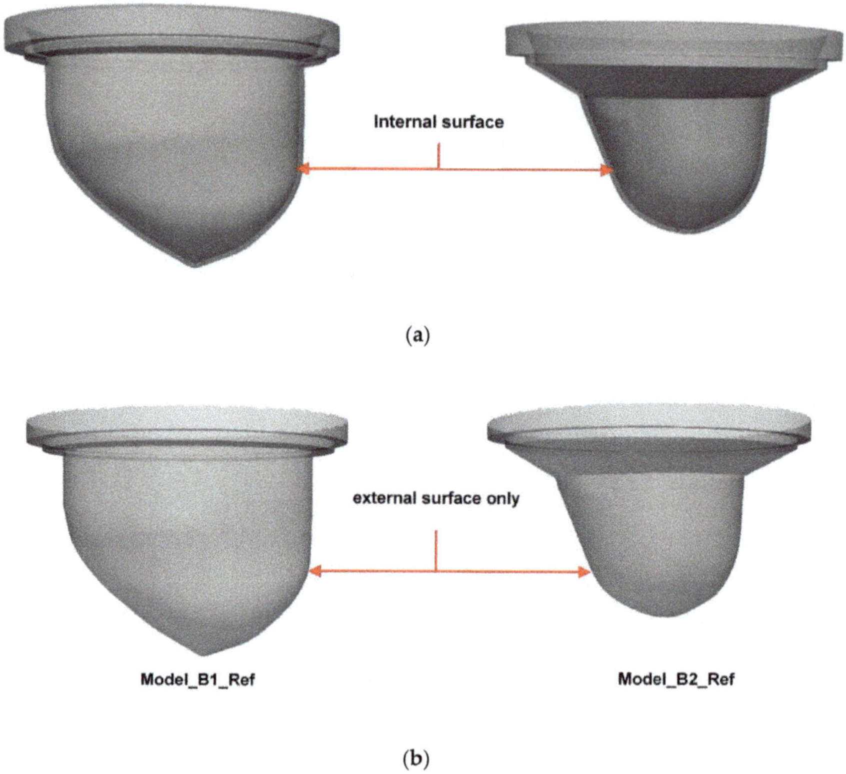


**Figure 28.** 3D breast phantom representations (**a**) before, (**b**) after application of the surface pre-processing step.

Model_B1_Ref and Model_B2_Ref have been further on used to assess the accuracy of the 3D breast surface reconstruction, the estimation of the vertical extent of the breast and the estimation of the volume of the breast, as performed by the Wavelia OBCD subsystem. Figure 29 shows two indicative results of the Wavelia 3D reconstruction method (see Section 2.3).

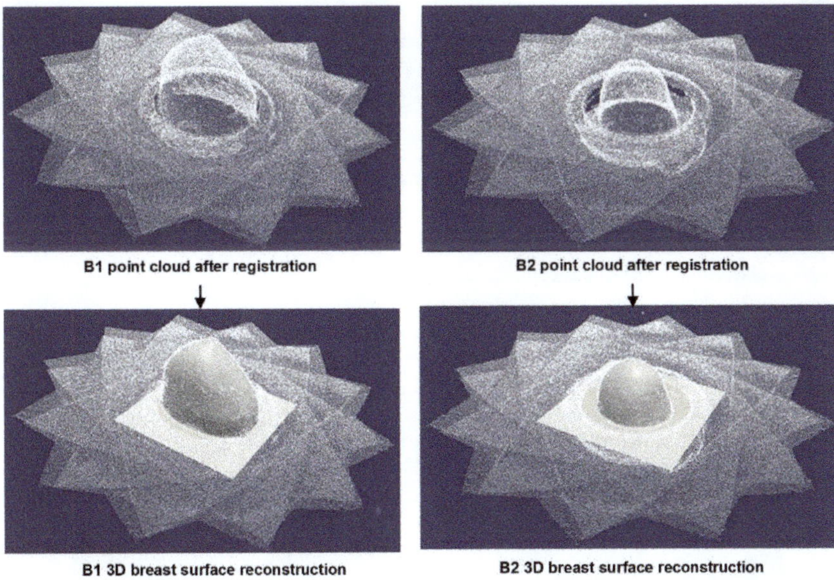


**Figure 29.** 3D breast reconstructions from point clouds using the Wavelia OBCD breast surface reconstruction method.

### 2.4.2. Procedure to Quantify the Accuracy of the 3D Breast Surface Reconstruction

To quantify the accuracy of the 3D breast surface reconstruction method and the breast volume computation, we first extract the part of the reconstructed surface that corresponds to the breast surface (removal of the flat horizontal surface corresponding to the examination table). To perform this operation, the Wavelia reference coordinate system is used, as presented in Section 2.1, to define the z value of the cutting plane that will be used to extract the breast surfaces (see Figure 30).

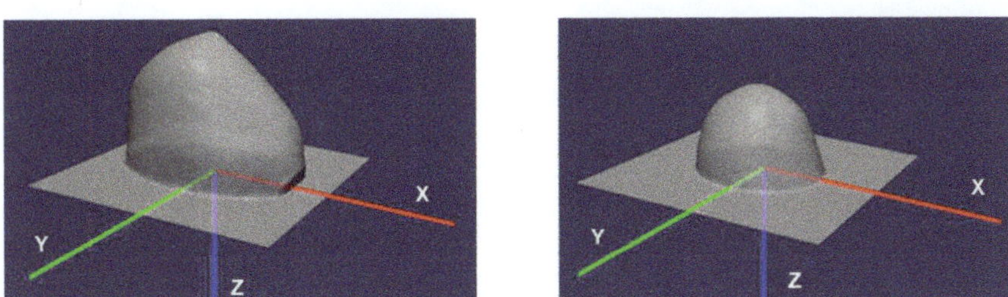


**Figure 30.** Wavelia reference coordinate system and 3D reconstructions of the breast phantoms B1 and B2.

The extraction procedure of the breast surfaces has been performed using the Paraview software [29]. After the breast surface has been extracted, a rigid registration is applied between this surface and its breast reference model. The Wavelia reference coordinate system is used as the main reference system for this operation. This procedure is performed by using the ICP (Iterative Closest Point) method, which is integrated in the MeshLab software [28]. The ICP transformation is applied to the breast reference model in order to obtain the registered breast reference model. The result of the registration procedure is shown in Figure 31a, where the extracted breast surface appears in blue and the registered breast reference model appears in semi-transparent grey. Then, the MeshLab software is used to compute the unsigned distance from each point of the extracted breast surface to the closest point of the registered breast reference model (distance error). We used the distance error as point color criterion to generate the breast distance color map of the extracted breast surface by using the MeshLab software as well (Figure 31b). Based on the distance data of the extracted breast surfaces, we also computed the following parameters to assess the accuracy of either 3D breast surface reconstruction:

- percentage of points with distance error less or equal to 0.5 mm;
- percentage of points with distance error less or equal to 1.0 mm;
- percentage of points with distance error less or equal to 1.5 mm;
- Root Mean Square (RMS) distance error in mm.

(**a**)

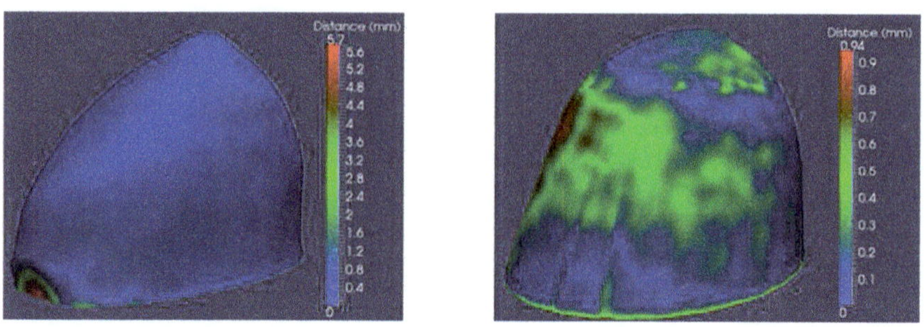


(**b**)

**Figure 31.** Breast phantom rigid registration: (**a**) example of results of the rigid registration of B1 and B2 reconstructed breast surfaces with the corresponding breast reference models; (**b**) example of breast distance color maps for the B1 and B2 extracted breast surfaces.

After the rigid registration has been performed, the breast reference surface is extracted from the registered breast reference model by using a cutting plane coinciding with the lower surface of the examination table. The breast reference surface is a closed surface.

Then, the vertical extent and the volume of the breast reference surface are computed by using the MeshLab software [28]. These parameters have been named: reference breast height ($H_{ref}$) and reference breast volume ($V_{ref}$). The following metrics are used to assess the accuracy of the vertical extent ($H_{comp}$) and the volume ($V_{comp}$) of the breast, as computed by the Wavelia OBCD subsystem:

- Breast vertical extent error [mm]: this metric is defined as: $H_{comp} - H_{ref}$.
- Breast volume error [mL]: this metric is defined as: $V_{comp} - V_{ref}$.
- Breast volume error in percentage: this metric is defined as: $\left(\frac{V_{comp}-V_{ref}}{V_{ref}}\right) \cdot 100$.

## 3. Results and Discussion

### 3.1. Calibration Evaluation

Several tests of the calibration procedure have been performed with the Wavelia OBCD subsystem. Indicative values of the calibration transform error, as evaluated during these tests, are presented in Table 1.

**Table 1.** Performance results of calibration procedure tests.

| Test ID | Calibration Transform Error (mm) |
|---|---|
| 1 | 1.40 |
| 2 | 0.58 |
| 3 | 0.70 |
| 4 | 0.97 |
| 5 | 0.88 |
| 6 | 0.80 |

Table 1 reports a value greater than 1 mm for the calibration transform error, in the case of Test 1. Such a relatively high value of the calibration transform error could be due to small movements of the cross-shaped reference object during the azimuthal 3D scan of the camera. To prevent these movements and to ensure sub-millimetric precision in the computation of the calibration transform, it is very important to avoid any contact with the OBCD subsystem during the calibration azimuthal scan of the camera.

Test 1 was one of the first tests of the calibration method with the Wavelia prototype, while it was still at the factory. All the other reported tests (Test 2 to Test 6) have been performed after installation of the Wavelia system at the Galway University Hospital, while carrying-out the calibration procedure after repeated training. It is demonstrated that sub-millimetric precision of the method has been consistently assured, after validation of the system for transfer to the clinical investigation site.

It is worthwhile noting that, this being a prototype version 1 of the Wavelia system, installed in the hospital for a first-in-human clinical investigation, the calibration is meant to be performed by trained technical staff during installation or maintenance of the system. It is not meant to be repeated by the daily operator, during the clinical investigation. The calibration measurements are used to define once the cumulative transformation operator, which is afterwards applied to align the multi-view camera measurements of all breast scans. The required accuracy for the calibration measurement (controlled environment) is higher than the effective accuracy of the breast scan measurements, such that the cumulative transformation operator is optimally defined.

### 3.2. D Breast Surface Reconstruction: Quantitative Evaluation Results

A series of eight 3D breast reconstruction tests have been performed. Six of the tests were performed on breast phantoms fully introduced into the opening of the examination table (four tests for B1 phantom and two tests for B2 phantom). To simulate a smaller breast phantom, the two remaining 3D breast reconstructions were performed with the B2 breast phantom partially introduced into the opening of the examination table. The results of the eight tests are shown in Figure 32 and Table 2.

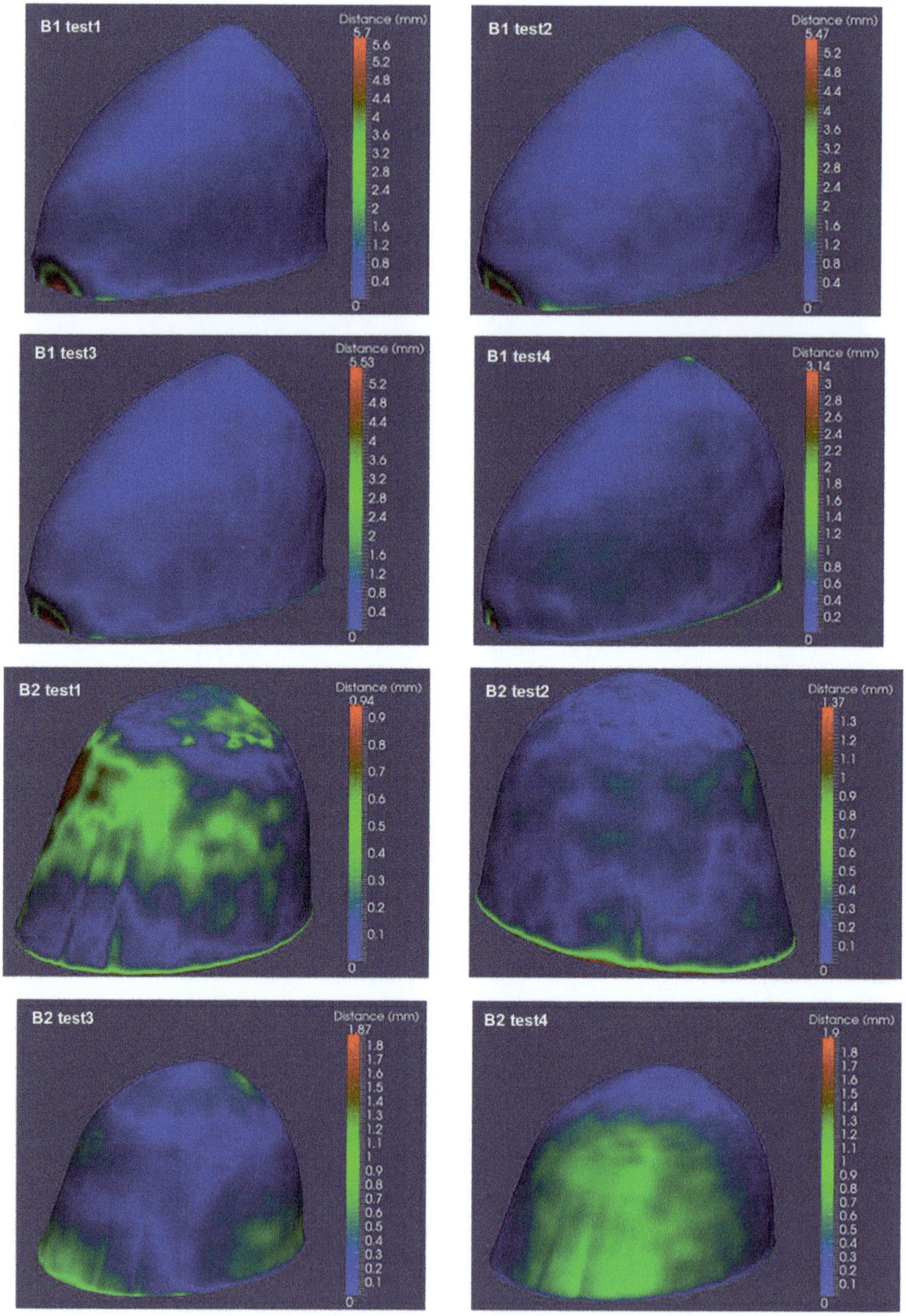


**Figure 32.** Breast distance color maps for B1 and B2 extracted breast surfaces: multiple scans used for the evaluation of the surface reconstruction of either of the two test breast phantoms.

**Table 2.** Summary of the accuracy results on eight 3D breast surface reconstruction tests involving the B1 and B2 breast phantoms.

| | B1 Test 1 | B1 Test 2 | B1 Test 3 | B1 Test 4 | B2 Test 1 | B2 Test 2 | B2 Test 3 | B2 Test 4 |
|---|---|---|---|---|---|---|---|---|
| Total number of points on the extracted breast surface | 177,228 | 178,570 | 178,562 | 179,896 | 134,808 | 137,000 | 119,758 | 119,748 |
| Percentage of points with distance error <= 0.5 mm (%) | 76.80 | 90.20 | 89.01 | 87.28 | 82.18 | 93.77 | 83.88 | 79.88 |
| Percentage of points with distance error <= 1.0 mm (%) | 96.26 | 95.19 | 94.58 | 97.45 | 100 | 99.14 | 98.12 | 99.77 |
| Percentage of points with distance error <= 1.5 mm (%) | 97.55 | 97.42 | 97.37 | 98.94 | 100 | 100 | 99.86 | 99.93 |
| Percentage of points with distance error >1.5 mm (%) | 2.45 | 2.58 | 2.63 | 1.06 | 0.00 | 0.00 | 0.14 | 0.07 |
| RMS distance error (mm) | 0.74 | 0.65 | 0.69 | 0.50 | 0.40 | 0.30 | 0.44 | 0.42 |
| Reference Vertical Extent of the Breast (mm) | 110.19 | 110.96 | 110.96 | 111.16 | 84.67 | 84.97 | 67.88 | 67.59 |
| Computed Vertical Extent of the Breast (mm) | 109.36 | 109.55 | 109.60 | 109.85 | 84.71 | 84.79 | 67.73 | 67.50 |
| Error in the Vertical Extent of the Breast (mm) | −0.83 | −1.41 | −1.36 | −1.31 | 0.04 | −0.19 | −0.15 | −0.09 |
| Reference Volume of the Breast (mL) | 915.57 | 924.57 | 924.57 | 928.28 | 498.92 | 502.51 | 349.48 | 350.01 |
| Computed Volume of the Breast (mL) | 909.81 | 921.25 | 921.25 | 921.97 | 494.27 | 502.21 | 349.51 | 347.48 |
| Breast Volume Error (mL) | −5.75 | −3.31 | −3.31 | −6.31 | −4.64 | −0.31 | 0.03 | −2.52 |
| Breast Volume Error (%) | −0.63 | −0.36 | −0.36 | −0.68 | −0.93 | −0.06 | 0.01 | −0.72 |

The breast distance maps shown in Figure 32 demonstrate the homogeneous spatial distribution of the points with a distance inferior or equal to 1.0 mm. Figure 32 also confirms the limited extension of the zones where the distance error is greater than 1.5 mm.

Table 2 shows that at least 76% of the extracted breast surface has a distance error inferior or equal to 0.5 mm, that at least 94% of this surface has a distance error inferior or equal to 1.0 mm and at least 97% of this surface has a distance error inferior or very close to 1.0 mm (greater than 1.0 mm and less or equal to 1.5 mm). It also shows that only approximately 3% of the extracted breast surface has a distance error greater than 1.5 mm. Based on these results, we can affirm that globally the accuracy of the extracted breast surface is sub-millimetric. This sub-millimetric accuracy may be also deduced from the RMS distance error which is overall less than 0.75 mm.

Concerning the computation of the vertical extent of the pendulous breast, Table 2 shows that the computed breast vertical extent is generally underestimated (breast vertical extent error is often negative) and that the absolute breast vertical extent error is very close to 1.0 mm (worst-case error of −1.41 mm for B1 test 2 in Table 2).

Table 2 also shows that the computed breast volume tends also to be underestimated (volume error is often negative) and that the uncertainty in the breast volume computation is less than 1%.

### 3.3. D Breast Surface Reconstruction: Indicative Results from the First Ten Patient Scans

In Figures 33 and 34, the reconstructed breast surfaces using the Wavelia OBCD subsystem are shown for the ten first patients enrolled in the on-going first in-human clinical investigation of the Wavelia system, at Galway University Hospital. These are preliminary clinical results which serve to demonstrate the efficacy of the presented calibration algorithm and 3D breast surface reconstruction method on real clinical data. It is demonstrated that meaningful breast surfaces of good quality have been achieved for all the ten patient scans.

In Figures 33 and 34, two views of the reconstructed surface of both breasts of each patient are shown. The two selected views for this illustration are: (a) a side view of the two pendulous breasts: looking at the patient from the front; (b) a supine-mimicking view of the two breasts: looking at the patient from the bottom. This comparative presentation of the breast anatomy for this group of ten patients, as captured during their Wavelia OBCD scans, highlights the expected variability in the positioning of the patients. Even though the same positioning of the arms and alignment of the patient's body with the bed is intended in all cases, unavoidable rotations of the breasts and/or twist of the torso seem to happen in the case of some patients.

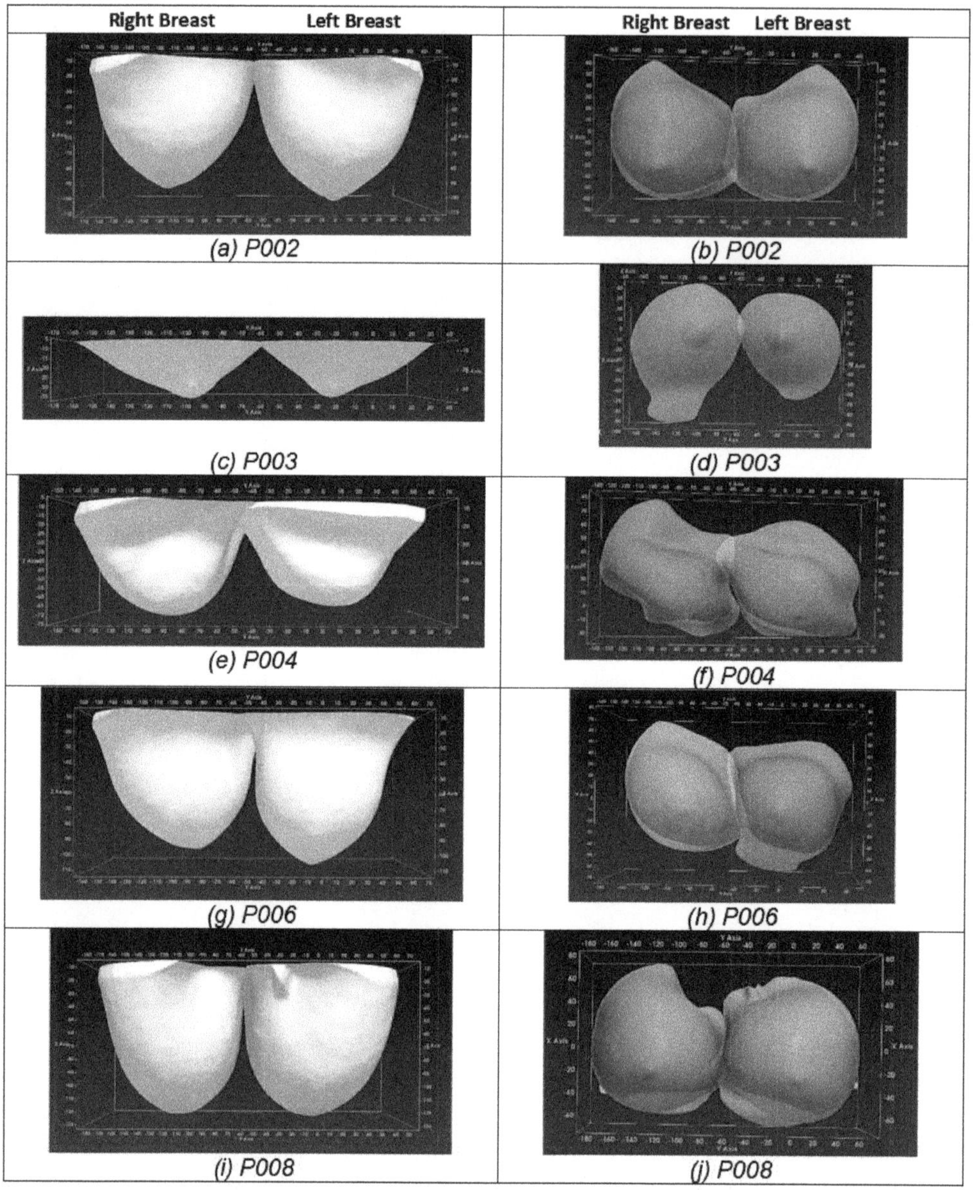


**Figure 33.** Breast Anatomy, as seen in the Wavelia OBCD scanner: Side view (**Left**), Supine position mimicking view (**Right**) - the first five patient scans.

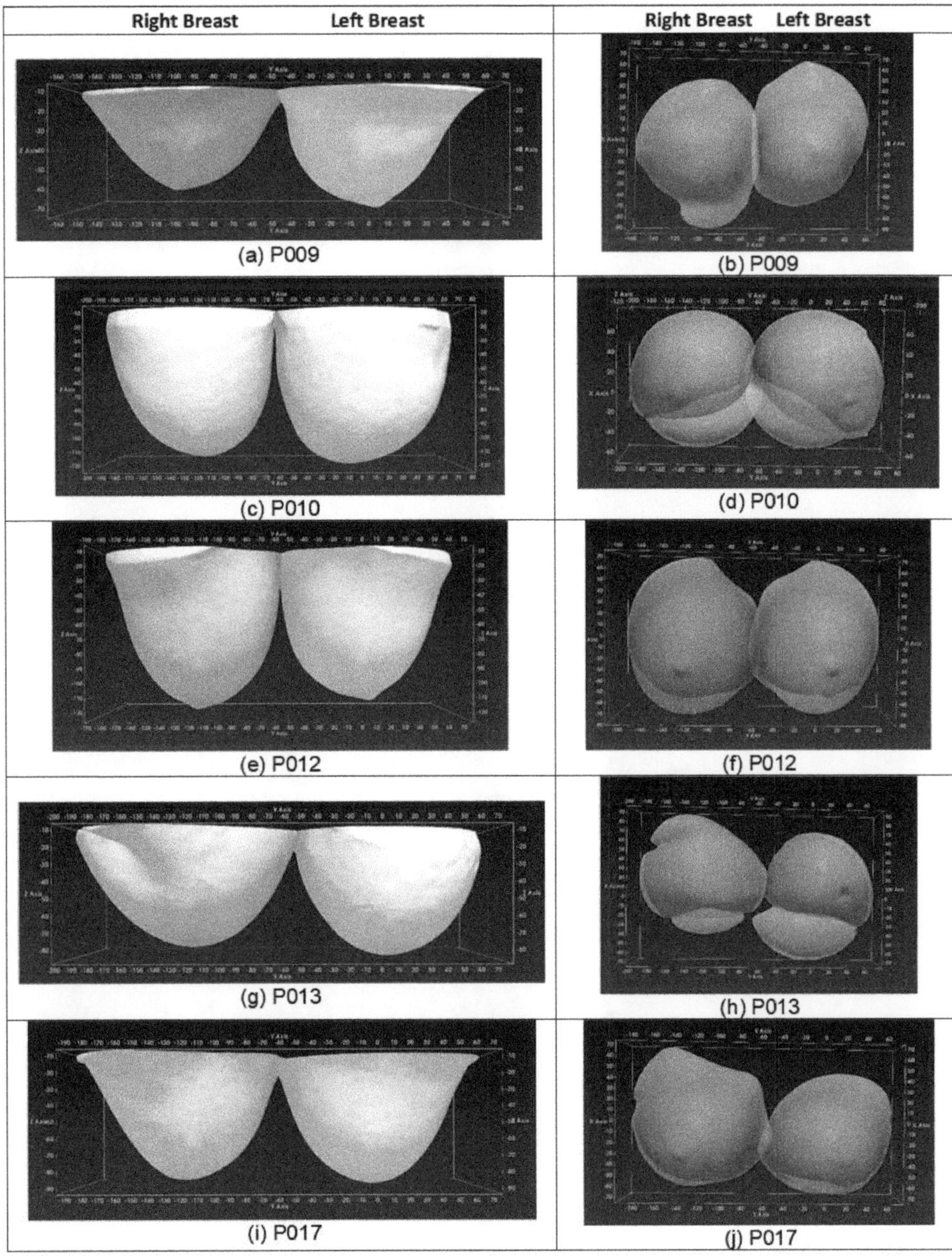


**Figure 34.** Breast Anatomy, as seen in the Wavelia OBCD scanner: Side view (**Left**), Supine position mimicking view (**Right**) - the second series of five patient scans.

While a quantitative evaluation of the reconstructed surfaces is not straightforward in the case of the patient scans, the computation of the breast volume, as performed using the Wavelia OBCD scan data, has been possible to be validated using as reference the X-ray mammography breast volume data, being made available for the same patients.

All the patients enrolled in the clinical study have had an X-ray mammogram performed no more than 6 weeks ahead of their scan with the Wavelia prototype medical device. The mammograms have been processed with the Volpara VDM (Volumetric Density Measurement) software package [30] and the breast volume has been computed. The Volpara VDM breast volume and density computation have been extensively validated, against MRI breast volume and density estimation [31–34] and can be safely considered as a valid reference. In [33] breast volume estimates obtained from 186 FFDM (Full Filed Digital Mammography) exams including mediolateral oblique (MLO) and cranial-caudal (CC) views to objective reference standard measurements obtained from breast MRI have been presented. The Pearson's correlation coefficient between the two computations of the breast volume has been found to be as high as 0.97. The Volpara commercial software was used to compute the breast volume and volumetric breast density, based on FFDM data, in this study. In [34], the agreement of three mammographic techniques (Single-energy X-ray Absorptiometry [SXA], Quantra, and Volpara) with MRI for percent fibro-glandular tissue volume, absolute fibro-glandular tissue volume, and total breast volume computation was assessed. In terms of breast volume computation, the squared Pearson's correlation coefficient was reported to be 0.91 between all the three mammographic techniques and the MRI, while the achieved RMSE (Root Mean Squared Error) was found to be 108 mL, 109 mL and 121 mL for the three mammographic methods correspondingly, when compared to MRI. This study involved data from 99 women. Both referenced studies have been used to validate volumetric breast density and also breast volume computation using FFDM data, the validation being performed against computation using 3D MRI data.

A similar approach has been adopted in our study to validate breast volume computation using the Wavelia OBCD 3D data, against the computation performed by the already validated Volpara software on the available FFDM data. Results on the 10 first patients that have participated in the first-in-human clinical investigation of the Wavelia system are presented. In Figure 35, the volume of both breasts of each of the ten patients is plotted, as computed from either:

- the Wavelia OBCD 3D scan data, or
- the 2D X-ray mammogram (Volpara software package, computation on 'For processing' DICOM files of the two standard mammographic views – Cranio-Caudal and Medio-Lateral Oblique).

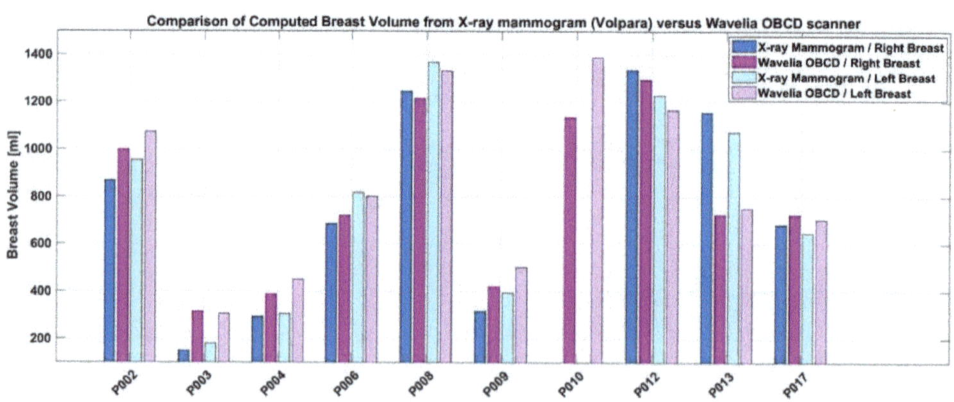


**Figure 35.** Comparative presentation of the breast volume, as computed by the Wavelia OBCD scanner (pendulous breast) and the X-ray mammogram (compressed breast - averaging over Cranio-Caudal and Medio-Lateral view captures).

Good level of correspondence between the breast volume computation, as performed by the two distinct imaging modalities, is demonstrated, thus validating the Wavelia OBCD breast surface reconstruction and volume computation. It is interesting to observe that the RMSE error computed between the Wavelia OBCD and FFDM (Volpara) breast volume computations based on this first small dataset of patients is 154.6 mL, thus in the same order of magnitude as the RMSE reported in [34] between FFDM and MRI.

Significant discrepancy between the X-ray mammography and the Wavelia OBCD breast volume computation can be noticed only in the case of Patient 013 (P013). This has most probably been due to the patient not being well positioned on the Wavelia scanner, resulting in part of her breasts not being scanned. Patient 003 had very small breasts, potentially difficult to be fully inserted in-between the X-ray mammography plates, resulting in under-estimation of the breast volume by the X-ray mammography in this case. The 'for processing' DICOM files, which are required for processing the mammograms with Volpara VDM, were not available in the case of Patient 010.

Finally, in order to evaluate the impact of potential movement of the patient's breast during the OBCD scan, a quality check has been developed and executed after each OBCD patient scan. For the purpose of the quality check, two versions of the breast surface are reconstructed using two interlaced and non-overlapping subsets of the full set of recorded azimuthal samples (Subset 1: camera at the azimuthal positions {0:60:300}°, Subset 2: camera at the azimuthal positions {30:60:330}°). The OBCD scan is validated, as long as the RMS distance between the two surfaces is inferior to 0.8 mm. The OBCD scans of all the patients participating in the first clinical investigation so far have been validated with this criterion, which indicates that if sub-millimetric precision has been demonstrated in the case of static breast phantoms, the same level of precision can be maintained during the OBCD scan of a real breast. Obviously, before the scan starts, the patient is instructed by the research assistant to stay still, avoid heavy breathing and maintain a relaxed position during the total duration of the scan, which lasts approximately 1 min.

### *3.4. Wavelia OBCD Breast Surface Reconstruction: Towards Development of a Real-Time Support Tool for the Wavelia MBI Examination and Breast Cancer Diagnosis*

As specified in the introduction, in the first prototype of the Wavelia breast imaging medical device, the OBCD and MBI subsystems are integrated in two separate examination tables on which the two scans are sequentially performed, while the patient positions herself in approximately the same way during both examinations. The OBCD scan is performed while the breast is hanging in the air, while the MBI scan is performed with the breast being immersed in a coupling (or transition) creamy liquid, which is opaque.

Given this implementation, at this stage of development of the Wavelia system, the breast surface reconstruction using the OBCD subsystem serves to define the amount of transition liquid to be used for the MBI scan and also to define the length of the vertical scan of the pendulous breast with the MBI system. In the context of the first-in-human clinical study of the Wavelia breast imaging medical device, it has also served to verify the correspondence of the breast anatomy, as retrieved using two distinct imaging modalities (Wavelia OBCD and MBI), tested for the first time on humans.

Rigid registration is performed to match the reconstructed breast surfaces, as computed during the two scans (OBCD and MBI). The ultimate objective of this operation is to exploit the sub-millimetric level of detail of the OBCD-derived breast surface, as a high-quality envelope for the MBI-derived breast abnormality detections for breast cancer diagnosis. It is recalled that due to the physically limited spatial resolution that can be achievable at the microwave frequency range, the exact orientation of the pendulous breast under scan, as well as details such as the nipple location, cannot be defined using the MBI-derived reconstruction of the breast surface alone. With no access to such level of detail, the annotation of the breast quadrant and clock-position at which an abnormality has been detected with MBI cannot be very accurate either, when using the MBI system stand-alone.

Given the potential deformation of the breast when submerged in the transition liquid, the exploitation of the OBCD breast surface for refined localization of the detected breast abnormalities with MBI is possible only in the case of minimally deformed breasts, in the actual implementation. In Figure 36a, the example of a young patient (37 years old) with medium-sized breasts, which have not been significantly deformed when immersed in the transition liquid, is shown. For such a patient, the registered OBCD contour can be safely used as envelope for the breast abnormalities, as detected with MBI. In Figure 36b, the example of an elderly patient (68 years old) with large breasts, which have been significantly deformed when immersed in the transition liquid, is shown. This second patient is a case in which the simple rigid registration of the OBCD-derived breast surface is not usable as an envelope for the breast abnormalities, as detected with MBI, in the actual implementation of the system.

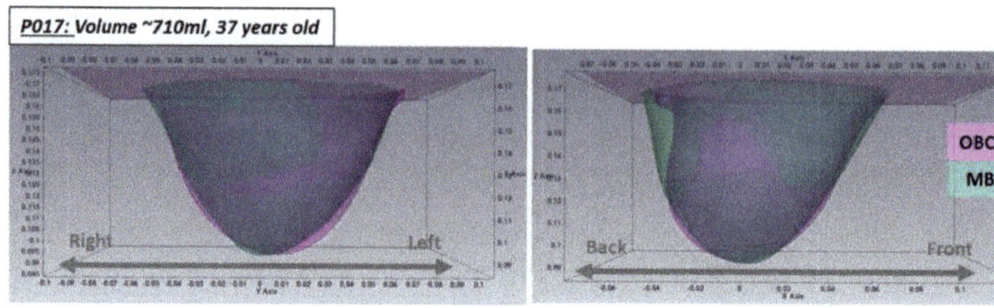


(**a**)

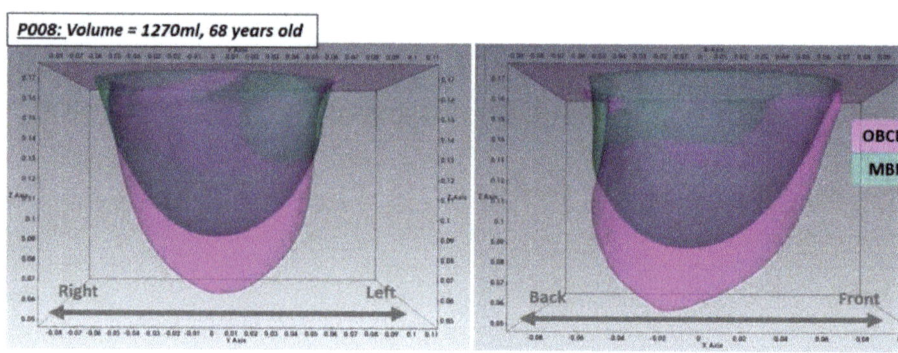


(**b**)

**Figure 36.** Illustration of the expected level of breast deformation due to immersion in the transition liquid of the Wavelia MBI scanner: (**a**) Patient 017: not significantly deformed breast, when immersed in the transition liquid of the Wavelia MBI subsystem; (**b**) Patient 008: significantly deformed breast, when immersed in the transition liquid of the Wavelia MBI subsystem.

In a future generation of the Wavelia breast imaging medical device prototype, it is planned to integrate both the OBCD and MBI subsystems in the same examination table, such that both scans are performed while the patient stays at the exact same position and the breast is in the exact same ambient conditions. A semi-transparent transition liquid would then need to replace the actual creamy and opaque transition liquid, such that the OBCD scan can be performed in the same conditions as the MBI scan and the breast is identically deformed during both scans. With such a future version of the Wavelia system, the OBCD-derived breast surface reconstruction is meant to enhance significantly the potential of the Wavelia Microwave Breast Imaging system in terms accurate localization of the detected breast abnormalities in the patient's breast, no matter what the level of breast deformation due to immersion in the transition liquid is. In addition, while being practically real-time available, the OBCD breast surface reconstruction is meant to drive the operator during the positioning of the patient (e.g., improve the centering of the breast in the system), such that the MBI scan can be performed in the best possible conditions.

## 4. Conclusions

In this paper, a new calibration algorithm for turntable-based 3D scanning systems, a new method for 3D surface reconstruction from point cloud based on a linear stretching function, and a breast volume computation method, which have been designed for and integrated in the OBCD subsystem of the Wavelia prototype breast imaging medical device, have been thoroughly presented. Indicative experimental results with two test breast phantoms and preliminary results from ten patient scans have demonstrated the efficacy of the system in reconstructing with high precision the external envelope of a female patient's breast hanging below the examination table, while the patient is in prone position. This consists in very valuable input information for the second counterpart of the Wavelia medical device, the MBI subsystem, which aims at detecting the presence of breast pathologies, based on the contrast, in terms of dielectric properties, between healthy and malignant breast tissue.

The envelope of the breast being scanned with the microwave breast imaging system is essential to be known for any detection in the microwave image to be meaningfully referenced in space. In addition, specifically for microwave breast imaging, as the dielectric properties of the interior of the breast are unknown, while the dielectric properties of the transition medium are known, the breast envelope which is extracted at first also serves as boundary between the two electromagnetic wave propagating media (dielectrically known exterior of the breast versus dielectrically unspecified interior of the breast), which is very important a priori information for any microwave imaging algorithm, further on. The spatial resolution of the OBCD (optical) subsystem is significantly finer than the one of the MBI (microwave) subsystem, but the optical subsystem cannot penetrate the breast skin, while the microwave subsystem can. The interest in fusing the optical and microwave data is exactly to profit from a highly detailed breast envelope, as spatial reference for any abnormalities of the breast tissues being detectable with microwaves.

The Wavelia prototype medical device has only recently been installed at the Galway University Hospital, Ireland, for its first-in-human pilot clinical test. To this day, the system has been tested on twenty patients in total (the aim being to test it on thirty patients by the end of this first-in-human clinical investigation). The OBCD subsystem, which has been the main focus of this paper, has been evaluated positively by the users, so far, and generated reliable data which are being efficiently exploited in the ongoing off-line processing of the MBI subsystem datasets. A first encouraging patient result, involving the detection of a palpable Invasive Ductal Carcinoma (IDC) of approximate size 15 mm in the breast of a 44-year old patient, using the Wavelia breast imaging medical device, has been presented in [35]. More complete associated clinical data results will be included in future publications on the Wavelia Microwave Breast Imaging system.

The proposed method for new 3D surface reconstruction from point clouds has been designed based on the shape of the breast when the patient is in prone position on the examination table (Wavelia MBI and OBCD subsystems). However, beyond this specific application, the method could be also used to reconstruct objects other than the breast, as long as the object is representable with the shape of a cone or a cylinder (e.g., the breast can be represented by a cone). The proposed calibration method could be used to calibrate any turntable-based 3D scanning system, in general.

**Author Contributions:** J.D.G.C. conceived and developed the calibration and breast surface reconstruction methods for the Wavelia OBCD subsystem, developed the piloting software for both Wavelia OBCD and MBI subsystems, including Graphical User Interface (GUI) appropriate for the clinical setting, wrote the original draft of the paper; A.F. designed the radar signal processing algorithms for microwave breast image formation and tumor detection with the Wavelia MBI subsystem, reviewed and edited the draft paper; L.D. supervised the hardware development of both Wavelia OBCD and MBI subsystems, designed the RF chain of the Wavelia MBI subsystem, reviewed and edited the draft paper; J.-G.B. participated to the design of the electrical architecture of both Wavelia OBCD and MBI subsystems and contributed to their installation at the hospital. All authors have read and agreed to the published version of the manuscript.

**Funding:** This research received no external funding.

**Conflicts of Interest:** The authors declare that they are employees of MVG Industries, which is the company developing the prototype of the Wavelia medical device described in this manuscript.

**Ethical Approval:** All procedures performed in the study involving human participants were in accordance with the ethical standards of the institutional and/or national research committee and with the 1964 Helsinki declaration and its later amendments or comparable ethical standards.

**Informed Consent:** Informed consent was obtained from all individual participants included in the study.

## Appendix A. Selection Procedure for the 3D Camera to be Integrated in the Wavelia OBCD Subsystem

For the development of the Wavelia OBCD subsystem, the 3D camera to be used needed to be carefully selected, first of all. Two 3D cameras using different acquisition technologies. have been tested and evaluated, before selection. For this purpose, 3D acquisitions were performed with both cameras around breast phantoms by using an azimuthal rotating camera system inside a closed environment with no light. The cameras were located at a distance of approximately 50 cm from the breast phantoms. The first 3D camera that was tested was the model O3D302 from IFM [36], which measures the distance between the sensor and the nearest point on the surface of the 3D object under scan, based on the time-of-flight (ToF) principle. The second 3D camera that was tested was the model Ensenso N10-804-18 from IDS [37], which works according to the "projected texture stereo vision" principle. This 3D camera has two integrated CMOS sensors and an infrared projector that projects high-contrast textures onto the object to be captured by using a pattern mask. Some technical data of the tested cameras are presented in Table A1. The experimental setups which have been used to test the two cameras are presented in Figure A1. A black curtain has been used to cover each setup, in order to obtain a closed obscure environment.

**Table A1.** Main technical data of the two tested cameras.

| | IFM O3D302 | Ensenso N10-804-18 |
|---|---|---|
| Image resolution [pixels] | 176 × 132 | 752 × 480 |
| Max. reading rate [Hz] | 25 | 30 |
| Operating distance [mm] | 300 ... 8000 | 450 ... 1600 |
| View field size X at 500 mm [mm] | 500 | 158.49 |
| View field size Y at 500 mm [mm] | 370 | 158.13 |
| Z-Accuracy [mm] | — | 0.452 |
| Baseline (Pupillary Distance) [mm] | — | 100 |
| Illumination | 850 nm, infrared | 850 nm, infrared |

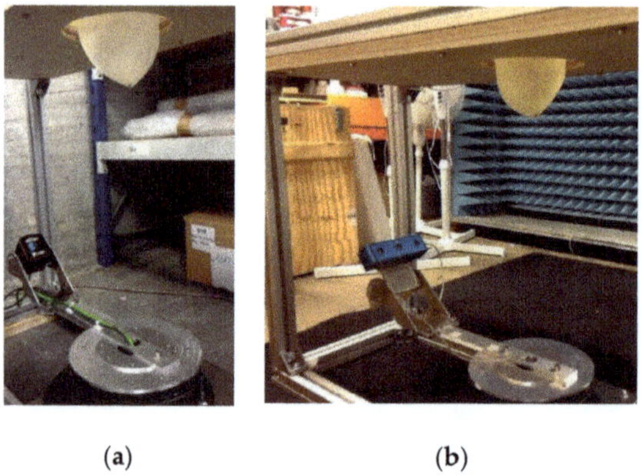

(a) (b)

**Figure A1.** Azimuthal rotating camera system: (**a**) Experimental test setup for IFM O3D302 camera; (**b**) Experimental test setup for Ensenso N10-804-18 camera.

As presented in Table A1, the resolution of the IFM camera is lower than the resolution of the Ensenso camera. Therefore, in order to compare both cameras with similar volumes of points, an azimuthal scan with an angular step of 10° has been performed with the IFM camera. This data has been compared with the data from an azimuthal scan with an angular step of 30° performed with the Ensenso camera. The two aforementioned 3D scans permitted to collect 36 point clouds with the first camera and 12 point clouds with the second one.

Mechanical data, such as the vertical inclination of the camera, the distance from the camera to the axis of azimuthal rotation, the 3D coordinates of the camera position and the azimuthal angle at which each point cloud is collected (0° to 350° for IFM camera and 0° to 330° for Ensenso camera), were used to approximatively align the acquired 3D point clouds. The results presented in Figure A2 correspond to the cumulated approximatively aligned 3D point clouds for both cameras.

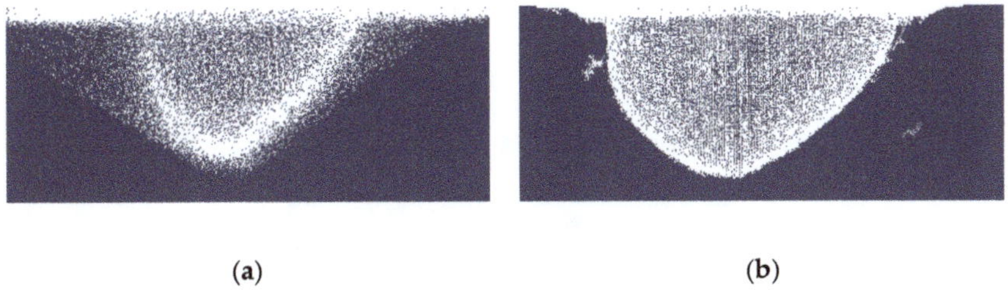


**Figure A2.** Cumulated approximatively aligned 3D point cloud: (**a**) From IFM O3D302 camera scan; (**b**) From Ensenso N10-804-18 camera scan.

By visually comparing the cumulative 3D point clouds from both cameras, it can be observed that the number of artifacts (ghost points) is significantly higher for the IFM camera than for the Ensenso camera.

To evaluate visually the accuracy of each cumulative 3D point cloud, a 3D reference surface of the breast phantom has been manually registered with the point cloud. The ground truth surface of the scanned breast phantom (red color) has been overlaid with the camera point clouds (white color) for the purpose of this illustration, in Figure A3.

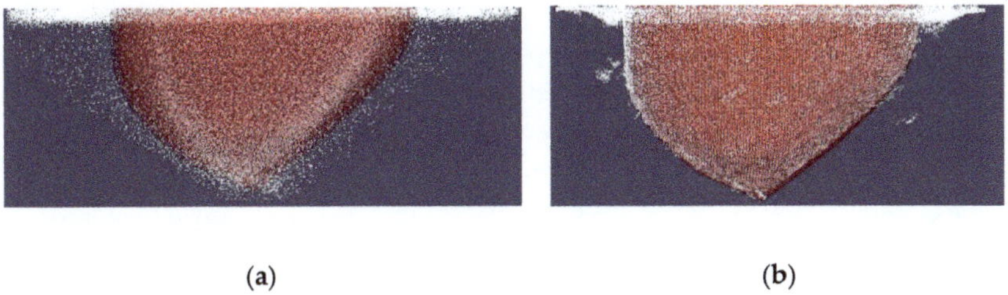


**Figure A3.** Cumulative approximately aligned 3D point cloud and registered 3D breast surface: (**a**) From IFM O3D302 Camera scan; (**b**) From Ensenso N10-804-18 camera scan.

Figure A3 shows than the cumulative point cloud collected with the Ensenso camera is significantly closer to the 3D breast reference surface than the cumulative point cloud collected with IFM camera is.

Based on these results, the 3D stereoscopic camera Ensenso N10-804-18 has been selected to be integrated in the Wavelia OBCD subsystem.

## Appendix B. Generation of the Input Point Cloud for the 3D Breast Surface Reconstruction Method

This procedure uses as input data the cumulative point cloud $Pc_{Acc}$ presented in Section 2.3. It consists in the following processing steps:

- extraction of the working point cloud;
- ring removal and verification.

Each of the above processing step are specified in a separate sub-section below.

### *Appendix B.1. Extraction of the Working Point Cloud*

In order to extract the points of the working point cloud, the z value of the lower plane of the examination table is required to be known. To compute this, a disk section points corresponding to this plane has been selected from $Pc_{Acc}$ (see illustration with the green point cloud in Figure A4). The mean of the z coordinates of the points located in this section have been further computed. The computed value is named $z_{ref}$.

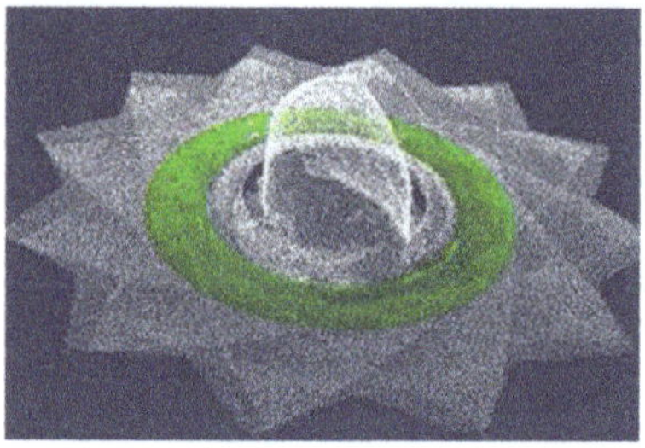

**Figure A4.** Disk section corresponding to the lower plane of the examination table.

To extract the working point cloud, the points where the z coordinate is lower than $z_{ref}$ are first selected from $Pc_{Acc}$.

The working point cloud is further extracted, by selecting the points that lie within an ellipsoid centered at the (0,0,0) position of the Wavelia reference coordinate system. The lengths of the semi-axes of the ellipsoid, *a* for the x and y axes and *b* for the z axis, have been selected such that all the points corresponding to the largest possible breast under scan lie, with sufficient margin, within the outer contour of this ellipsoid. To select the points inside the ellipsoid, the condition $\frac{{p_x}^2}{a^2} + \frac{{p_y}^2}{a^2} + \frac{{p_z}^2}{b^2} \leq 1$ is applied.

After the ellipsoidal point selection, an additional operation step is applied to remove remaining outlier points, if any, lying far from the breast surface. An illustration of this operation is provided in Figure A5 below.

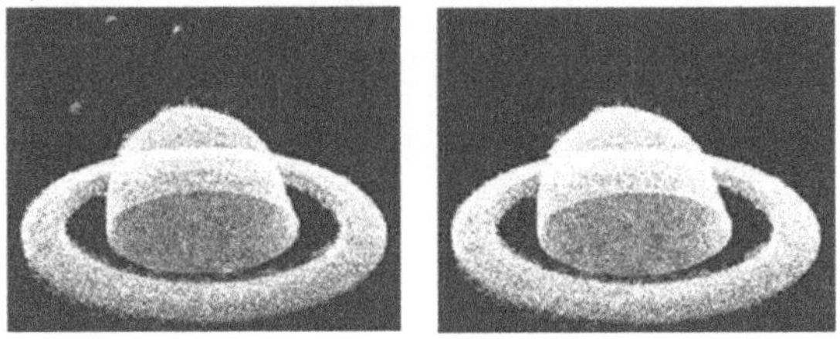

(a) (b)

**Figure A5.** Example of outlier removal, after ellipsoidal point selection: (**a**) before, (**b**) after outlier removal: extracted point cloud for the test breast phantom B2.

Finally, the point (0,0, $z_{ref}$) is set as the origin of the extracted working point cloud.

*Appendix B.2. Ring Removal and Verification*

As mentioned in the introduction, the opening of the examination table of the Wavelia OBCD subsystem is fitted with a plastic ring. The ring removal procedure consists in identifying the points of the working point cloud corresponding to the ring and setting to zero their z coordinate values. To do this, a disk section of points corresponding to the flat surface of the ring is selected, then those points are fitted to a plane. The signed distance to this plane is used as a criterion for the removal of points (positive distance for points above the plane and negative distance for points below the plane). To apply the point removing criterion, a parameter is defined:

- *inner radius (InRad)*: this parameter is defined as the physical radius of the ring minus an estimate of the expected variation due to reflections on the vertical wall of the ring during the OBCD scan of the breast.

The removing criterion is applied to the points with Euclidean distance to the xy plane greater than *InRad (see inner radius* in Figure A6a). The complete ring removal procedure, as applied to a test breast phantom, is shown in Figure A6b.

The new point cloud, as obtained after ring removal, is named $Pc_B$.

Some points corresponding to the ring may still remain. To verify this, the following parameters are used:

- *ring height*: it is the physical height of the ring, including an estimate of the noise along the z axis on the flat ring section (see Figure A6b) during the OBCD scan;
- *mesh resolution (Dxy)*: it corresponds to the desired resolution on the xy plane for the 3D breast reconstructed mesh;
- *remaining reference factor*: it is a factor in the range [0–1], defining whether processing to remove remaining ring points is required, or not. This factor has been determined empirically, based on observations from a series of available scan data, involving both human and phantom breasts.

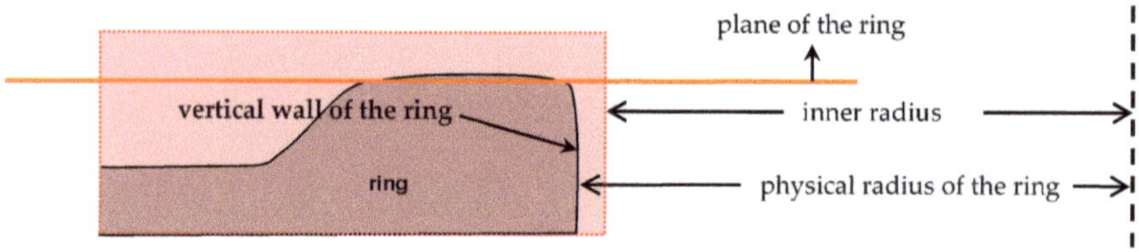


(a)

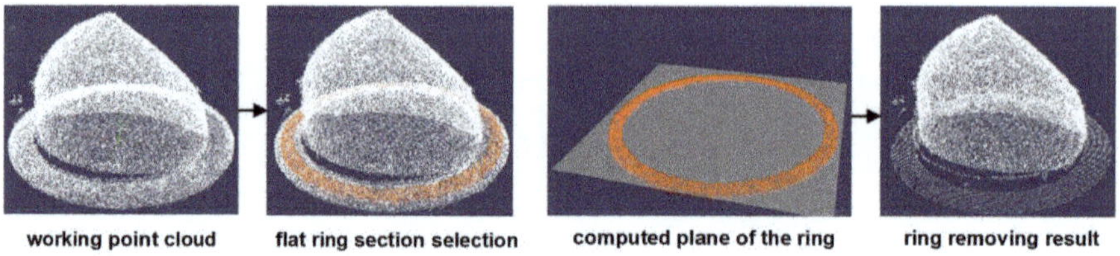


(b)

**Figure A6.** Ring removing procedure: (**a**) point removing criterion; (**b**) full ring removal procedure, as applied to the Test breast phantom B1.

The points are selected with an absolute z coordinate value lower than *ring height*. A rectangular flat grid is generated with a resolution corresponding to *Dxy* by using the minimum and maximum values for the x and y coordinates of the selected points. The z coordinate values of this flat grid are zero. Then a disk section is defined on that grid between radii *InRad* and *InRad - 2·Dxy*. Then, the selected points on the flat grid are projected, and the points of the grid which correspond to the disk section where at least one projected point is present are set to 1. The other points, which do not correspond to the disk section, are set to 0 (see illustration in Figure A7).

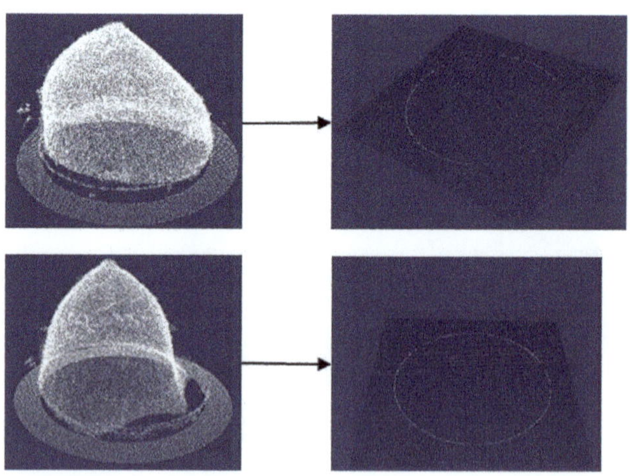

**Figure A7.** Verification procedure after ring removal. The white points on the flat grid (the right-hand images) correspond to the remaining ring points.

The remaining ring factor is computed by dividing the number of points equal to 1 present within the disk section on the flat grid (white points in the right-hand images in Figure A7) by the maximal number of points which may be present in the disk section. If this ratio is greater than *remaining reference factor*, a specific process to remove remaining ring points will be necessary to be applied. Otherwise, the regular process for artifact removal will be sufficient. Both processes are defined in Appendix C.1.

**Appendix C. Final Denoising**

This procedure performs 2D image processing steps to finalize the denoising process (see Section 2.3.2). It consists in the following processing steps:

- removal of the artifacts and/or removal of ring residuals;
- removal of remaining artifacts.

Each of the above processing steps is specified in a separate sub-section below.

*Appendix C.1. Removal of the Artifacts And/or Removal of Ring Residuals*

In this step, $Pcg_A$ (see Section 2.3.2) is processed as a 2D image to perform the artifact removal processing and/or the ring residuals removal processing. The selection of one of these processes has been described in Appendix B.2.

The removal procedures are based on morphological operations, so a binary image needs to be generated from $Pcg_A$, first of all. Then, erosion and dilation operations are performed, the result being a binary mask, which is applied to $Pcg_A$ by a multiplication operation. The resulting point cloud is named $Pcg_B$; this point cloud is further considered clean from any significant artifacts related to ring residuals.

The number of iterations and the structuring elements used in the erosion and dilation operations were empirically determined, based on the so far cumulated experience on both human and phantom breast scans.

Figure A8 shows the procedure to remove ring residuals in the case of Patient 004.

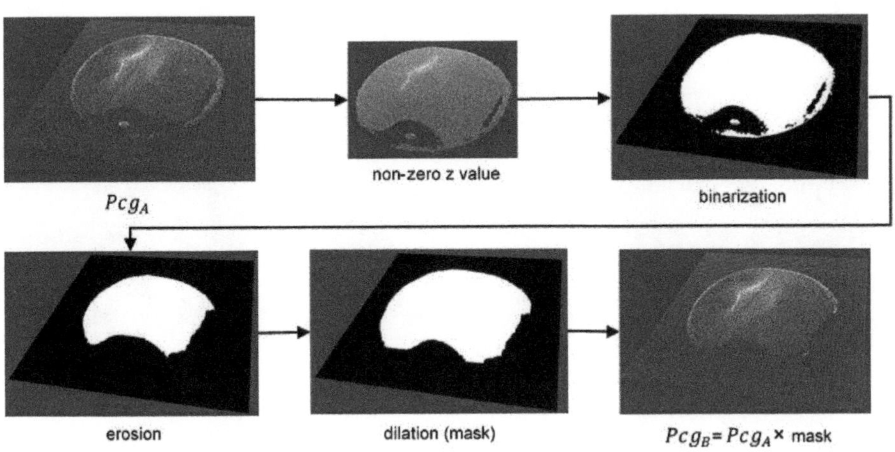


**Figure A8.** Illustration of the erosion-dilation procedure, which is applied to remove ring residuals. Illustration on Patient 004.

*Appendix C.2. Removal of Remaining Artifacts*

At this stage, residual artifacts from the previous steps may remain. The aim of this procedure is to further reduce the remaining artifacts along the z axis in $Pcg_B$, while preserving the details in the vicinity of the nipple. To do this, $Pcg_B$ is processed as a 2D image. $Pcg_B$ is divided in four vertical sections (cyan, magenta, blue and red in Figure A9). Median filtering using distinct kernel sizes for the four vertical sections of the breast is further applied.

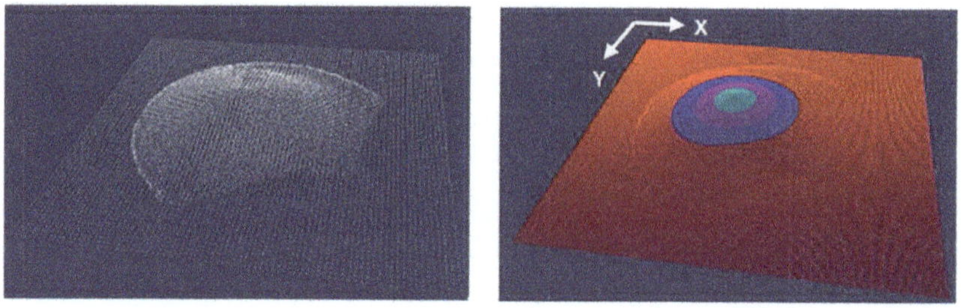


**Figure A9.** Illustration of the division of the stretched breast into four sections along the z axis. Illustration on Patient 002.

To divide $Pcg_B$ in four vertical sections Equation (16) is applied, with the first section corresponding to the nipple, the fourth section corresponding to the zone of the scanned breast the closest to the chest wall, while the second and third are middle sections of the scanned breast.

At this stage, the median filter is applied with a variable kernel size, which is defined as follows:

- a (1 × 1) kernel size for the nipple section (this kernel size allows to best preserve the breast surface details around the nipple);
- a (5 × 5) kernel size for the second breast section;
- a (7 × 7) kernel size for the third breast section;
- a (9 × 9) kernel size for the fourth breast section;

The specific combination of kernel sizes has been empirically defined and allowed to preserve a good continuity between the four sections of the reconstructed breast surfaces (see Figure A9), in the case of all the experimental and patient scans performed with the Wavelia OBCD subsystem, so far. The result of the median filter with variable kernel sizes is a point cloud named $Pcg_C$, as illustrated in Figure A10 for Patient 002.

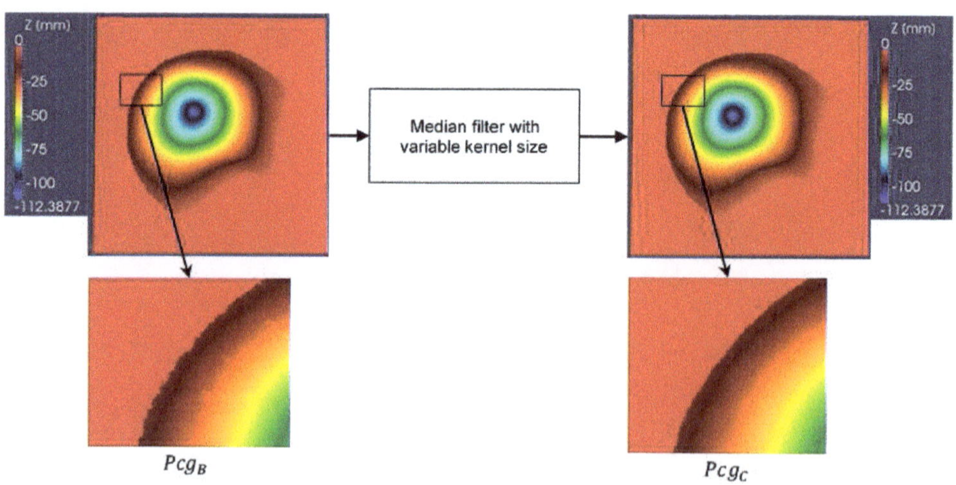


**Figure A10.** Result of the median filter with variable kernel size (the intensity of the images corresponds to the z coordinate of the corresponding pixels). Illustration for Patient002.

## References


1. Fear, E.C.; Meaney, P.M.; Stuchly, M.A. Microwaves for Breast Cancer Detection? *IEEE Potentials* **2003**. [CrossRef]
2. O'Loughlin, D.; O'Halloran, M.; Moloney, B.M.; Glavin, M.; Jones, E.; Elahi, M.A. Microwave breast imaging: Clinical advances and remaining challenges. *IEEE Trans. Biomed. Eng.* **2018**, *65*, 2580–2590. [CrossRef] [PubMed]
3. Fasoula, A.; Duchesne, L.; Cano, J.G.; Lawrence, P.; Robin, G.; Bernard, J.-G. On-Site Validation of a Microwave Breast Imaging System, before First Patient Study. *Diagnostics* **2018**, *8*, 53. [CrossRef] [PubMed]
4. Iversen, P.O.; Garreau, P.; Burrell, D. Real-time spherical near-field handset antenna measurements. *IEEE Antennas Propag. Mag.* **2001**. [CrossRef]
5. Fasoula, A.; Anwar, S.; Toutain, Y.; Duchesne, L. Microwave vision: From RF safety to medical imaging. In Proceedings of the 2017 11th European Conference on Antennas and Propagation, EUCAP, Paris, France, 19–24 March 2017. [CrossRef]
6. Fasoula, A.; Bernard, J.; Robin, G.; Duchesne, L. Elaborated breast phantoms and experimental benchmarking of a microwave breast imaging system before first clinical studyTitle. In Proceedings of the 12th European Conference on Antennas and Propagation, London, UK, 9–13 April 2018.
7. Registered Cinical Trial Protocol. Available online: https://clinicaltrials.gov/ct2/show/NCT03475992 (accessed on 11 February 2020).
8. Lawrence, P.; Fasoula, A.; Duchesne, L. RF-based Breast Surface Estimation—Registration with Reference Imaging Modality. In Proceedings of the 2018 IEEE International Symposium on Antennas and Propagation & USNC/URSI National Radio Science Meeting, Boston, MA, USA, 8–13 July 2018.
9. Winters, D.W.; Shea, J.D.; Madsen, E.L.; Frank, G.R.; van Veen, B.D.; Hagness, S.C. Estimating the breast surface using UWB microwave monostatic backscatter measurements. *IEEE Trans. Biomed. Eng.* **2007**. [CrossRef]
10. Sakamotov, T.; Sato, T. A target shape estimation algorithm for pulse radar systems based on boundary scattering transform. *IEICE Trans. Commun.* **2004**. [CrossRef]
11. Endo, F.; Kidera, S. Accuracy enhanced beamforming method based on envelope surface extraction for non-contact UWB breast cancer radar. In Proceedings of the ISAP 2016—International Symposium on Antennas and Propagation, Okinawa, Japan, 24–28 October 2016.
12. Sarafianou, M.; Preece, A.W.; Craddock, I.J.; Klemm, M.; Leendertz, J.A. Evaluation of Two Approaches for Breast Surface Measurement Applied to a Radar-Based Imaging System. *IEEE Trans. Antennas Propag.* **2016**. [CrossRef]
13. Williams, T.C.; Bourqui, J.; Cameron, T.R.; Okoniewski, M.; Fear, E.C. Laser surface estimation for microwave breast imaging systems. *IEEE Trans. Biomed. Eng.* **2011**. [CrossRef]
14. Kurrant, D.; Bourqui, J.; Fear, E. Surface estimation for microwave imaging. *Sensors* **2017**, *7*, 1658. [CrossRef]
15. Lee, W.Y.; Kim, M.J.; Lew, D.H.; Song, S.Y.; Lee, D.W. Three-dimensional surface imaging is an effective tool for measuring breast volume: A validation study. *Arch. Plast. Surg.* **2016**. [CrossRef]
16. Chae, M.P.; Rozen, W.M.; Spychal, R.T.; Hunter-Smith, D.J. Breast volumetric analysis for aesthetic planning in breast reconstruction: A literature review of techniques. *Gland. Surg.* **2016**. [CrossRef]
17. Choppin, S.B.; Wheat, J.S.; Gee, M.; Goyal, A. The accuracy of breast volume measurement methods: A systematic review. *Breast* **2016**. [CrossRef]
18. O'Connell, R.L.; Khabra, K.; Bamber, J.C.; deSouza, N.; Meybodi, F.; Barry, P.A.; Rusby, J.E. Validation of the Vectra XT three-dimensional imaging system for measuring breast volume and symmetry following oncological reconstruction. *Breast Cancer Res. Treat.* **2018**. [CrossRef] [PubMed]
19. Khawaldeh, S.; Aleef, T.A.; Pervaiz, U.; Minh, V.H.; Hagos, Y.B. Complete End-to-End Low Cost Solution to a 3D Scanning System with Integrated Turntable. *Int. J. Comput. Sci. Inf. Technol.* **2017**. [CrossRef]
20. Song, T.; Zhou, L.; Ding, X.; Yi, W. 3D Surface Reconstruction Based on Kinect Sensor. *Int. J. Comput. Theory Eng.* **2013**. [CrossRef]
21. Ye, Y.; Song, Z. An accurate 3D point cloud registration approach for the turntable-based 3D scanning system. In Proceedings of the 2015 IEEE International Conference on Information and Automation, ICIA, Lijiang, China, 8–10 August 2015. [CrossRef]

22. Xiao, W.; Deng, M.; Li, C. A surface reconstruction algorithm based on 3D point cloud stratified sliced. *Sens. Transducers* **2014**, *173*, 197.
23. Rankin, J.R. *Computer Graphics Software Construction*; Prentice Hall of Australia: Sydney, Australia, 1989; ISBN 0-7248-0194-9.
24. Tomasi, C.; Manduchi, R. Bilateral filtering for gray and color images. In Proceedings of the IEEE International Conference on Computer Vision, Bombay, India, 7 January 1998. [CrossRef]
25. Zhang, C.; Chen, T. Efficient feature extraction for 2D/3D objects in mesh representation. In Proceedings of the IEEE International Conference on Image Processing, Thessaloniki, Greece, 7–10 October 2001. [CrossRef]
26. Klamkin, M.S. On the Volume of a Class of Truncated Prisms and Some Related Centroid Problems. *Math. Mag.* **1968**. [CrossRef]
27. Zastrow, E.; Davis, S.K.; Lazebnik, M.; Kelcz, F.; van Veen, B.D.; Hagness, S.C. Development of anatomically realistic numerical breast phantoms with accurate dielectric properties for modeling microwave interactions with the human breast. *IEEE Trans. Biomed. Eng.* **2008**. [CrossRef]
28. Cignoni, P.; Callieri, M.; Corsini, M.; Dellepiane, M.; Ganovelli, F.; Ranzuglia, G. MeshLab: An open-source mesh processing tool. In Proceedings of the 6th Eurographics Italian Chapter Conference, Salerno, Italy, 2–4 July 2008.
29. Ahrens, J.; Geveci, B.; Law, C. ParaView: An end-user tool for large-data visualization. In *Visualization Handbook*; Elsevier: Amsterdam, The Netherlands, 2005; ISBN 978-0123875822.
30. Highnam, R.; Brady, M.; Yaffe, M.J.; Karssemeijer, N.; Harvey, J. Robust breast composition measurement—Volpara$^{TM}$. In *International Workshop on Digital Mammography*; Springer: Berlin, Germany, 2010. [CrossRef]
31. Brandt, K.R.; Scott, C.G.; Ma, L.; Mahmoudzadeh, A.P.; Jensen, M.R.; Whaley, D.H.; Wu, F.F.; Malkov, S.; Hruska, C.B.; Norman, A.D.; et al. Comparison of Clinical and Automated Breast Density Measurements: Implications for Risk Prediction and Supplemental Screening. *Radiology* **2016**. [CrossRef]
32. Ko, S.Y.; Kim, E.-K.; Kim, M.J.; Moon, H.J. Mammographic density estimation with automated volumetric breast density measurement. *Korean J. Radiol.* **2014**. [CrossRef]
33. Gubern-Mérida, A.; Kallenberg, M.; Platel, B.; Mann, R.M.; Martí, R.; Karssemeijer, N. Volumetric breast density estimation from full-field digital mammograms: A validation study. *PLoS ONE* **2014**. [CrossRef] [PubMed]
34. Wang, J.; Azziz, A.; Fan, B.; Malkov, S.; Klifa, C.; Newitt, D.; Yitta, S.; Hylton, N.; Kerlikowske, K.; Shepherd, J.A. Agreement of mammographic measures of volumetric breast density to MRI. *PLoS ONE* **2013**. [CrossRef] [PubMed]
35. Fasoula, A.; Moloney, B.M.; Duchesne, L.; Cano, J.G.; Oliveira, B.L.; Bernard, J.G.; Kerin, M.J. Super-resolution radar imaging for breast cancer detection with microwaves: The integrated information selection criteria. In Proceedings of the 41st Annual International Conference of the IEEE Engineering in Medicine & Biology Society (EMBC), Berlin, Germany, 23–27 July 2019.
36. O3D Camera. Available online: https://www.ifm.com/fr/en/product/O3D302 (accessed on 11 February 2020).
37. Ensenso N10 Camera. Available online: https://en.ids-imaging.com/ensenso-n10.html (accessed on 11 February 2020).

*applied sciences*

MDPI

*Article*

# Low-Cost, Compact, and Rapid Bio-Impedance Spectrometer with Real-Time Bode and Nyquist Plots

**Didik R. Santoso [1,*], Bella Pitaloka [1], Chomsin S. Widodo [2] and Unggul P. Juswono [2]**

1. Division of Instrumentation, Department of Physics, Brawijaya University, Jl. Veteran Malang 65145, Indonesia; pitalokabella@student.ub.ac.id
2. Division of Medical Physics & Biophysics, Department of Physics, Brawijaya University, Jl. Veteran Malang 65145, Indonesia; chomsin@ub.ac.id (C.S.W.); unggul-pj@ub.ac.id (U.P.J.)

* Correspondence: dieks@ub.ac.id

Received: 31 December 2019; Accepted: 20 January 2020; Published: 28 January 2020

**Abstract:** Bioelectric impedance spectroscopy (BIS) has been widely used to study the electrical properties of biological tissue based on the characteristics of the complex electrical impedance dispersions. One of the problems in using the BIS method is the length of time required for the data acquisition process and possibly data analysis as well. In this research, a compact and work rapidly BIS instrumentation system has been developed at a low cost. It is designed to work in the frequency range of 100 Hz to 100 kHz, which is generally used in the fields of biophysics and medical physics. The BIS instrumentation system is built using several integrated modules. The modules are an AC current source to produce a selectable injection current; a data acquisition system to measure voltage, current, and phase difference rapidly and simultaneously; and software to calculate and display measurement results in the form of Bode and Nyquist plots in real time. The developed BIS system has been validated using a simple RC circuit as the sample being tested. The average time needed in the process of data acquisition and analysis until the formation of impedance dispersion curves in the form of Bode and Nyquist plots, for 54 sample frequencies, is less than one minute. The system is able to identify R and C values of the sample with a maximum error of 1.5%. In addition, some simple application examples are also presented in this paper.

**Keywords:** BIS-instrumentation; biological-tissue; gain-phase detector; Bode plot; Nyquist plot

## 1. Introduction

Bioelectrical impedance spectroscopy (BIS) has been currently used in broad fields of studies, such as biophysics, medical physics, agriculture, and other fields of life-sciences. BIS is a non-invasive method for characterizing biological materials based on their electrical properties, which is expressed in terms of complex electrical impedance dispersion curves over a certain frequency range [1,2]. In here, there are three regions of bioelectrical impedance dispersion, $\alpha$, $\beta$, and $\gamma$. The $\alpha$-dispersion (10 Hz to 10 kHz) is related to the phenomena of ionic diffusion of the cell membrane and the counterion effects, the $\beta$-dispersion (10 kHz to 100 MHz) is related to the polarization phenomenon of cell membranes, and the $\gamma$-dispersion (in GHz order) is related to the polarization of water molecules [3]. Properties of electrical impedance have been reported to understand the nature and behavior of cell [4–8], bacteria detection and identification [9,10], identification of beverages and food ingredients [11–16], and so on. In the medical field, BIS is used to characterizations of serum and blood [17–19], identification of body composition and disease [20–22], and to assist the process of diagnosis and functional electrical stimulation of the body [23–25]. Research on BIS, as mentioned, mostly covers $\alpha$ and $\beta$ dispersion, and is carried out using a frequency of 10 Hz to 1 MHz.

In principle, identification of the electrical impedance of biological tissue performed by injecting alternating current, $I$, into the sample through a pair of current electrodes, and measuring the voltage,

$\boldsymbol{V}$, which arises through a pair of voltage electrodes. Then, the impedance, $\boldsymbol{Z}$, is calculated by using formula $\boldsymbol{Z} = \boldsymbol{V}/\boldsymbol{I}$. It should be noted that $\boldsymbol{V}$, $\boldsymbol{I}$, and $\boldsymbol{Z}$ is complex variables, therefore a BIS instrumentation system must not only be able to measure the magnitudes of current and voltage, but must also be able to measure their phase difference. Therefore, a BIS instrumentation system is at least having three functional units. First is an AC current source, which must be stable within working frequency range; second is a data acquisition system to record voltage, current, and their phase difference; and third is software for computing and displaying the impedance dispersion curves, which is commonly in the Bode plot or Nyquist plot.

Furthermore, in an effort to provide a BIS instrumentation system in good performance, some researchers have offered the new design. Starting from precision AC current sources [26], the high accuracy data acquisition system [27,28], even the whole system device [29–33]. Commonly, the offered BIS instrumentation system is arranged in modular and work separately. It has weaknesses in compactness and synchronization of the hardware system. Besides that, most of the offered BIS instrumentation cannot perform automatically frequency scans. The measurement process is carried out manually starting from the lowest to the highest frequency, then the measurement results are plotted using MS-Excel or others data processing software by entering the recorded measuring data. These methods will require a long time, so it is inefficient to produce an impedance dispersion curve.

In this research, we propose a low-cost, rapid, and compact BIS instrumentation system with Bode and Nyquist plot in real-time. There are several new things we offer:

- The system contains three main modules in an integrated (compact) ones, so synchronization among modules can be maintained/controlled by PC, and it can work automatically.
- The system uses programmable AC current source, so that the value of the current to be injected into biological samples can be selected/regulated through software via PC.
- The data acquisition system is built based on gain phase detector (GPD-AD8302), so that the value of $\boldsymbol{V}$, $\boldsymbol{I}$, and phase difference can be identified rapidly.
- We developed computational software (in Delphi) to calculate the complex impedance values of the samples, and also software to draw Bode and Nyquist plots. We do not need another application program (e.g., MS-Excel) to do the calculations separately. Therefore, the time needed for computation and data analysis to be reduced very significantly.

## 2. Materials and Methods

A block diagram of BIS instrumentation system which is proposed in this research is given in Figure 1. It is built in several integrated modules, and can be grouping into three main units, i.e., AC current source, data acquisition, and software. The unit of AC current source is a stable current source with selectable frequency within a certain frequency range.

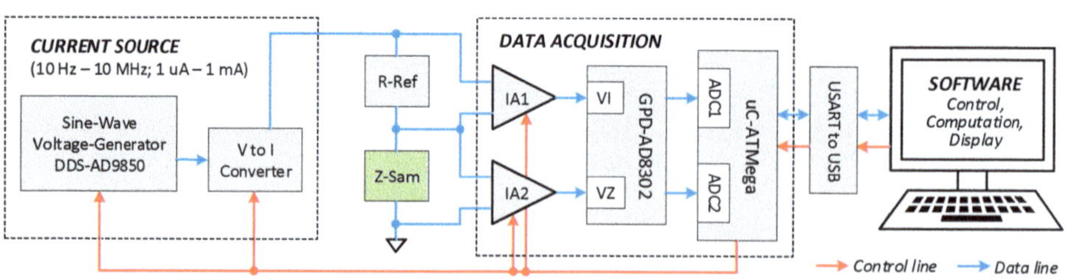


**Figure 1.** Block diagram of the developed BIS instrumentation system.

In the implementation, an AC current source with a certain frequency is injected into the sample through a reference resistor (R-Ref). Function of R-Ref is to guarantee that the current which is recorded by the data acquisition system is the actual current received by the sample. In this case, the actual current value is the voltage value read by IA1 divided by R-Ref, while the sample voltage arising due to injection current is read by IA2. Next, the outputs of IA1 and IA2 are forwarded to the gain phase detector module (GPD-AD8302). GPD performs an internal process to produce the gain in dB of (*V/I*) and the phase difference, the both in form of the equivalent voltages. These two signals are then converted into digital data by the ADC contained in the ATmega microcontroller, before being sent to the computer (PC). On the PC, the data is calculated to produce an impedance value, and plotted directly as single point on the Bode and Nyquist charts. This process is repeated for other frequencies until one full frequency cycle is completed.

Here, the software is developed and installed on the microcontrollers and PC, used for three purposes. The first is to control the working function of the hardware as a whole, specifically the selection of the current value and the desired working frequency range. The second is to calculate the complex impedance of the sample based on data from GPD, and the third is to present the complex impedance dispersion of the sample in the form of Bode and Nyquist plots in real time.

### 2.1. Unit of AC Current Source

The BIS instrumentation system requires a pure sine current signal with a selectable frequency and amplitude. For this purpose, the unit of AC current source is developed using two modules, i.e., the voltage generator and the voltage to current (*V/I*) converter. The voltage generator is serving to produce a sine voltage in the frequency range of 10 Hz to 10 MHz, while the *V/I* converter functions to change the voltage source into current source at the related frequency, with selectable current amplitude of the 1 μA, 10 μA, 100 μA, and 1 mA.

#### 2.1.1. Voltage Generator

The AC voltage generator is built based on the AD9850 module, as shown in Figure 2b. The module is widely available in the commercial market with price around $12. The AD9850 is a highly integrated device that uses advanced DDS technology, digitally programmable frequency synthesizer, and clock generator function. It has capabilities to produce sine and square waves signals 1-volt peak to peak ($V_{pp}$) in amplitude and frequency range from DC to 40 MHz. The frequency tuning, control, and phase modulation words are loaded into the AD9850 via a parallel byte or serial loading format. The parallel load format consists of five iterative loads of an 8-bit control word, while serial loading is accomplished via a 40-bit serial data stream on a single pin [34].

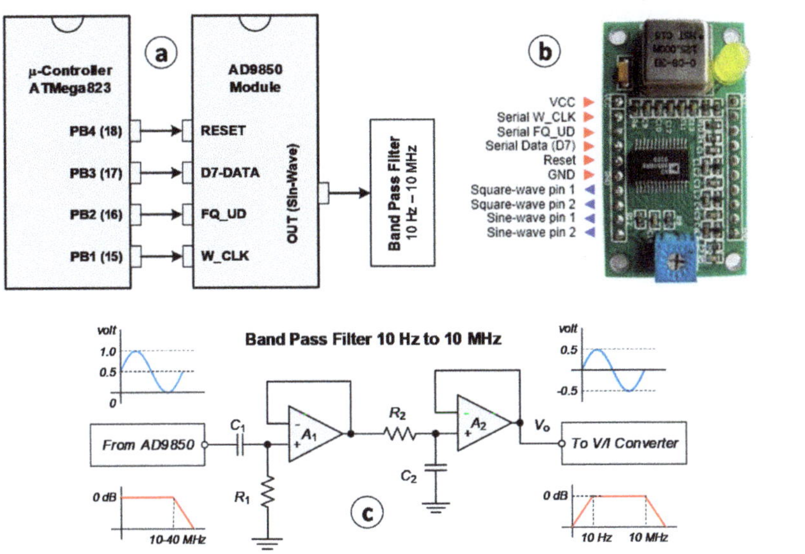


**Figure 2.** AD9850-based voltage generator module with frequency range of 10 Hz to 10 MHz.

In this project, the AD9850 module is used to generate sine-wave voltage signals. The determining frequency value is controlled by software that installed on the microcontroller ATMega823 via serial loading format. Figure 2a shows pins connectivity between the AD9850 and the ATMega823. W_CLK is a clock signal for 40-bits serial data streaming, FQ_UD is frequency update signal for new data streaming, D7-Data is 5-bytes data streaming, and RESET is signal to reset the AD9850 to the original state. In here, simple op-amp circuit as shown in Figure 2c is perform a band pass filter (BPF), is used to conditioning the output of AD9850 in 'pure' sine wave signal within frequency range of 10 Hz to 10 MHz and amplitude of 0.5 volt.

2.1.2. Voltage to Current (*V/I*) Converter

As explained above, in this project we use a current source (not a voltage source) to be injected into biological samples. Therefore, the voltage that has been generated by the voltage generator must be converted to an equivalent current. For this purpose, we use a voltage to current (*V/I*) converter. There are several types of *V/I* converter, one of the most popular is Howland current source (HCS). The HCS is a voltage controlled current source (VCCS) with loads connected to the ground. The HCS has advantages in the stability, grounded mode, and simple application method. In this project HCS with buffered feedback path is used, as given in Figure 3. The use of buffered feedback path can provide advantages in high impedance measurements.

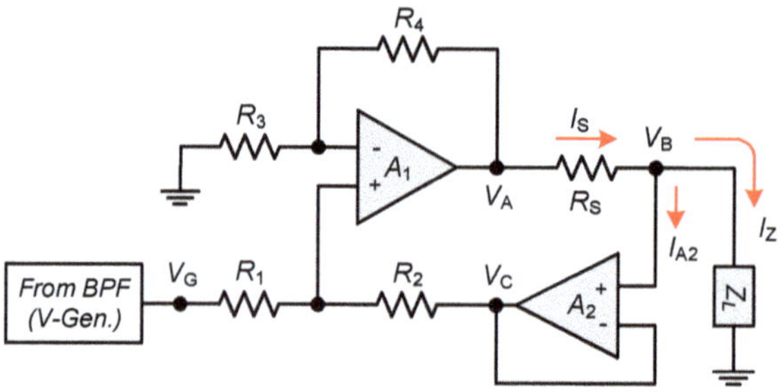


**Figure 3.** Howland current source with buffered feedback.

Considering the op-amp role and Kirchhoff law, the output voltage at the point of the circuit in Figure 3 can be obtained

$$V_{-A1} = V_{+A1}; V_C = V_B \tag{1}$$

$$\left(\frac{R_3}{R_3+R_4}\right)V_A = \left(\frac{R_1}{R_1+R_2}\right)V_G + \left(\frac{R_2}{R_1+R_2}\right)V_C \tag{2}$$

$$V_A = \left(\left(\frac{R_1}{R_1+R_2}\right)V_G + \left(\frac{R_2}{R_1+R_2}\right)V_C\right)\left(\frac{R_3+R_4}{R_3}\right) \tag{3}$$

For $R_1 = R_2 = R_3 = R_4 = R$, then

$$V_A = \left(\left(\frac{R}{2R}\right)V_G + \left(\frac{R}{2R}\right)V_B\right)\left(\frac{2R}{R}\right) = V_G + V_B \tag{4}$$

$$I_S = \frac{V_A - V_B}{R_S} = \frac{V_G + V_B - V_B}{R_S} = \frac{V_G}{R_S} \tag{5}$$

Current injected to the material load $\mathbf{Z}_L$ is $\boldsymbol{I}_Z$. According to Kirchhoff's Current Law (KCL),

$$I_Z = I_S - I_{A2} \tag{6}$$

$I_{A2}$ is current entering the A2 op-amp, and for high input impedance op-amp, $I_{A2} \approx 0$, so that

$$I_Z = I_S = \frac{V_G}{R_S} \tag{7}$$

Based on Equation (7), and with $V_G$ set constant at 0.5 volts, then the current value is only determined by $R_S$. By selecting $R_S$ of 500 kΩ, 50 kΩ, 5 kΩ, and 500 Ω, an $I_Z$ current value of 1 μA, 10 μA, 100 μA, and 1 mA will be obtained, respectively. In addition, in this research we use LF357 for op-amp A1 and LF351 for op-amp A2. The LF357 has GBW 20 MHz and input impedance in order TΩ, while the LF351 has GBW of 4 MHz and input impedance in order of TΩ. Both are low-cost JFET input op-amp. In addition, to provide current selector is used an analog multiplexer CD4051, and it is controlled by the microcontroller.

### 2.2. Unit of Data Acquisition System

Unit of the data acquisition system is tasked to read and record data of current, voltage and phase difference along measuring process. Figure 4 provides a technical illustration of how the data is recorded. $R_{Ref}$ is reference resistor, use to ensure that the current measured by the data acquisition system is the actual current that is injected into the sample. The value of the actual current $Iz$ is the voltage on the reference resistor divided by the value of the reference resistor, $R_{Ref}$.

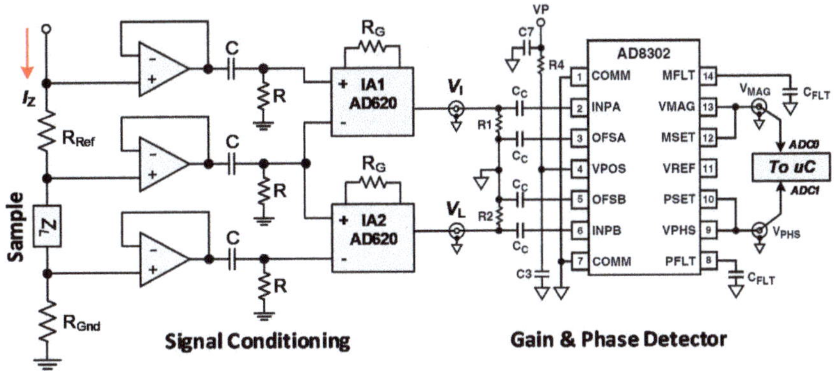


**Figure 4.** Data acquisition system: signal conditioning (SC) and gain-phase detector (GPD).

There are two parts in data acquisition system device, i.e., signal conditioning circuit (SC) and gain phase detector (GPD). The SC part perform high impedance buffering, filtering, and amplifying, as required by the GPD. The buffer is an isolation circuit, it serves to ensure that the current and voltage signals are not affected by the circuit afterwards. Each buffer circuit is performed by LF353 op-amp. The RC circuit after buffer forms a high pass filter (HPF). Each HPF is set at 10 Hz cut-off frequency, is to eliminate dc-offset voltage that might arise. Finally, the AD620 instrumentation amplifier (IA) is used in two functions, first is to convert the differential signal to single ended, and second is to amplify the signals ($G_1$ and $G_2$), if needed. To provide voltage gain control through $R_G$ selection, is used an analog multiplexer CD4051, and it is controlled by the microcontroller. In addition, the AD620 is low-cost IA with working frequency up to 1 MHz. However, it should be noted that a greater gain will cause the maximum working frequency will also decrease.

Based on the explanation, then the complex impedance values of the measured sample can be calculated as

$$Z_L = |Z_L|\angle\theta_L = \frac{V_L}{I_L} = \frac{|V_L|\angle\theta_{VL}}{|I_L|\angle\theta_{IL}};\ \textit{where}\ |Z_L| = |V_L|/|I_L|\ \textit{and}\ \theta_L = (\theta_{VL} - \theta_{IL}) \tag{8}$$

$$V_Z = \frac{V_L}{G_2} = \frac{|V_L|\angle\theta_L}{G_2} = |V_Z|\angle\theta_L;\ where\ |V_Z| = \frac{|V_L|}{G_2} \quad (9)$$

$$I_Z = \frac{V_R}{G_1 * R_{Ref}} = \frac{|V_R|\angle\theta_R}{G_1 * R_{Ref}} = |I_Z|\angle\theta_R;\ where\ |I_Z| = \frac{|V_R|}{G_1 * R_{Ref}} \quad (10)$$

so that,

$$Z_L = |Z_L|\angle\theta_L; where\ |Z_L| = |V_L|/|I_L|\ and\ \theta_L = (\theta_{VL} - \theta_{IL}) \quad (11)$$

Furthermore, in the development of a BIS instrumentation system that can work quickly, the use of GPD devices such as AD8302 as part of a data acquisition system is the right choice. The AD8302 has a very good ability in terms of identifying the gain and phase difference between two signals, from low frequencies up to 2700 MHz [35]. Common configuration the use of AD8302 for gain and phase difference measurement, especially for low frequency operation is shown if Figure 4 above. In the figure, two $C_C$ capacitors connected to the OFSA and OFSB pins are the useful components to set high pass corner frequency ($f_{CH}$) of the measured signals. According to the application note of the AD8302 at low frequency operation [36], formula to calculate $f_{CH}$ is given by

$$f_{CH}(\text{kHz}) = 2/C_C\ (\mu\text{F}) \quad (12)$$

Working principle of the AD8302 is compares the two input voltage signals i.e., $V_{INPA}$ and $V_{INPB}$, and generates two output voltage signals i.e., $V_{MAG}$ and $V_{PHS}$. Where the $V_{MAG}$ is output voltage proportional to the decibel ratio of voltage signals applied to INPA and INPB, while the $V_{PHS}$ is output voltage proportional to the phase difference between INPA and INPB. Relationship between input signals ($V_{INPA}$ and $V_{INPB}$) and output signals ($V_{MAG}$ and $V_{PHS}$) is as shown in Figure 5. The AD8302 has magnitude measurement range from −30 dB to +30 dB with a sensitivity of 30 mV/dB, and phase measurement range from (−180 to 0) degrees with a sensitivity of 10 mV/degree or (0 to +180) degrees with a sensitivity of −10 mV/degree.

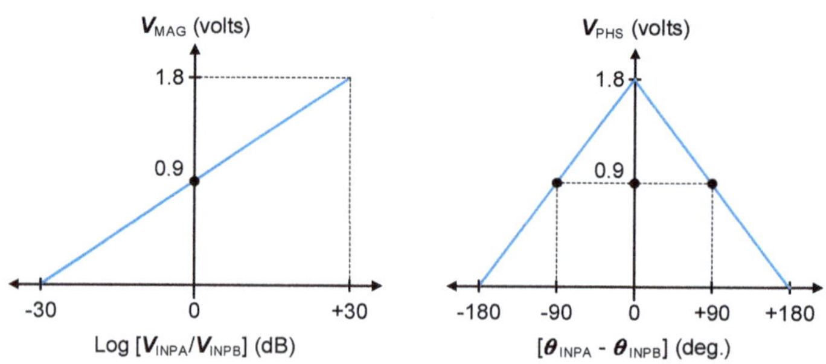


**Figure 5.** Relationship between magnitude (dB) to the $V_{MAG}$, and between phase difference (deg.) to the $V_{PHS}$.

Based on the graphs in Figure 5, then the formulas for calculating $V_{MAG}$ and $V_{PHS}$ are

$$V_{MAG} = (30\ \text{mV}/\text{dB}) * (\log(V_{INPA}/V_{INPB})) + 900\ \text{mV} \quad (13)$$

$$V_{PHS} = (10\ \text{mV}/\text{degree}) * ((\theta_{INPA} - \theta_{INPB}) - 90^\circ) + 900\ \text{mV} \quad (14)$$

From Equations (13) and (14), the magnitude (dB) and phase angle (degree) can be written as,

$$mag\ (\text{dB}) = \log\left(\frac{V_{INPA}}{V_{INPB}}\right) = \frac{V_{MAG} - 900\text{mV}}{30\text{mV}} \quad (15)$$

$$\theta\ (\text{degree}) = [\theta_{INPA} - \theta_{INPB}] = \frac{1800\ \text{mV} - V_{PHS}}{10\ \text{mV}} \quad (16)$$

Then, the real and imaginary parts of complex impedance can be written as,

$$\begin{aligned} |Z_L| &= 10^{\frac{mag}{20}} * R_{Ref} \\ Z_{Re} &= |Z_L| \cos\,\theta \\ Z_{Im} &= |Z_L| \sin\,\theta \end{aligned} \tag{17}$$

### 2.3. Software

The software was developed in Delphi, and is used for two purposes. First is for controlling the hardware system, and second is for data analysis and display.

Software for controlling the hardware system is responsible to:

- Determine the value of the current used (1 μA, 10 μA, 100 μA, or 1 mA). This is related to the selection of the $\boldsymbol{R_S}$ value that stated in Equation (7).
- Determine low frequency and high frequency limits, and the number of data samples. This is related to the DDS-AD9850 voltage generator programming.
- Determine the voltage gain value of the reference signal and sample signal. This is related to the selection of the $\boldsymbol{R_G}$ value to produce $\boldsymbol{G_1}$ and $\boldsymbol{G_2}$ that stated in Equation (10).

Whereas, software for data analysis and display is responsible to:

- Calculation of values of the current, voltage, phase angle, and impedance (using Equations (15) and (16))
- Calculation of real and imaginary parts of impedance (using Equation (17))
- Plot the impedance value in the Bode (magnitude and phase as a function of frequency)
- Plot the impedance value in Nyquist (real part vs. imaginary parts)

## 3. Results and Discussion

### 3.1. Prototype of the BIS Instrumentation System

The prototype of the developed BIS instrumentation system is given in Figure 6, where Figure 6a is photo of the hardware system and Figure 6b is main display of the software system. The main unit of the hardware system is current source and data acquisition system. The hardware system is equipped with connectors for biological samples and a connector to the PC as the main control device and data processing and display unit. To inject the current from the BIS device into the biological sample, two types of connectors can be chosen—i.e., female USB and BNC. The female USB connector is used for interdigitated electrodes (IDT), while BNC connectors is used for parallel plates or needle electrodes. In the software section, it provides several pre-acquisition menus such as selecting low and high frequency limits as well as determining the frequency samples, selecting the current value, and selecting the voltage amplification value. At the time of data acquisition, the measurement results are displayed in the form of Bode plots (magnitude and phase-angle), Nyquist plots, and also in MS-Excel table data for further purposes. In addition, the software system also displays the results of calculating the resistance and capacitance values of the samples that are carried out automatically by the program.

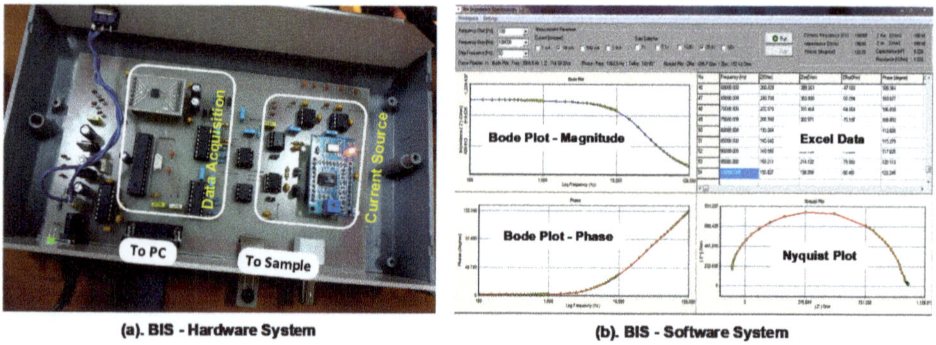


**Figure 6.** Prototype of developed BIS instrumentation system: (**a**) hardware, and (**b**) software.

Furthermore, it is important to state the validation results on the BIS instrumentation system that has been made, specifically the performance of the current source and the data acquisition system. Figure 7a shows the characteristics of the voltage generator, that stated in Figure 2, from frequencies of 1 Hz to 10 MHz obtained from the test results. As we can see, the signal has a good and perfect sine wave shape at frequencies of 1 kHz, 1 MHz, and 10 MHz. At low frequencies up to 2 MHz, the signal has an almost constant amplitude of 0.5 volts. However, starting at 2 MHz, the signal amplitude starts to decrease with increasing frequency. So that the voltage generator module can work well in low frequencies up to 2 MHz.

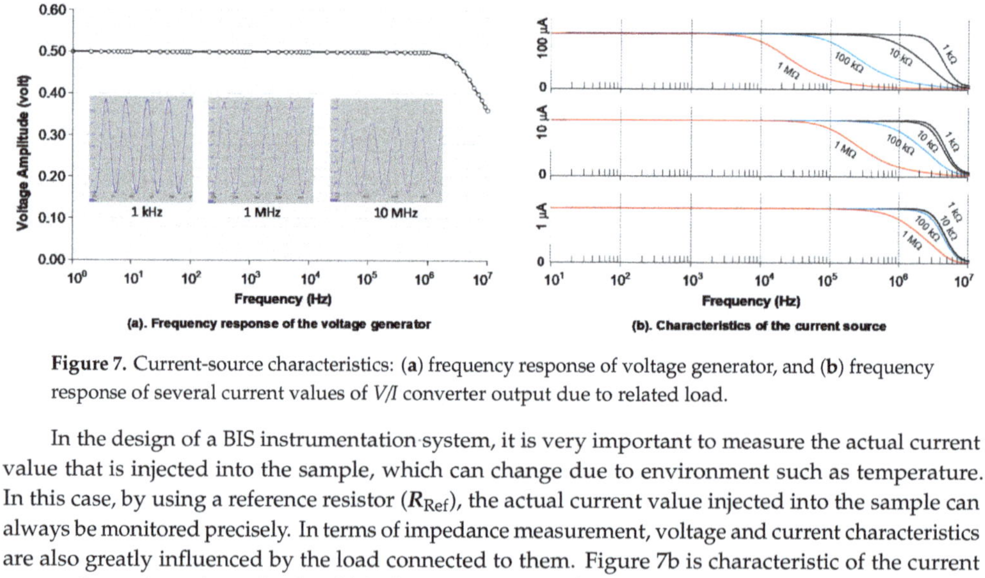


**Figure 7.** Current-source characteristics: (**a**) frequency response of voltage generator, and (**b**) frequency response of several current values of *V/I* converter output due to related load.

In the design of a BIS instrumentation system, it is very important to measure the actual current value that is injected into the sample, which can change due to environment such as temperature. In this case, by using a reference resistor ($R_{Ref}$), the actual current value injected into the sample can always be monitored precisely. In terms of impedance measurement, voltage and current characteristics are also greatly influenced by the load connected to them. Figure 7b is characteristic of the current source due to impedance loads within frequency range of 10 Hz to 10 MHz, for current setting of 1 µA, 10 µA, and 100 µA. We can see, although the voltage source to be changed has a flat shape at frequencies up to 2 MHz, but the current will decrease at high frequencies, along with the greater load. In here, the choice of a smaller current will provide benefits in terms of high frequency operations. However, keep in mind that biological tissue commonly has a small impedance value, so the use of a small current will also produce very small voltage. This is where a voltage amplifier device is needed. In addition, with regard to the operating frequency, the BIS system has effective performance in the frequency operation of 100 Hz to 100 kHz. Although the frequency range of 10 Hz to 1 MHz still can be achieved, but it will provide relatively large error, especially in the lower and upper frequency borders.

Moreover, it is important to mention here how to make Bode and Nyquist plots in the BIS system that was developed. In here, Bode and Nyquist plots is carried out per single frequency, and repeated until full frequency cycle has been reached. The procedure for measuring impedance using the developed BIS system including the estimated length of time required is described as below.

(a) The first step is preparing for pre-acquisition; such as determining the current value, the frequency limit, number of sampling frequencies, and voltage gain.
(b) Initialization and generation of sine-voltage at one frequency by the AD9850. The time required, including providing a delay time so that the signal is stable is around 200 ms.
(c) The microcontroller takes 50 pieces of $V_{MAG}$ and $V_{PHS}$ from AD8302 for single-frequency operation and directly sent to the PC. On the PC the data is averaged to get the best data. The time required is around 600 ms.
(d) Calculation the value of 'mag (dB)' using Equation (15) and the value of '$\theta$ (degree)' using Equation (16). Calculation the value of complex impedance magnitude, real part and imaginary part of complex impedance using Equation (17). Next, plotting the results on the Bode and Nyquist graphs provided. The time required is around 200 ms.
(e) Return (looping) to point (b), until all frequencies stated in point (a) have been completed.

For example, in our experiments we used the frequency range from 100 Hz to 100 kHz, and the sample frequencies were: 100 Hz, 150 Hz, 200 Hz, … , 1 kHz, 1.5 kHz, 2 kHz, … , 10 kHz, 15 kHz, 20 kHz, … 100 kHz (there are 54 sample frequencies). The time needed to measure complex impedances at a single frequency is around 1000 ms (1 s). Thus, the time needed to get the experimental data measured in one full frequency cycle is 54 s (less than 1 min).

### 3.2. Application Examples

In order to verify the reliability of the developed BIS system, we carry out two kinds experiments with different samples. First experiment used a parallel RC circuit, and the second experiment used liquid solution—i.e., mixed water and milk with different concentration.

#### 3.2.1. Parallel RC Circuit

A parallel RC circuit is the simplest equivalent electrical model of biological tissue. In this experiment, four RC circuit configurations were used, with the R value being fixed at 5.6 kΩ and the C values set different at values of 1 nF, 10 nF, 33 nF, and 47 nF. These values are those listed on the components body. Measurement by using Precision LCR meters ST2830, the following values are obtained R fix is 5.45 kΩ, and C are 0.93 nF, 9.41 nF, 34.15 nF, and 51.46 nF simultaneously.

In the experiment, measurements were made using a BNC connector, with both components soldered directly to the ends of the connector. We use 10 μA current value and several choices voltage amplification. Result of the experiment for each configuration is displayed at the Bode plot, Nyquist plot, and Excel's data format in real-time. Cumulative results is given in Figure 8 as Bode plot, both in magnitude (ohm) and phase difference (degree). Based on the data in Figure 8, the calculation of component values performed by the program gets the values as shown in Table 1. The comparison of the 'true values' and measured values of components give maximum error of 1.5%.

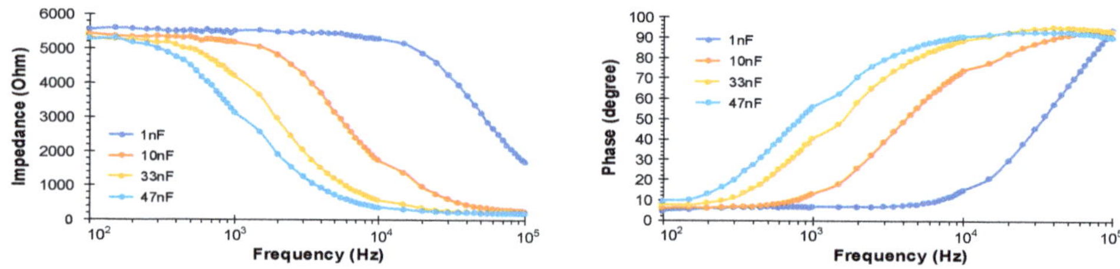


(a) (b)

**Figure 8.** Bode plot of RC experiments: (**a**) impedance magnitude (ohm), and (**b**) phase (degree).

**Table 1.** Measurement result and error calculation

| Prec. RLC Meter | | Dev. BIS | | Error (%) | |
|---|---|---|---|---|---|
| R (kΩ) | C (nF) | R (kΩ) | C (nF) | C (%) | R (%) |
| 5.45 | 0.93 | 5.472 | 0.938 | 0.40 | 0.86 |
| 5.45 | 9.41 | 5.489 | 9.341 | 0.72 | 0.73 |
| 5.45 | 34.15 | 5.377 | 34.653 | 0.64 | 1.47 |
| 5.45 | 51.46 | 5.415 | 52.602 | 1.34 | 1.17 |
| | | Average Error (%) | | 0.78 | 1.06 |

3.2.2. Mixed Water and "Yakult" with Different Concentration

In this experiment we use interdegitated electrode (IDT). The IDT electrode is very suitable for the measurement of impedance of small amounts of liquid material. More about the use of IDT electrodes can be seen in several references, and not discussed in this paper. The liquid sample used in this experiment is a mixture of mineral water and milk ("Yakult") in several types of concentrations. The experiment results are given in Figure 9.

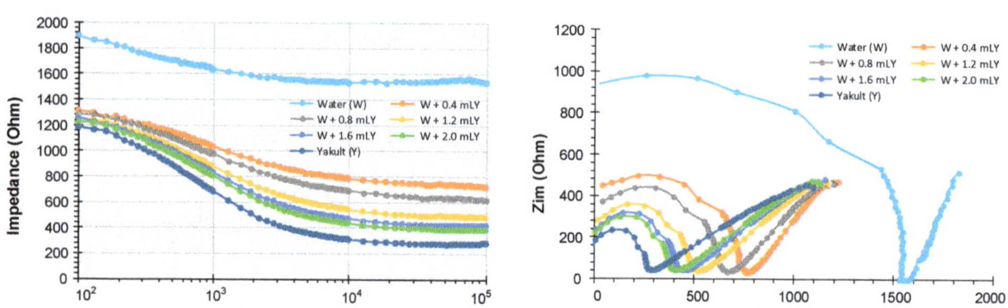


(a) (b)

**Figure 9.** Impedance values of Yakult solution: (**a**) in Bode plot, and (**b**) in Nyquist plot.

Figure 9a presents impedance values of samples in magnitude Bode plot, and Figure 9b is in Nyquist plot. It appears that the electrical impedance of mineral water without Yakult addition has the highest value, this is because water is a material with weak electrolyte properties so the impedance value is relatively high. The addition of Yakult material of 0.4 mL, 0.8 mL, 1.2 mL, 1.6 mL, and 2.0 mL to the water causes the electrical impedance of the mixture liquids to decrease. In here, the greater the volume of Yakult added to the water, the greater the decrease in impedance that occurs. This is because Yakult is an electrolyte solution, so the conductivity of the sample will increase, which means the total resistance or impedance will decrease. Furthermore, using the Nyquist plot given in Figure 9b, we can study the phenomenon of ion transport and bulk electrolyte resistance, which is very important in material characterization by using electrical impedance spectroscopy method.

## 4. Conclusions

In this research, a low-cost, concise, and fast bioelectrical impedance spectrometer (BIS) system has been successfully developed. This system works with current sources that can be selected through software, with values of 1 μA, 10 μA, 100 μA, and 1 mA. Working frequency that can be achieved is from 10 Hz to 1 MHz, and is optimal in the frequency range of 100 Hz to 100 kHz. This is related to the working ability of GPD-AD8302 at low frequencies, and also the limitations of op-amps in high frequency operations. The results of measurement of complex impedance dispersion of the measured material are displayed in real-time in the form of Bode and Nyquist plots. The use of automatic frequency scanning techniques, the AD8302 GPD device, and the calculation of complex impedance values carried out automatically by software, is the key to this equipment being able to work quickly. To measure the full cycle in a certain frequency range, including displaying the results in the Bode and Nyquist plots, it only takes no more than one minute. The developed BIS system has been tested to measure the complex impedance of an RC circuit, and is able to detect component values with a maximum error of 1.5%. It has also been used to measure the impedance distribution of a solution in different concentrations, and is able to distinguish its concentration very well.

**Author Contributions:** Conceptualization, D.R.S. and C.S.W.; Methodology, D.R.S., B.P., and U.P.J.; Software, B.P. and D.R.S.; Validation, D.R.S. and C.S.W.; Formal analysis, D.R.S.; Investigation, B.P. and D.R.S.; Resources, D.R.S., C.S.W., and U.P.J.; Writing, B.P. and D.R.S.; Funding acquisition, D.R.S. All authors have read and agreed to the published version of the manuscript.

**Funding:** Directorate of Research and Community Service, Directorate General of Strengthening for Research and Development, Ministry of Research, Technology and Higher Education, Republic of Indonesia, grant number (167/SP2H/LT/DRPM/2019).

**Acknowledgments:** Thanks to the Institute of Research and Community Services (LPPM), Brawijaya University, which has supported this research (330.20/N10.C10/PN/2019).

**Conflicts of Interest:** The authors declare no conflict of interest.

## References

1. Pérez, P.; Maldonado, A.; Yúfera, A.; Huertas, G.; Rueda, A.; Huertas, J.L. Towards Bio-impedance Based Labs: A Review. *J. Electr. Eng.* **2016**, *4*, 116–127. [CrossRef]
2. Bera, T.K. Bioelectrical impedance methods for noninvasive health monitoring: A review. *J. Med. Eng.* **2014**, *2014*. [CrossRef] [PubMed]
3. Naranjo-Hernández, D.; Reina-Tosina, J.; Min, M. Fundamentals, recent advances, and future challenges in bioimpedance devices for healthcare applications. *J. Sens.* **2019**, *2019*. [CrossRef]
4. Ho, J.S.; Low, J.H.; Sim, L.N.; Webster, R.D.; Rice, S.A.; Fane, A.G.; Coster, H.G.L. In-situ monitoring of biofouling on reverse osmosis membranes: Detection and mechanistic study using electrical impedance spectroscopy. *J. Membr. Sci.* **2016**, *518*, 229–242. [CrossRef]
5. Sarró, E.; Lecina, M.; Fontova, A.; Solà, C.; Gòdia, F.; Cairó, J.J.; Bragós, R. Electrical impedance spectroscopy measurements using a four-electrode configuration improve on-line monitoring of cell concentration in adherent animal cell cultures. *Biosens. Bioelectron.* **2012**, *31*, 257–263. [CrossRef] [PubMed]
6. Xu, Y.; Xie, X.; Duan, Y.; Wang, L.; Cheng, Z.; Cheng, J. A review of impedance measurements of whole cells. *Biosens. Bioelectron.* **2016**, *77*, 824–836. [CrossRef] [PubMed]
7. Mansor, M.A.; Takeuchi, M.; Nakajima, M.; Hasegawa, Y.; Ahmad, M.R. Electrical impedance spectroscopy for detection of cells in suspensions using microfluidic device with integrated microneedles. *Appl. Sci.* **2017**, *7*, 170. [CrossRef]
8. Zhang, F.; Jin, T.; Hu, Q.; He, P. Distinguishing skin cancer cells and normal cells using electrical impedance spectroscopy. *J. Electroanal. Chem.* **2018**, *823*, 531–536. [CrossRef]
9. Asami, K. Low-frequency dielectric dispersion of bacterial cell suspensions. *Colloids Surf. B Biointerfaces* **2014**, *119*, 1–5. [CrossRef]
10. Pal, N.; Sharma, S.; Gupta, S. Sensitive and rapid detection of pathogenic bacteria in small volumes using impedance spectroscopy technique. *Biosens. Bioelectron.* **2016**, *77*, 270–276. [CrossRef]

11. Jinyang, L.; Meiqing, L.; Hanping, M.; Wenjing, Z. Diagnosis of potassium nutrition level in Solanum lycopersicum based on electrical impedance. *Biosyst. Eng.* **2016**, *147*, 130–138. [CrossRef]
12. El Khaled, D.; Castellano, N.N.; Gazquez, J.A.; García Salvador, R.M.; Manzano-Agugliaro, F. Cleaner quality control system using bioimpedance methods: A review for fruits and vegetables. *J. Clean. Prod.* **2017**, *140*, 1749–1762. [CrossRef]
13. Soares, C.; Tenreiro Machado, J.A.; Lopes, A.M.; Vieira, E.; Delerue-Matos, C. Electrochemical impedance spectroscopy characterization of beverages. *Food Chem.* **2020**, *302*, 125345. [CrossRef] [PubMed]
14. Chowdhury, A.; Singh, P.; Bera, T.K.; Ghoshal, D.; Chakraborty, B. Electrical impedance spectroscopic study of mandarin orange during ripening. *J. Food Meas. Charact.* **2017**, *11*, 1654–1664. [CrossRef]
15. Grossi, M.; Lecce, G.D.; Toschi, T.G.; Riccò, B. Fast and accurate determination of olive oil acidity by electrochemical impedance spectroscopy. *IEEE Sens. J.* **2014**, *14*, 2947–2954. [CrossRef]
16. Aitzaz, A.M.; Kim, J.; Kim, T.; Park, K.D.; Cho, S. Electrical characterization of pork tissue measured by a monopolar injection needle and discrete fourier transform based impedance measurement. *Appl. Sci.* **2019**, *9*, 4049. [CrossRef]
17. Pradhan, R.; Mitra, A.; Das, S. Impedimetric characterization of human blood using three-electrode based ECIS devices. *J. Electr. Bioimpedance* **2012**, *3*, 12–19. [CrossRef]
18. Addabbo, T.; Fort, A.; Mugnaini, M.; Parri, L.; Pinzi, M.; Vignoli, V.; Mvemba, P.K.; Becatti, M.; Barygina, V.; Taddei, N.; et al. On the Suitability of Low-Cost Compact Instrumentation for Blood Impedance Measurements. *IEEE Trans. Instrum. Meas.* **2019**, *68*, 2412–2424. [CrossRef]
19. Bohli, N.; Chammem, H.; Meilhac, O.; Mora, L.; Abdelghani, A. Electrochemical Impedance Spectroscopy on Interdigitated Gold Microelectrodes for Glycosylated Human Serum Albumin Characterization. *IEEE Trans. Nanobiosci.* **2017**, *16*, 676–681. [CrossRef]
20. Sanchez, B.; Vandersteen, G.; Martin, I.; Castillo, D.; Torrego, A.; Riu, P.J.; Schoukens, J.; Bragos, R. In vivo electrical bioimpedance characterization of human lung tissue during the bronchoscopy procedure. A feasibility study. *Med. Eng. Phys.* **2013**, *35*, 949–957. [CrossRef]
21. Zarafshani, A.; Bach, T.; Chatwin, C.R.; Tang, S.; Xiang, L.; Zheng, B. Conditioning Electrical Impedance Mammography System. *Meas. J. Int. Meas. Confed.* **2018**, *116*, 38–48. [CrossRef]
22. Salazar-Anguiano, J.; Chávez-López, M.G.; Zúñiga-García, V.; Camacho, J.; Elías-Viñas, D. Resistive Part of Impedance as a Possible Indicator of Hepatocellular Carcinoma. *Arch. Med. Res.* **2018**, *49*, 89–93. [CrossRef] [PubMed]
23. Zink, M.D.; Weyer, S.; Pauly, K.; Napp, A.; Dreher, M.; Leonhardt, S.; Marx, N.; Schauerte, P.; Mischke, K. Feasibility of bioelectrical impedance spectroscopy measurement before and after thoracentesis. *BioMed Res. Int.* **2015**, *2015*. [CrossRef] [PubMed]
24. Freeborn, T.; Fu, B. Fatigue-Induced Cole Electrical Impedance Model Changes of Biceps Tissue Bioimpedance. *Fractal Fract.* **2018**, *2*, 27. [CrossRef]
25. Ben Hamed, K.; Zorrig, W.; Hamzaoui, A.H. Electrical impedance spectroscopy: A tool to investigate the responses of one halophyte to different growth and stress conditions. *Comput. Electron. Agric.* **2016**, *123*, 376–383. [CrossRef]
26. Bertemes-Filho, P.; Felipe, A.; Vincence, V.C. High Accurate Howland Current Source: Output Constraints Analysis. *Circuits Syst.* **2013**, *04*, 451–458. [CrossRef]
27. Chen, T.A.; Wu, W.J.; Wei, C.L.; Darling, R.B.; Liu, B. Da Novel 10-Bit Impedance-To-Digital Converter for Electrochemical Impedance Spectroscopy Measurements. *IEEE Trans. Biomed. Circuits Syst.* **2017**, *11*, 370–379. [CrossRef]
28. Mohamadou, Y.; Momo, F.; Theophile, L.; Njike Kouekeu Landry, C.; Fabrice, T.; Emmanuel, S. Accuracy enhancement in low frequency gain and phase detector (AD8302) based bioimpedance spectroscopy system. *Meas. J. Int. Meas. Confed.* **2018**, *123*, 304–308. [CrossRef]
29. Jiang, Z.; Yao, J.; Wang, L.; Wu, H.; Huang, J.; Zhao, T.; Takei, M. Development of a Portable Electrochemical Impedance Spectroscopy System for Bio-Detection. *IEEE Sens. J.* **2019**, *19*, 5979–5987. [CrossRef]
30. Bouchaala, D.; Fendri, A. Handheld Bioimpedance Spectrometer for the Total Frequency Range of the E -Dispersion. *Proc. Sens.* **2013**, *2013*, 175–180. [CrossRef]
31. Mylott, E.; Kutschera, E.; Widenhorn, R. Bioelectrical impedance analysis as a laboratory activity: At the interface of physics and the body. *Am. J. Phys.* **2014**, *82*, 521–528. [CrossRef]

32. Grassini, S.; Corbellini, S.; Angelini, E.; Ferraris, F.; Parvis, M. Low-cost impedance spectroscopy system based on a logarithmic amplifier. *IEEE Trans. Instrum. Meas.* **2015**, *64*, 1110–1117. [CrossRef]
33. Jiang, H.; Sun, A.; Venkatesh, A.G.; Hall, D.A. An audio jack-based electrochemical impedance spectroscopy sensor for point-of-care diagnostics. *IEEE Sens. J.* **2017**, *17*, 589–597. [CrossRef] [PubMed]
34. Analog Devices. CMOS 125 MHz Complete DDS Synthesizer; AD9850 Datasheet. 2004. Available online: https://www.analog.com/media/en/technical-documentation/data-sheets/AD9850.pdf (accessed on 22 January 2020).
35. Analog Devices. LF-2.7 GHz RF/IF Gain and Phase Detector; AD8302 Datasheet. 2018. Available online: https://www.analog.com/media/en/technical-documentation/data-sheets/ad8302.pdf (accessed on 22 January 2020).
36. Pilotte, M. *Operation of RF Detector Products at Low Frequency*; AN-691 Application Note; Analog Devices: Norwood, MA, USA, 2005; Available online: https://www.analog.com/media/en/technical-documentation/application-notes/AN-691.pdf (accessed on 22 January 2020).

*applied sciences*

MDPI

*Article*

# Real-Time Localization of Epileptogenic Foci EEG Signals: An FPGA-Based Implementation

**Jose V. Frances-Villora [1,*], Manuel Bataller-Mompean [1], Azeddine Mjahad [1]; Alfredo Rosado-Muñoz [1], Antonio Gutierrez Martin [2], Vicent Teruel-Marti [3], Vicente Villanueva [4], Kevin G. Hampel [4] and Juan F. Guerrero-Martinez [1]**

1 Processing and Digital Design Group, Department of Electronic Engineering, University of Valencia, 46100 Burjassot, Spain; Manuel.Bataller@uv.es (M.B.-M.); Mjahad.Azeddine@uv.es (A.M.); Alfredo.Rosado@uv.es (A.R.-M.); Juan.Guerrero@uv.es (J.F.G.-M.)

2 Functional Neurosurgery Unit. La Fe Hospital, 46009 Valencia, Spain; antoniogutierrez100@gmail.com

3 Neural Circuits Laboratory, Department of Anatomy and Embryology, University of Valencia, 46010 Valencia, Spain; Vicent.Teruel@uv.es

4 Refractory Epilepsy Unit. La Fe Hospital, 46026 Valencia, Spain; vevillanuevah@yahoo.es (V.V.); hampel_kev@gva.es (K.G.H.)

* Correspondence: Jose.V.Frances@uv.es

Received: 11 December 2019; Accepted: 18 January 2020; Published: 23 January 2020

**Abstract:** The epileptogenic focus is a brain area that may be surgically removed to control of epileptic seizures. Locating it is an essential and crucial step prior to the surgical treatment. However, given the difficulty of determining the localization of this brain region responsible of the initial seizure discharge, many works have proposed machine learning methods for the automatic classification of focal and non-focal electroencephalographic (EEG) signals. These works use automatic classification as an analysis tool for helping neurosurgeons to identify focal areas off-line, out of surgery, during the processing of the huge amount of information collected during several days of patient monitoring. In turn, this paper proposes an automatic classification procedure capable of assisting neurosurgeons online, during the resective epilepsy surgery, to refine the localization of the epileptogenic area to be resected, if they have doubts. This goal requires a real-time implementation with as low a computational cost as possible. For that reason, this work proposes both a feature set and a classifier model that minimizes the computational load while preserving the classification accuracy at 95.5%, a level similar to previous works. In addition, the classification procedure has been implemented on a FPGA device to determine its resource needs and throughput. Thus, it can be concluded that such a device can embed the whole classification process, from accepting raw signals to the delivery of the classification results in a cost-effective Xilinx Spartan-6 FPGA device. This real-time implementation begins providing results after a 5 s latency, and later, can deliver floating-point classification results at 3.5 Hz rate, using overlapped time-windows.

**Keywords:** electroencephalogram; epileptic EEG signal classification; epilepsy; epileptogenic focus; real-time implementation; FPGA

---

## 1. Introduction

Epilepsy is a common neurological disorder usually described by seizures which are recurrent in nature. This disorder can be produced by different brain disorders, such as brain tumors, intracranial hemorrhages and brain malformations [1], and depending on the affected area, a disorder may generate, apart from epileptic seizures, malfunctions in motion and patient perception [2].

An epileptic seizure is a period of time where the patient experiences a set of symptoms with different levels of severity: uncontrolled shaking movements of the body with loss of consciousness (generalized tonic-clonic seizure), shaking movements of a specific part of the body with different levels of consciousness (focal seizure), or short moments of focal seizures with impairment of awareness (absence seizure). Epileptic seizures can be originated by abnormal, synchronous, or even excessive brain neural activity, causing a temporary disruption to the way that the brain normally works. Anyway, what happens to someone during a seizure depends on the affected part of the brain and how far the seizure activity disseminates in the brain.

Epilepsy can be classified primarily into two types: generalized and partial (or focal) epilepsy [3]. Generalized onset seizures affect, at the same time, both sides of the brain or groups of cells on both sides of the brain. On the other hand, focal onset seizures (the term focal is used instead of partial to be more accurate when talking about where seizures begin) usually start in one area or group of cells on one side of the brain.

The activity of the brain is usually registered using either electroencephalography (EEG) or functional magnetic resonance imaging (fMRI). Although fMRI has better spatial resolution, the use of the multidimensional time series generated from electroencephalogram EEG is more popular, as it allows high precision time measurements, is functionally fast and is relatively cheap.

The EEG epileptogenic source's localization has been studied for decades [4,5]; however, the methods were not implemented in clinical practices until recently. Nowadays, the EEG is considered a noninvasive and useful test to assess whether a pharmacoresistant patient can benefit from the resective epilepsy surgery [6]. As the resective surgery aims to remove surgically the brain sections involved in the focal onset epilepsy, it is important to distinguish precisely between "focal signals," those recorded in brain areas where first ictal signals are detected, and "non-focal signals," those registered from brain areas not related to the seizure onset [7]. Many patients with epilepsy may require EEG signals to be recorded from deep structures of the brain using intracranial electrodes.

Usually, the focus localization is essentially made using registers acquired monitoring the patient 24 h a day, during a stay of several days in an epilepsy monitoring unit (EMU). In these kind of units, apart from scalp EEG (and intracranial electrodes to record signals from deep structures of the brain, in many patients), the epilepsy patients are recorded on video, along with their speech and movements. Thus, all data are collected targeting the evaluation of its seizure disorder, seeking to gather data before a seizure starts, during one and during recovery. The evaluation of this information can be used to locate candidate areas for the epileptogenic focus, although in some cases it is not enough to locate the epileptogenic focus precisely prior to the surgery.

The visual analysis of the EEG recordings of seizures with intracranial electrodes can help in locating the seizure. However, visual inspection is a hard and time consuming process that can be affected by the clinician subjectivity. In addition, it is not easy to determine the seizure source by a direct visual inspection of the EEG signal recordings.

To select candidate areas from EEG signals could be helpful a computerized analysis of the EEG [8,9]. As with other pathologies [10–12], machine learning has been applied in epilepsy at many works [13–15] to classify EEG signals as normal versus epileptic or seizure versus inter-ictal. However, the most challenging classification problem is focal (F) versus non-focal (NF). The classification of seizure from normal and seizure-free signals has achieved a 100% classification accuracy. However, this goal has not been achieved to date for the classification of focal and non-focal EEG signals. Neurosurgeons have difficulties determining the brain region responsible of the initial seizure discharge, so this kind of classification may serve as a tool to help epileptologists to resect the epileptogenic area. Compared with signals of the epileptogenic areas, the signals from non-epileptogenic areas are more nonlinear, less random and more nonstationary.

Many machine learning systems have been developed to classify and detect the epileptogenic source signals. Sharma et al. [13] used entropies derived from the coefficients of the wavelet transform of the EEG signals to feed a least squares-support vector machine (LS-SVM) model to distinguish focal and non-focal EEG signals. In [16], Sharma et al. also used the LS-SVM classifier to feed the entropies derived from some subbands decomposed using tunable-Q wavelet transform (TQWT).

In [17], Sharma et al. utilized empirical mode decomposition (EMD) and entropy for the classification of focal and non-focal EEG signals. In this work, intrinsic mode functions (IMFs) from focal and non-focal EEG signals were extracted using EMD, and then the entropies were fed the input of a LS-SVM classifier. Das et al. [18] also used entropy-based features from the EMD, DWT (discrete wavelet transform) and EMD-DWT domains, along with a k-nearest neighbor (k-NN) classifier model. In turn, Zeng et al. [19] used features derived from euclidean measures obtained from the phase space reconstruction (PSR) of several IMFs, obtained using EMD as well. Bhattacharyya et al. [20] proposed the decomposition of the EEG signal into rhythms using the empirical wavelet transform (EWT), and then used some area measures from them as input for a LS-SVM classifier model, to recognize focal and non-focal EEG signals. Another work of Bhattacharyya et al. [15] also used multivariate subband entropy measures from TQWT along with multivariate fuzzy entropy in combination with a LS-SVM classifier model. Chatterjee et al. [21] also used SVM and k-NN classifiers fed by multifractal, detrended fluctuation analysis (MFDFA) based feature sets. Singh et al. [22] used features derived from DFT-based rhythms of the EEG to fed the LS-SVM classifier. Taran et al. [23] proposed the use of spectral moment based features extracted from the modes of the clustering VMD (CVMD) and extreme learning machine (ELM) classifiers. Deivasigamani et al. [24] utilized features extracted from the dual tree complex wavelet transform (DT-CWT) to fed an adaptive neurons fuzzy interference system (ANFIS).

However, these works tend to require a considerable computational load, especially the most recent ones. As an example, San-Segundo et al. [25] proposed a deep neural network (DNN) made up of two convolutional layers for feature extraction and three fully connected layers for classification. In this work, authors increased the classification accuracy a little at the expense of increasing, considerably, its computational needs. In turn, Daoud et al. [26] used both a deep convolutional autoencoder and an unsupervised learning scheme merging a deep convolutional variational autoencoder and a K-means algorithm.

This progressive increase of the computational demand could impede the jump of using the epileptogenic source localization during surgery. This application needs a real-time implementation, and could be used as a help-decision tool by neurosurgeons to refine the localization of an epileptogenic area during resective epilepsy surgery. Note that the recent technology is mature enough to implement machine learning processes in real-time [27,28].

Thus, the goals of this work were to assess the possibility of locating the epileptogenic focus in real-time, and study the simplification of the classification process to reduce its computational needs as much as possible while maintaining similar classification accuracy to previous works. We also studied the resource usage and performance of the real-time application on a recent Xilinx FPGA reconfigurable device.

The main contributions of this work are:

- The proposal of an automatic classification procedure optimized to classify in real-time the location of the epileptogenic focus from EEG inter-ictal signals. It is conceived to be used in a portable device as a decision-assisting tool by neurosurgeons during surgery.
- The proposed feature set and the classifier model have been selected to minimize both the number of features and the computational cost, while preserving the classification accuracy at a level similar to that in previous works.
- The classification procedure has been implemented using a reconfigurable logic FPGA device. This hardware implementation computes the whole procedure, accepts the EEG raw input signal and delivers the classification result. Two designs have been implemented, using single and double floating point precision following the IEEE 754 standard for floating-point arithmetic.
- The analysis of the resource usage of this kind of implementation, its accuracy with respect a Matlab implementation and how fast the device can deliver results (maximum frequency of operation).

The rest of the paper is organized as follows. Section 2 introduces the dataset, and the analyzed features and classifier models. The details of the hardware implementation and the proposed computational method are described in Section 3. Results of the analysis and discussion are presented in Sections 4 and 5, respectively. Finally, Section 6 concludes the paper.

## 2. Methodology, Materials and Methods

This section presents an overview of the dataset utilized, and introduces the analyzed features and classifier models.

### 2.1. Dataset

In this work, the publicly available Bern-Barcelona database [7] was used. This is an open source EEG dataset that has been used for a large number of epilepsy studies [16–20]. This dataset collects intracranial EEG recordings from five pharmacoresistant epilepsy patients, including two classes of EEG signals: focals and non-focals. *Focal* signals (F) are those captured from an epileptogenic area (acquired from those channels that detected ictal EEG signal changes first, as decided by at least two neurologists via visual inspection) and *non-focal* signals (NF) are captured in channels out of this area. Each class contains 3750 pairs of simultaneously acquired signals "x" and "y," all of them randomly selected, and consisting of 20 s windows of simultaneous recording, sampled at 512 Hz. Each focal pair consists of one of the focal EEG channels for the $x$ signal, and one of this channel's neighboring focal channels for the $y$ signal, both simultaneously acquired from the same patient. The non-focal pairs were selected from nonfocal EEG channels in the same way [7,29]. All EEG signal were band-pass filtered by an fourth order Butterworth (0.5 Hz–150 Hz). In addition, before being included into the database, signal pairs were visually inspected to discard prominent measurement artifacts.

Note that all recordings of seizure activity, and three hours after the last seizure, were excluded. Thus, this database contains neither ictal nor postictal stage activity.

### 2.2. Preprocessing

The 50 Hz of the EEG signals was filtered using a moving average of order 5. In addition, EEG signals were filtered using a Butterworth low pass IIR filter with $f = 80$ Hz and order $N = 6$.

### 2.3. Feature Extraction

This is one of the most important steps in classification problems. Table 1 lists the 39 features considered in this work for each segment of the signal dataset. All these have been used succesfully as features in previous EEG seizure detection works [14,30–37].

The features considered are from different domains, such as time, frequency, information theory and entropy. But note that all of them are univariate and imply low or medium computational load to extract them.

Statistical parameters such as mean, variance, skewness and kurtossis have been used to extract information on changes in the distribution and amplitude of the EEG data. Those parameters have been considered on the first and second derivative too. Frequency parameters have been calculated by means of the DFT transform, as spectral power or relative energy between different bands. In addition, some nonlinear features have been calculated, such as fractal dimension, used to compare rhythms in the self-similarity present in the signals; entropy of the signal; and spectral entropy, to depict randomness of the EEG in the frequency domain.

However, note than other features having greater computational complexity were not considered, even if these had been used successfully in other works. The reason is obvious; this work aimed to select the set of features having the lowest computational load while providing similar classification accuracy to other works in the bibliography. Thus, the more computationally complex features were discarded from the beginning, such as the calculation of certain entropy measures from IMF signals calculated using EMD, wavelet transform or other time-frequency domain features.

**Table 1.** List of extracted features.

| EEG Feature | Description |
|---|---|
| Log energy entropy (LogEn) [30] | Non-normalized energy based entropy |
| Median frequency (MDF) [31] | Division of the EEG power spectrum into two regions |
| Mean frequency (MNF) [31] | Mean normalized frequency of the power spectrum |
| Katz fractal dimension (KFD) [31] | Index characterizing the fractal pattern complexity |
| Lower quartile 1 (Q1) [31] | 25% of the EEG signal |
| Upper quartile 3 (Q3) [31] | 75% of the EEG signal |
| Inter quartile range (IQR) [31] | Difference between Q3 and Q1 |
| Semi inter quartile deviation (SID) [31] | Measure of spread |
| Skewness (Sk) [32] | Measure of the degree of symmetry |
| Kurtosis (Kr) [32] | Measure of tailedness of the probability distribution |
| Root mean square (RMS) [33] | Root mean square of the EEG signal |
| Band power (PB) [33] | Average power of the EEG signal (0 to $f_s/2$) |
| Zero crossing (ZC) [33] | Number of times that the signal changes of sign |
| Complexity (Comp) [33] | Hjorth parameter |
| Mobility (Mob) [33] | Hjorth parameter |
| Activity (Act) [33] | Hjorth parameter |
| Spurious free dynamic range (SFDR) [34] | Length along a EEG signal |
| Curve length (CL) [34] | Length along a EEG signal |
| Teager energy (TE) [34] | Non linear energy |
| Variance (Var) [34] | Variance of the EEG signal |
| Standard deviation (Std) [34] | Standard deviation of the signal |
| Mean (Mean) [34] | Mean of the EEG signal |
| 1st derivative variance (Var1) [34] | Variance of the first derivative |
| 1st derivative standard deviation (Std1) [34] | Standard deviation of the first derivative |
| 1st derivative mean (Mean1) [34] | Mean of the first derivative |
| 2nd derivative variance (Var2) [34] | Variance of the second derivative |
| 2nd derivative standard deviation (Std2) [34] | Standard deviation of the second derivative |
| 2nd derivative mean (Mean2) [34] | Mean of the second derivative |
| Derivative variance ratio (RatioVar) [36] | Ratio of derivative respect absolute of derivative variances |
| Power (P) [35] | Power of the signal window |
| 1st difference (1d) [35] | Feature extraction and selection for emotion recognition from eeg |
| Normalized 1st difference (N1d) [35] | Normalization of 1st difference |
| 2nd difference (2d) [35] | Feature extraction and selection for emotion recognition from eeg |
| Normalized 2nd difference (N2d) [35] | Normalization of 2nd difference |
| Normalized Length Density (NLD) [35] | Quantifies self-similarities |
| Higher order crossings (HOC) [38] | Describes the oscillatory pattern of a signal |
| Band power (Pu) [35] | Spectral power in certain spectral band |
| Recursive Efective Efficiency (REE) [35] | Energy ratio of spectral bands |
| Relative Energy Ratio (rE) [37] | Relative energy between bands |

### *2.4. Feature Reduction*

Feature reduction reduces the computational complexity of the classifier and also avoids the possibility of redundancy. In this study, we obtained several discriminatory features for the two class classification process (Section 4.1). The number of features was reduced using the RelieF feature selection technique [39], as is explained in Subsection 4.1.

#### 2.4.1. Log Energy Entropy

Entropy is a concept handling predictability and randomness, with higher values of entropy always being related to a lesser system order and more randomness. The entropy of an EEG channel is a measure of uncertainty, where the EEG signals are considered a random variable. The log energy entropy of a x EEG signal is defined as [40]:

$$H_{LogEn}(x) = -\sum_{i=0}^{N-1} (log_2(p_i(x)))^2, \qquad (1)$$

$p(x)$ being the probability density function. With this entropy calculated regarding the signal power spectrum as a probability distribution, the log energy spectral entropy is obtained.

#### 2.4.2. Skewness

It is a higher-order statistical attribute of a time series. Skewness is a measure of the asymmetry of the probability distribution (pdf) of a real-valued random variable around its mean.

$$Skew[x] = E[(\frac{x-\mu}{\sigma})^3] = \frac{E[x^3] - 3\mu\sigma^2 - \mu^3}{\sigma^3}. \tag{2}$$

#### 2.4.3. Root Mean Square

Is the square root of the mean square, the arithmetic mean of the squares of a set of numbers, also known as the quadratic mean:

$$RMS = \sqrt{\frac{1}{N}\sum_{i=1}^{N} x_i^2}. \tag{3}$$

#### 2.4.4. Derivative Variance Ratio

This is the quotient between the variance of the derivative of the signal and the variance of the absolute value of said derivative. It is a derivative variance ratio (called RatioVar) [36]:

$$Ratiovar = \frac{E[\frac{dx^2}{dt}] - E[\frac{dx}{dt}]^2}{E[\frac{d|x|^2}{dt}] - E[\frac{d|x|}{dt}]^2}. \tag{4}$$

#### 2.4.5. Relative Energy Ratio

It is used to observe the changes in EEG frequency bands due to the stressors. When stress occurs, the energy of Alpha band, $HF$, will reduce. Meanwhile, energy of lower bands will increase [37]:

$$rE = \frac{Total\ Energy\ In\ HF}{Total\ Energy\ In\ LF}. \tag{5}$$

### *2.5. Classifier Models*

Several classifiers have been used in this analysis (Section 4.1), each one having its own specific strengths and weaknesses. All them are briefly outlined below.

#### 2.5.1. Support Vector Machine (SVM)

It is a supervised classification technique that constructs a separating hyperplane maximizing the margin between the input data classes that are viewed in an n-dimensional space (n is the number of features used as inputs). Essentially, this involves orienting the separating hyperplane to be perpendicular to the shortest line separating the convex hulls of the training data for each class, and locating it midway along this line.

In addition to performing linear classification, SVMs can efficiently perform a non-linear classification using what is called the kernel trick, implicitly mapping their inputs into high-dimensional feature spaces.

2.5.2. K-Nearest Neighbor (KNN)

KNN is a supervised learning technique where a new instance is classified based on the closest training samples present in the feature space. It does not use any model to fit, and is only based on memory. When a test data is entered, it is assigned to the class that is most common amongst its k nearest neighbors.

2.5.3. Decision Tree

It is a method that creates a model that enables one to predict the target value of an item (represented in leaves) based on several input variables (represented as branches). In the case of using a classification tree analysis, the predicted outcome is the class (discrete) to which the data belongs.

2.5.4. Logistic Regression

It is a classification algorithm used to assign observations to a discrete set of classes. Unlike linear regression, which outputs continuous values, the logistic regression transform its output using the logistic sigmoid function, to return a probability value that can be mapped to two or more discrete classes.

2.5.5. Discriminant Analysis

Linear and quadratic discriminant analysis were used. Linear discriminant analysis (LDA) is a generalization of the Fisher's linear discriminant that finds a linear combination of features characterizing or separating two or more classes. In turn, quadratic discriminant analysis separates input features into two or more classes of objects by a quadratic surface, becoming a general version of the linear version.

2.5.6. Ensembles

Ensemble classification improves results by combining several models. It can be used with any learning method. Thus, this approach allows for better predictive performance compared to single models. The number of classifier components has a great impact on the classification accuracy. In this work, trees, discriminant and k-NN classifier components were used.

2.5.7. Neural Network Classifier (NN)

A neural network consists of a series of units (neurons) arranged in layers. This arrangement converts an input vector into some output. To do so, each neuron takes its inputs and calculates the output by applying a usually nonlinear function (the activation function), to later pass the output to the next layer. Generally, neural networks are defined as feed-forward: a unit feeds its output to all the units on the next layer, but there is no feedback to the previous layer. Signals are weighted when fed the input of a unit neuron. The weights are tuned in the training phase of the classifier.

Neural networks are considered to be good classifiers due to their inherent features, such as adaptive learning, robustness, self-organization and generalization capability.

*2.6. Performance Analysis*

In order to evaluate the performance of the proposed method, the performance of the classifiers are expressed in terms of classification accuracy (Acc), defined as follows [41]:

$$Accuracy(\%) = \frac{TP + TN}{TP + FN + TN + FP} \times 100 \quad (6)$$

where $TP$, $TN$, $FP$ and $FN$ denote true positives, true negatives, false positives and false negatives, respectively.

## 3. Hardware Implementation and Computational Method

This section discusses the implementation details. We will leave for Section 4 the reasons for and how this computational procedure was chosen. At this point, all that matters to know that the classifier model used was a perceptron with 25 neurons in the single hidden layer, 5 neurons in the input layer and 2 neurons in the output layer (Algorithm 2). The 5 extracted features (Table 2) were computed following Algorithm 1. Details about feature selection and classifier model selection are left for Section 4.

The proposed hardware implementation was conceived as an intellectual property (IP) core using Xilinx Vivado HLS and the Xilinx Vivado Design Suite 2016.2 [42]. It provides a signal interface definition that enables it as a standalone module, being able to also be used as a peripheral of a more complex system on chip (SoC), embedded microprocessor, etc. Further, this approach offers the capability of customization for specific needs in many different hardware applications.

### 3.1. Working Modes

The IP core perform two different working modes:

- *Initialization mode.* The initialization of the IP core consists of the load of the matrices $\mathbf{x}_{nmax}, \mathbf{x}_{nmin}, \mathbf{W}_h, \mathbf{b}_h, \mathbf{W}_o, \mathbf{b}_o, \mathbf{x}_{dmax}, \mathbf{x}_{dmin}, y_{min},$ and $y_{max}$, using the external signal interface. These matrices are essentially weights and bias of the neural network, along with normalization and denormalization values and vectors. All together, these matrices allow the IP core to perform a proper classification. Note that the calculation of these matrices is achieved out of this IP core, and the results are transferred to it during this initialization process. Once the initialization is complete, the core can change to another working mode, never before.
- *Run mode* or *on-line mode*. In this mode, the input data $\mathbf{x}$ is fed into the IP core. Then, several features are extracted, and in turn, fed into the neural network system. The output is computed according to the initialized network topology. The IP core, when running in this mode, computes and serves the corresponding output before accepting a new input.

### 3.2. IP Core Signal Interface

In Xilinx FPGAs, external core signal interfaces are used to follow proprietary protocol specifications, such as AXI4 [43,44], AXI4-Lite or AXI4-Stream [44,45]. In this work the AXI4 protocol was selected to permit specifying arrays as arguments. However, note that the protocol interface has almost no influence on performance when the core is running in the on-line mode. The reason is that the reported performance refers to the complete epileptic focus classification task, while the load of one input vector by iteration implies a negligible overhead. Therefore, for replication purposes, it can be expected to achieve similar results, for the running mode, regardless of the protocol specification implemented for the core signal interface (e.g., AXI4-Lite, AXI4 or AXI4-Stream).

Figure 1 outlines the external signal interface of the IP core, where signal lines are represented by thin black arrows and buses, and bunches of signals are represented by white thick arrows.

Signal START indicates when the core can start processing data, the READY signal indicates when the core is ready to accept new inputs, the IDLE signal indicates when the core is idle and the DONE signal indicates when the core operation has been completed. Altogether, these signals constitute the block-level interface, controlling the core independently of the port-level I/O protocol.

On the other hand, the input and output data ports implement a handshake data flow protocol. Lines A_TVALID, A_TREADY and the bus A_TDATA integrate the input data port, while B_TVALID, B_TREADY and the bus B_TDATA integrate the output data port. The TDATA bus is the payload, while TVALID and TREADY lines signal when the information pass across the interface. These signals integrate a two-way flow control mechanism that enable master and slave to control the data transmission rate rate across the interface.

Finally, the MODE signal is used as application-signaling. It requests the entering on initialization mode or on-line mode.

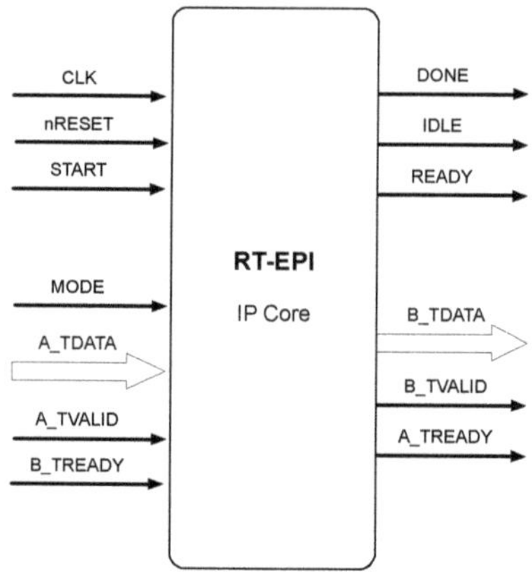


**Figure 1.** Interface signals used by the RT-EPI IP core. White arrows represents buses.

### 3.3. *System Parameterization*

The definition of the SLFN neural network topology was conceived parametrically to achieve a flexible design with minimal code modifications. The main parameters are:

- L: Is the length, in samples, of the input signal window.
- IN: Is the number of neurons in the input layer.
- $\tilde{N}$: Is the number of neurons in the hidden layer.
- ON: Is the number of neurons in the output layer.
- FT: Boolean determining the type of IEEE 754 floating-point arithmetic precision: single or double.

### 3.4. *Computation of the RT-EPI Algorithm*

The proposed IP hardware core implements a real-time epileptogenic focus classification of a different EEG input window signal each iteration.

#### 3.4.1. Algorithm Description

Algorithm 1 shows the steps in which perform the feature computation. This computational procedure follows the expressions described in Table 2 to extract features from the input signal, and then, to compose the input data pattern to be used as input to the neural network.

Algorithm 2 shows the computational procedure implementing the classification. Note that it is mandatory to the previous IP core so that the implementation may be fully functional and begin accepting any input data. The initialized structures are the normalization and denormalization parameters ($\mathbf{x}_{nmax}, \mathbf{x}_{nmin}, \mathbf{x}_{dmax}, \mathbf{x}_{dmin}, y_{min}, y_{max}$), plus the weights and biases of the hidden and output layers ($\mathbf{W}_h, \mathbf{b}_h, \mathbf{W}_o, \mathbf{b}_o$). Further, the first step of the algorithm is the normalization of the data input pattern. Then, the outputs of the hidden and output layers are calculated. Although the hyperbolic tangent sigmoid function has been used as an activation function, the proposed algorithm allows the use of a wide range of activation functions (including piece-wise linear activation functions). The results of the output layer are then denormalized to obtain the final output.

The pseudocode in Algorithms 1 and 2 show all the matrices and vectors involved in each step of the computation, with their respective dimensions.

The computational procedure was implemented using a sequential architecture. This architecture minimizes both the usage of memory and arithmetic slices. And, although the throughput results can be improved using a parallel architecture, the use of a sequential architecture enabled us to establish a standard machine to be used as a reference in subsequent works.

**Algorithm 1** Feature extraction pseudocode.

**Input:** $\mathbf{x}_{(1\times L)} \rightarrow$ Input signal.

**Output:** $\mathbf{x}'_{(1\times 5)} \rightarrow$ Output features vector.

$\bar{\mathbf{x}}_{(1\times 1)} = \frac{1}{L}\sum_{i=1}^{L} x_i.$

$\mathbf{x}_{(1\times L)} = \mathbf{x} - \bar{\mathbf{x}}.$

$\mathbf{x}_{(1\times L)} = 50Hz\_removal(\mathbf{x}).$

$\mathbf{x}_{(1\times L)} = low\_pass\_filtering(\mathbf{x})\ ^{(*)}.$

$\bar{\mathbf{x}}_{(1\times 1)} = \frac{1}{L}\sum_{i=1}^{L} x_i.$

$\mathbf{x}_{(1\times L)} = \mathbf{x} - \bar{\mathbf{x}}.$

$RMS_{(1\times 1)} = \sqrt{\frac{1}{L}\sum_{i=1}^{L} x_i^2}.$

$skew_{(1\times 1)} = \dfrac{\frac{1}{L}\sum_{i=1}^{L}(x_i - \bar{\mathbf{x}})^3}{\frac{1}{L}\sum_{i=1}^{L}(x_i - \bar{\mathbf{x}})^2)^{\frac{3}{2}}}.$

$\mathbf{d}_{(1\times L-1)} \mid d_{i-1} = x_i - x_{i-1}\ for\ i = \{2, ..., L\}.$

$\mathbf{a}_{(1\times L-1)} \mid a_i = |d_i|\ for\ i = \{1, ..., L-1\}.$

$RatioVar_{(1\times 1)} = \frac{\sigma_{\mathbf{d}}}{\sigma_{\mathbf{a}}}.$

$\mathbf{tmp}_{(1\times L)} = |FFT(\mathbf{x})|.$

$\mathbf{p}_{(1\times 1+L/2)} \mid p_i = k \cdot tmp_i,\ k = 2\ for\ i = \{2, ..., L/2\}\ and\ k = 1\ for\ i = \{1, 1 + L/2\}.$

$rE_{(1\times 1)} = \dfrac{\sum_{i=49}^{321} p_i}{\sum_{j=5}^{48} p_j} \quad (\ p_i \subset [12Hz, 80Hz],\ p_j \subset [1Hz, 12Hz]\ ).$

$LogEn_{(1\times 1)} = \sum_i log2(p_i^2).$

$\mathbf{x}'_{(1\times 5)} = \left[ rE\ \ LogEn\ \ \sigma\ \ RatioVAr \right].$

*(*) Filter coefficients: B[0] = 0.0001; B[1] = 0.0006; B[2] = 0.0014; B[3] = 0.0019; B[4] = 0.0014; B[5] = 0.0006; B[6] = 0.0001; A[0] = 1.0000; A[1] = -4.1069; A[2] = 7.2450; A[3] = -6.9795; A[4] = 3.8569; A[5] = -1.1558; A[6] = 0.1464.*

**Algorithm 2** Classification of an input pattern. pseudocode

**Input:** $\mathbf{x}'_{(1\times IN)}$ → Input data pattern.
$\mathbf{x}_{nmax\,(1\times IN)}$, $\mathbf{x}_{nmin\,(1\times IN)}$ → Input normalization vectors.
$y_{min\,(1\times 1)}$, $y_{max\,(1\times 1)}$ → Range normalizing input.
$\mathbf{W}_{h\,(IN\times \tilde{N})}$ → Hidden layer weights matrix.
$\mathbf{b}_{h\,(1\times \tilde{N})}$ → Hidden layer bias.
$\mathbf{W}_{o\,(\tilde{N}\times ON)}$ → Output weights matrix.
$\mathbf{b}_{o\,(1\times ON)}$ → Output layer bias.
$\mathbf{x}_{dmax\,(1\times ON)}$, $\mathbf{x}_{dmin\,(1\times ON)}$ → Output denormalization vectors.

**Output:** $\mathbf{y}_{(1\times ON)}$.

1: $$\mathbf{x}_{norm\,(1\times IN)} = \frac{(y_{max} - y_{min})\cdot(\mathbf{x}' - \mathbf{x}_{nmin})}{\mathbf{x}_{nmax}) - \mathbf{x}_{nmin}} + y_{min}.$$

2: $$\mathbf{tmp1}_{(1\times \tilde{N})} = Tansig(\ \mathbf{x}_{norm}\cdot \mathbf{W}_h + b_h\ ).$$

3: $$\mathbf{tmp2}_{(1\times ON)} = Tansig(\ \mathbf{tmp1}\cdot \mathbf{W}_o + b_o\ ).$$

4: $$\mathbf{y}_{(1\times ON)} = \frac{(\mathbf{tmp1} - y_{min})\cdot(\mathbf{x}_{dmax} - \mathbf{x}_{dmin})}{y_{max} - y_{nmin}} + \mathbf{x}_{dmin}.$$

*(Ñ: number of hidden neurons, IN: number of input neurons, ON: number of output neurons.)*

### 3.4.2. Design Considerations

The proposed design allows the definition of IEEE 754 floating point units using single or double data type precision. This selection together with the parametric definition of the SLFN neural network permits one to test the design in different conditions with few code modifications.

The activation function implemented in this design was the hyperbolic tangent sigmoid.

On the other hand, the proposed design uses pipelining. The main reasons for that is that pipelining alleviates the great latency involving the use of floating point operations, and the suitability of the algorithm for its use, since most of the steps of Algorithms 1 and 2 can be implemented with *for* loops or nested *for* loops.

The pipelining technique helps to optimize the initiation interval, defined as the number of clock cycles that must occur before a new input can be applied. Thus, the initiation interval becomes the parameter to optimize, and the effort must focus on approximating it to one as much closely as possible. To carry this out, we used the PIPELINE optimization pragma directive in each step of the computational procedure implemented using *for* loops. This generates a pipeline design with an initiation interval as low as possible, which dramatically reduces the total latency of the loop implementation.

In addition, the clock period target was set to 4 ns. This forces the compiler to obtain the fastest hardware implementation.

Take into account that both the above design considerations and the computational procedure described in Algorithms 1 and 2 must be followed to replicate the implementation results in this work.

## 4. Results

The analysis described in this work was carried out using the Bern-Barcelona dataset (Section 2.1). The first 50 focal and non-focal register pairs of the dataset were used. As all considered features are univariate, and 40 s of EEG signals are available in each register pair (20 s $\times$ 2), a total of 4000 s of EEG signals were considered in this study.

In order to better compare our results with the results provided by previous works, we used five-fold cross validation in our experimental procedure. In this approach, the EEG signals are divided randomly into five equal portions. Four out five portions were considered for training and the rest, half for validation and half for testing.

All the feature extraction algorithms and classification models were implemented using Matlab R2018a and the "Statistics and Machine Learning Toolbox."

The first goal was to find the classification procedure with the lowest computing cost, because of the real-time implementation aim. Thus, we had to be find (1) a set of features with the minimum number of features and minimum computing cost, using (2) the simplest classifier model, and (3) the optimum segment length. A minimum classification accuracy of 95% was required, a value above the average of related works in the bibliography.

To do so, all considered features, Table 1, were ranked by discrimination capacity and by computational cost. Then, different classification procedures were checked in a loop. At each iteration of the loop, a combination of features were selected (taking into account the ranks); then, fed to 24 classification models, Table 3; and finally the classification accuracies for different segment lengths were calculated. The loop finished when the smallest set of features reached 95% of classification accuracy on at least one of the classification models.

Once the classification procedure was defined, it was coded in C using Xilinx Vivado HLS and the Xilinx Vivado Design Suite 2016.2 [42]. These tools were also used to carry out synthesis, simulations and cosimulations. The Xilinx Virtex-7 XC7VX1140T FPGA device was selected for synthesis and implementation, because it is a biggest Virtex-7 device that permits implement the application without resource restrictions.

The coded design is parameterizable, and follows a pipelined and sequential architecture that computes (Algorithms 1 and 2) all the classification procedure from accepting the raw signal to the delivery of the classifier output.

The reported analyses were conducted using two different arithmetic precisions: a 32-bit floating-point algorithm ("single" design), and 64-bit floating-point algorithm ("double" design). Both implementation designs used the IEEE 754 standard.

### 4.1. Set of Features and Classification Model

The RelieF algorithm was used as feature extraction algorithm. RelieF returns a rank of features and its weights to represent the discrimination capacity of these features. These ranks and weights were used to select the relevant set of features, along with the computational cost criteria, optimizing the real-time implementation of the application. The number of selected features was defined as the minimum set of features that allowed us to obtain a minimum threshold of 95% classification accuracy on at least one of the classification models (Table 3).

To determine the best classification model, 24 classification models were tested. Table 3 lists all these models along with their prediction speed and memory usage characteristics. Note that only those classification models with low or medium speed and memory usage were chosen, due to the importance of minimizing these parameters in the real-time implementation.

All classification models were trained with several sets of features, seeking to determine the feature set and classification model at the same time. However, the evaluation of the classification models depends on a third parameter: the window length. Figure 2 shows the accuracy results for the selected sets of features (Table 2), for all the classification models and the variations of the window length from 1 to 10 s. Note that this figure only shows the best score for the neural network classification model. Figure 3 details the classification accuracy obtained using different numbers of neurons for the hidden layer.

Thus, five features were finally selected, those in Table 2, to be used with a neural network as the classification model. As it can be seen, the computational cost of the five selected features is low (case of features derived from the temporal domain) or medium (in the case of frequency domain features, where it is necessary to compute the Fourier transform).

This minimum set of features achieves a classification accuracy of 95% using a Neural Network and five seconds of segment length. In addition, the optimum number of neurons in the hidden layer of the neural network is 25, as it can be shown in the analysis of Figure 3.

Thus, from that point we will assume that the implementation is done by a neural network of type perceptron with just one hidden layer of 25 hidden neurons and two output neurons in the output layer (provided the two classes of this classification problem). In addition, the neural network will have five input neurons, because five is the dimensionality of the selected number of features, which, in turn, would be computed from window lengths of 5fiveseconds of the input signal.

**Table 2.** Set of selected features.

| **Features** | |
|---|---|
| Root Mean Square | $RMS_x = \sqrt{\frac{1}{N}\sum_{i=1}^{N} x_i^2}$ |
| Skewness | $s = \frac{E(x-\mu)^3}{(E(x-\mu)^2)^{\frac{3}{2}}}$ |
| Derivative Variance Ratio | $RatioVar = \frac{\sigma_\Delta}{\sigma_{\lvert\Delta\rvert}}$ $\quad with\ \Delta(i){=}x(i){-}x(i{-}1)$ |
| Relative Energy Ratio | $rE = \frac{\sum_{12-80Hz} P(x)}{\sum_{1-12Hz} P(x)}$ $\quad with\ P(x)\ the\ Spectral\ Power$ |
| Log Energy Spectral Entropy | $LogEn = \sum log2(P(x)^2)$ |

**Table 3.** Acronyms, prediction speed and memory usage characteristics of all the classification models used in this work.

| Acronym | Classifier | Prediction Speed | Memory Usage |
|---|---|---|---|
| [*TreeCoarse*] | Decision Tree (Coarse Tree) | Fast | Small |
| [*TreeMedium*] | Decision Tree (Medium Tree) | Fast | Small |
| [*TreeFine*] | Decision Tree (Fine Tree) | Fast | Small |
| [*DiscrLin*] | Discriminat Analysis (Linear Discriminant) | Fast | Small |
| [*DiscrQuad*] | Discriminat Analysis (Quadratic Discriminant) | Fast | Large |
| [*RegrLog*] | Logistic Regresion | Fast | Medium |
| [*SVMLin*] | Support Vector Machine (Linear SVM) | Fast[1] | Medium |
| [*SVMQuad*] | Support Vector Machine (Quadratic SVM) | Fast[2] | Medium[3] |
| [*SVMCubic*] | Support Vector Machine (Cubic SVM) | Fast[2] | Medium[3] |
| [*SVMFineGaus*] | Support Vector Machine (Fine Gaussian SVM) | Fast[2] | Medium[3] |
| [*SVMMedGaus*] | Support Vector Machine (Medium Gaussian SVM) | Fast[2] | Medium[3] |
| [*SVMCoarseGaus*] | Support Vector Machine (Coarse Gaussian SVM) | Fast[2] | Medium[3] |
| [*KNNFine*] | Nearest Neighbor (Fine KNN) | Medium | Medium |
| [*KNNMedium*] | Nearest Neighbor (Medium KNN) | Medium | Medium |
| [*KNNCoarse*] | Nearest Neighbor (Coarse KNN) | Medium | Medium |
| [*KNNCosine*] | Nearest Neighbor (Cosine KNN) | Medium | Medium |
| [*KNNCubic*] | Nearest Neighbor (Cubic KNN) | Slow | Medium |
| [*KNNWeighted*] | Nearest Neighbor (Weighted KNN) | Medium | Medium |
| [*BoostTrees*] | Ensemble (Boosted Trees) | Fast | Low |
| [*BaggTrees*] | Ensemble (Bagged Trees) | Medium | High |
| [*SubspaceDiscr*] | Ensemble (Subspace Discriminant) | Medium | Low |
| [*SubspaceKNN*] | Ensemble (Subspace KNN) | Medium | Medium |
| [*RUSBoostTrees*] | Ensemble (RUSBoost Trees) | Fast | Low |
| [*NeuralNetwork*] | Neural Network (3-Layer Perceptron) | Fast[6] | Low[6] |

Fast for binary classification, medium for multiclass classification. Fast for binary classification, slow for multiclass classification. Medium for binary classification, high for multiclass classification. Slow for high-dimensional data. Medium for high-dimensional data. When a moderate number of neurons are used.

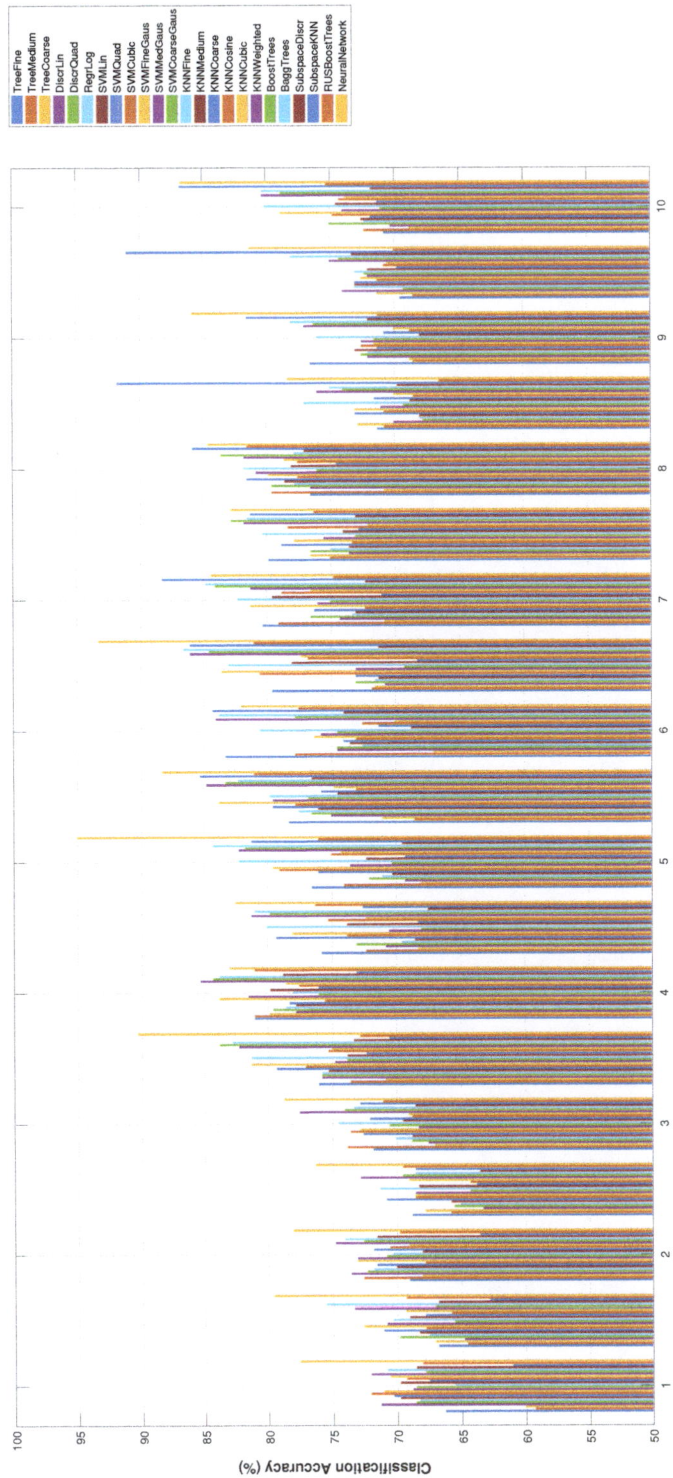


**Figure 2.** Classification accuracy for all the classification models (using the selected set of features) in front of the window length. As can be seen, the neural network model enabled us to achieve 95% accuracy for a window of five seconds.

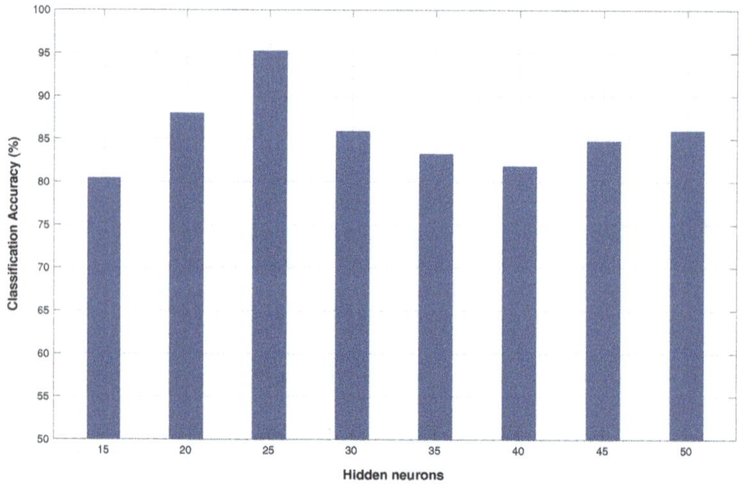


**Figure 3.** Classification accuracy of the neural network as a function of the number of hidden neurons in the hidden layer.

## 4.2. Hardware Resources Analysis

Table 4 gathers all the resource analysis results. It also shows resource usage as a percentage of occupation of the Xilinx XC7VX1140T FPGA, intending to provide an idea of the design occupancy in current FPGAs.

It can be seen that the design demands near the same DSP48E slices for both designs, 133 for single precision and 137 for double precision. Obviously, this slight variation has been achieved thanks to the pipelined design.

**Table 4.** Resource usage, performance and precision of the FPGA implementation as a function of the data type ("float" is for 32-bit floating-point arithmetic precision, and "double" for 64 bit floating-point arithmetic precision). Resource usage is indicated by the number of required slices and the percentage of occupation of a Xilinx XC7VX1140T FPGA.

| | Data Type | |
|---|---|---|
| **Resources** | **Double** | **Float** |
| DSP48E | 137 | 133 |
| BRAM | 185 | 95 |
| FF | 45,240 | 39,921 |
| LUT | 52,571 | 46,302 |
| DSP48E | 4.0% | 4.1% |
| BRAM | 4.9% | 2.5% |
| FF | 3.2% | 2.8% |
| LUT | 7.4% | 6.5% |
| Clock Period (ns) | $5.35 \pm 0.5$ | $5.21 \pm 0.5$ |
| Max. Clock Frequency (MHz) | 186.9 | 191.9 |
| Clock cycles | 73,999,901 | 53,996,354 |
| Max. Classification Frequency (Hz) | 2.53 | 3.55 |
| Accuracy (MAE) | $3.9428 \times 10^{-15}$ | $1.0005 \times 10^{-6}$ |

The required number of flip-flops (FF) and look up tables (LUT) were not high, presenting only small variations between designs.

On the other hand, as is natural, the amount of block RAM doubled for double arithmetic precision with respect to single arithmetic precision. That is due to the use of 64-bit representation of double precision and 32-bit of single precision. Then, block RAM requirement halves according the amount of memory needed for its representation.

### 4.3. Hardware Performance and Accuracy

Mean absolute error (MAE) was used to measure the accuracy of the results of each design with respect to the Matlab implementation. MAE measures the average magnitude of the errors without considering the direction of its deviation, taking into account the absolute differences between them.

The accuracy shown in Table 4 is the maximum of the accuracy obtained for the results of each output neuron. As it can be seen, the accuracy for the single design is very low, but the accuracy for the double design is so low that, in practice, it indicates that results in this case are similar to the results obtained in its Matlab implementation. It is natural, because both implementations use double precision data types.

In turn, the number of clock cycles shows a strong dependency of the data type (Table 4), and the numbers of minimum allowable cycles reported for both designs were similar.

The maximum frequency of operation is obtained from the minimum clock period and the required number of clock cycles. It is represented in Table 4 for both data type designs. As it can be seen, the computation of all the classification procedure, from accepting raw signals to the delivery of the classification output can be done at a rhythm of 2.53 Hz for the "double" design and 3.55 Hz for the "single" designs (3.55 classification outputs by second).

## 5. Discussion

The analysis described in this paper was carried out using the Bern-Barcelona dataset. Thus, all related works in the bibliography, selected for the sake of comparison, use the same database. That provides a more comparable framework, given that the use of works using other datasets may expose significant differences when performing the same classification method. As an example, San-Segundo et al. [25] shows that the focal-nonfocal (F-NF) classification accuracy may differ more than 20% when the same methods applies to the Bern-Barcelona dataset [7] and the Epileptic Seizure Recognition dataset [46]. In this example, the nature of the signals in the dataset, mainly the difference between signal lengths (only 1s for the latter), makes the difference. Thus, note that the Bern-Barcelona dataset is the logical selection when facing just the F-NF problem, provided its longer signal length (20 s) and its specialization in inter-ictal signals (recordings of seizure activity and three hours after the last seizure activity are excluded).

Table 5 compares our results with previous works for the task of classifying the focal and non-focal signals using the Bern-Barcelona EEG dataset. It details the obtained classification accuracy, summarizes the extracted features, and indicates the type of classifier used in each work.

Note that the purpose of this work was not to beat the accuracy results of previous works in the bibliography; our goal was to obtain, at the same time, the best classification procedure having the lowest possible computational load (for feature extraction and classification), aiming at its real-time implementation. Despite this, we obtained pretty good accuracy results. The 95.5% classification accuracy obtained in this work surpasses many other related works [13,15–18,20–22,47], while some other works [19,23,24,48] surpasses this result by a maximum of 1.5% classification accuracy ( [19] achieved 97% classification accuracy). Thus, when not considering [25], the comparison with the other related works can be considered pretty good in light of the great simplification achieved for the feature extraction process. In turn, San-Segundo et al. [25] obtained to 98.6% classification accuracy (3.1% more than this work), but at the expense of using a computationally intensive tool, a deep neural network, which is far away from the simplicity sought in this work; that is the reason to exclude this work from comparison from this point forward.

**Table 5.** Comparison of focal and non-focal classification results with previous works.

| Authors | Classifier Type | Extracted Feature | Accuracy |
|---|---|---|---|
| Sharma et al. (2015) [17] | LS-SVM | EMD with IMFs | 87.0% |
| Das et al. (2016) [18] | k-NN | EMD-DWT, log-energy entropy | 89.4% |
| Chatterjee et al. (2017) [21] | SVM, k-NN | MFDFA | 92.2% |
| Singh et al. (2017) [22] | LS-SVM | DFT based filter bank | 89.7% |
| Sharma et al. (2017) [16] | LS-SVM | TQWT [1] | 95.0% |
| Sharma et al. (2017) [13] | LS-SVM | Wavelet based entropies | 94.3% |
| Bhattacharyya et al. (2017) [15] | LS-SVM | Fuzzy entropy of TQWT | 84.7% |
| Bhattacharyya et al. (2018) [20] | LS-SVM | Rhythm separation from EWT | 90.0% |
| Taran et al. (2018) [23] | ELM | Clustering VMD | 96.0% |
| Deivasigamani et al. (2018) [24] | ANFIS | Extracted from CWT | 96.0% |
| Zeng et al. (2019) [19] | RBF-NN | EMD and PSR | 97.0% |
| San-Segundo et al. (2019) [25] | DNN | Fourier transform | 95.5% |
| San-Segundo et al. (2019) [25] | DNN | Wavelet transform | 97.4% |
| San-Segundo et al. (2019) [25] | DNN | Raw data | 98.6% |
| San-Segundo et al. (2019) [25] | DNN | 6 IMFs from EMD | 98.9% |
| Rahman et al. (2019) [47] | Stacked SVM | VMD, DWT , and others [2] | 95.2% |
| Sharma et al. (2019) [48] | SVM | LSDA from bispectrum | 96.2% |
| *This work* | *Neural Network* | *5 low computational complexity features extracted in real-time* | 95.5% |

[1] With different kernel function. [2] Other features: refined composite multiscale dispersion entropy (RCMDE), refined composite multiscale fuzzy entropy (RCMFE) and autoregresive model coefficients (AR).

As it can be seen in Table 5, extracted features were used to proceed from computationally intensive processes, such as the decomposition of EEG signals using empirical mode decomposition (EMD) to extract intrinsic mode functions (IMFs). Thus, Sharma et al. [17] obtained 87.0% classification accuracy using five entropy features extracted from IMFs; Das et al. [18] also used entropy-based features from the EMD, DWT (discrete wavelet transform) and EMD-DWT domains, achieving 89.4% classification accuracy; and Zeng et al. [19] arrived to a 97% classification accuracy using features derived from Euclidean measures obtained from the phase space reconstructions (PSRs) of several IMFs obtained using EMD.

The wavelet transform has been also a computationally intensive process used in related works as the basis of the feature extraction process. Thus, Sharma et al. [16] obtained several entropy features from the tunable-Q wavelet transform (TQWT), reporting a 95.0% classification accuracy; Bhattacharyya et al. [20] obtained 90.0% classification accuracy using as features, projections of the reconstructed phase space (RPS) from the rhythm separation achieved using the empirical wavelet transform (EWT); Sharma et al. [13] obtained 94.25% classification accuracy from various wavelet based entropies; Bhattacharyya et al. [15] obtained 84.67% classification accuracy using TQWT based multivariate sub-band fuzzy entropy with LS-SVM classifiers; and Deivasigamani et al. [24] obtained 96.0% classification accuracy based on a set of features extracted from the dual tree complex wavelet transform (DT-CWT) and using an adaptive neuron fuzzy interference system (ANFIS).

Other works use variational mode decomposition as the basis of feature extraction, such as Rahman et al. [47], who obtained a 95.2% classification accuracy using features such as refined composite multi scale dispersion entropy (RCMSDE), refined composite multiscale fuzzy entropy (RCMSFE) and autoregressive model (AR) coefficients extracted from variational mode decomposition (VMD), DWT and VMD-DWT domains; or Taran et al. [23], who obtained 96.0% classification accuracy using spectral moment based features extracted from the modes of the clustering VMD (CVMD) and extreme learning machine classifiers.

Note that the computational cost of the feature extraction process in previous works is greater than the computational cost of the feature set proposed in our work. The only work in the bibliography having a computational cost comparable to that of our work is Singh et al. [22], that obtained a 89.7% of classification accuracy deriving features from DFT-based rhythms of the EEG. In the same way, we have to compute the DFT too. Nevertheless, we obtain a better classification accuracy (5.8% more).

Thus, despite its simplicity, a classification procedure that performs better than most of related works, or, in the worst case, got surpassed by a maximum of 1.5% of classification accuracy (not considering [25]) was achieved.

However, this work does not propose just an optimum feature set, but an optimum classification procedure, as a whole. Thus, the proposed feature set, Table 2, can be combined optimally with a neural network classifier model when five seconds of segment length are used (Section 4.1). Our analysis indicates that this, altogether, guarantees the best accuracy performance with a minimum computational cost. The proposed neural network is of perceptron type, with just one hidden layer of 25 hidden neurons, five input neurons (the dimensionality of the proposed feature set) and two output neurons in the output layer.

On the other hand, the FPGA real-time implementation of the classification procedure, following Algorithms 1 and 2, has been done using a sequential architecture. The benefits of using this architecture are the minimization of the memory usage and the number of arithmetic hardware blocks. Anyway, to improve the throughput results this computation can be easily parallelized.

The proposed hardware design allows the definition of floating point arithmetic units of single or double data type precision (following the IEEE 754 standard for floating-point arithmetic). As it was expected, both designs offer a great MAE accuracy. "Double" design achieves an accuracy similar to the Matlab environment implementation: $3.94 \times 10^{-15}$, while the "float" design offers a great accuracy: $1.00 \times 10^{-6}$. MAE accuracy was measured using the Matlab implementation as a reference.

The analysis demonstrates that the proposed hardware implementation does not uses many resources. Both designs need no more than 137 DSP slices, while BRAM usage is 95 MB for the "single" design (doubling to 185 MB the requirements for the "double" design, provided that, obviously, the 64-bit double precision data types doubles the memory needs of the single 32-bit data type).

Note that the computational needs of the implementation in a Virtex 7 Xilinx FPGA device requires a reduced portion of its total resources; see Table 4. In fact, this application can be executed even on a small and cost-effective Xilinx XC6SLX100 Spartan-6 FPGA, assuring a low-cost of implementation.

Regarding the performance, it has been shown that the proposed implementation can perform all computation tasks at a maximum of 3.55 Hz when using the single data types, or 2.53 Hz when using double data types. That means that the single design can deliver outputs at a rhythm of 3.55 times by second.

However, note that this 3.55 Hz of classification frequency (2.53 Hz for double design) is only effective after the first 5 s of acquisition time due to the 5 s segmentation. This implies a minimum latency time to achieve the first result from the beginning of an acquisition without artifacts of the EEG signal. From this 5 s latency, the proposed implementation is capable of handling overlapped time windows, delivering results at the maximum classification frequency. Thus, for the single design, a result will be provided each $1/3.55 = 0.28$ seconds after the first 5 s window length.

Thus, we have shown that an adequate selection of the set of features, classifier model and length of the window segment, allows one to obtain good classification accuracy results (above the average of previous related works) while maintaining a low computational load for the whole classification procedure. It enables us to move the classification procedure to the real-time field, embedded in a logic-reconfigurable FPGA.

The proposed implementation can be carried out on a small portable device embedding a fast classification engine of epileptogenic focus. This device can serve as a help decision tool to assist neurosurgeons to refine the localization of the epileptogenic area during the resective epilepsy surgery in those cases where greater precision or confirmation were needed.

## 6. Conclusions

The locating of the epileptogenic focus using interictal EEG signals is generally a computerized analysis carried out off-line by neurosurgeons and epileptologists to determine the brain regions responsible for the initial seizure discharge. However, previous works tend to propose more computationally costly procedures the more recent they are.

This work shows that an adequate selection of the set of features, classifier model and length of the window segment, allows one to obtain good classification accuracy results (above the average) while maintaining a low computational load. It enables the real-time implementation of the whole classification procedure, on an FPGA reconfigurable device, from accepting the raw EEG signals to the delivery of the classification outputs at a rhythm of up to 3.55 Hz. It opens the door to the use of the automatic classification as a decision-assisting tool during surgery, enabling neurosurgeons to refine the localization of the epileptogenic area during the resective epilepsy surgery.

Concluding, it has been shown that the proposed hardware implementation of the epileptogenic foci locator can be embedded on a small portable device, embedding, thus, a fast classification engine of epileptogenic signals in epilepsy.

**Author Contributions:** Individual contributions: conceptualization, J.F.G.-M., A.G.M., V.V., K.G.H. and V.T.-M.; methodology, J.F.G.-M., J.V.F.-V. and M.B.-M.; software, J.F.G.-M., A.M. and J.V.F.V.; formal analysis, J.F.G.-M., J.V.F.-V., A.M., M.B.-M. and A.R.-M.; writing—original draft preparation, J.V.F.-V.; writing—review and editing, J.V.F.-V., A.R.-M. and M.B.-M. All authors have read and agreed to the published version of the manuscript.

**Funding:** This research received no external funding.

**Conflicts of Interest:** The authors declare no conflict of interest.

## Abbreviations

The following abbreviations are used in this manuscript:

| | |
|---|---|
| EMD | Empirical mode decomposition |
| IMF | Intrinsic mode function |
| DWT | Discrete wavelet transform |
| MFDFA | Multifractal detrended fluctuation analysis |
| DFT | Discrete Fourier transform |
| TQWT | Tuneable-Q wavelet transform |
| EWT | Empirical wavelet transform |
| ELM | Extreme learning machine |
| VMD | Variational mode decomposition |
| ANFIS | Adaptive neuro fuzzy interference system |
| RBF-NN | Radial basis function neural network |
| PSR | Phase state reconstruction |
| DNN | Deep neural network |
| LSDA | Locality sensitive discriminant analysis |

## References and Notes

1. Fisher, R.S.; Acevedo, C.; Arzimanoglou, A.; Bogacz, A.; Cross, J.H.; Elger, C.E.; Engel, J., Jr; Forsgren, L.; French, J.A.; Glynn, M.; et al. ILAE Official Report: A practical clinical definition of epilepsy. *Epilepsia* **2014**, *55*, 475–482. [CrossRef] [PubMed]
2. Falco-Walter, J.J.; Scheffer, I.E.; Fisher, R.S. The new definition and classification of seizures and epilepsy. *Epilepsy Res.* **2018**, *139*, 73–79. [CrossRef] [PubMed]
3. Daroff, R.B.; Jankovic, J.; Mazziotta, J.C.; Pomeroy, S.L. *Bradley's Neurology in Clinical Practice E-Book;* Elsevier Health Sciences, 2016; ISBN 9780323287838.
4. Ray, A.; Tao, J.X.; Hawes-Ebersole, S.M.; Ebersole, J.S. Localizing value of scalp EEG spikes: A simultaneous scalp and intracranial study. *Clin. Neurophysiol.* **2007**, *118*, 69–79. [CrossRef] [PubMed]

5. Bast, T.; Boppel, T.; Rupp, A.; Harting, I.; Hoechstetter, K.; Fauser, S.; Schulze-Bonhage, A.; Scherg, M. Noninvasive source localization of interictal EEG spikes: effects of signal-to-noise ratio and averaging. *J. Clin. Neurophysiol.* **2006**, *23*, 487–497. [CrossRef] [PubMed]
6. Dworetzky, B.A.; Reinsberger, C. The role of the interictal EEG in selecting candidates for resective epilepsy surgery. *Epilepsy Behav.* **2011**, *20*, 167–171. [CrossRef] [PubMed]
7. Andrzejak, R.G.; Schindler, K.; Rummel, C. Nonrandomness, nonlinear dependence, and nonstationarity of electroencephalographic recordings from epilepsy patients. *Phys. Rev.* **2012**, *86*, 046206. [CrossRef]
8. Wilson, S.B.; Turner, C.A.; Emerson, R.G.; Scheuer, M.L. Spike detection II: automatic, perception-based detection and clustering. *Clin. Neurophysiol.* **1999**, *110*, 404–411. [CrossRef]
9. Liu, D.; Wang, Q.; Zhang, Y.; Liu, X.; Lu, J.; Sun, J. FPGA-based real-time compressed sensing of multichannel EEG signals for wireless body area networks. *Biomed. Signal Process. Control.* **2019**, *49*, 221–230. [CrossRef]
10. Mjahad, A.; Rosado-Muñoz, A.; Bataller-Mompeán, M.; Francés-Víllora, J.; Guerrero-Martínez, J. Ventricular Fibrillation and Tachycardia detection from surface ECG using time-frequency representation images as input dataset for machine learning. *Comput. Methods Programs Biomed.* **2017**, *141*, 119–127. [CrossRef]
11. Mjahad, A.; Rosado-Muñoz, A.; Guerrero-Martínez, J.F.; Bataller-Mompeán, M.; Francés-Villora, J.V.; Dutta, M.K. Detection of Ventricular Fibrillation Using the Image from Time-Frequency Representation and Combined Classifiers without Feature Extraction. *Appl. Sci.* **2018**, *8*. [CrossRef]
12. Mjahad, A.; Muñoz, A.R.; Mompeán, M.B.; Víllora, J.V.F.; Martínez, J.F.G. Detección de Fibrilación Ventricular Mediante Tiempo-Frecuencia y Clasificador KNN sin Extracción de Parámetros. *Rev. Iberoam. AutomáTica InformáTica Ind.* **2017**, *15*, 124–132. [CrossRef]
13. Sharma, M.; Dhere, A.; Pachori, R.B.; Acharya, U.R. An automatic detection of focal EEG signals using new class of time–frequency localized orthogonal wavelet filter banks. *Knowledge-Based Syst.* **2017**, *118*, 217–227. [CrossRef]
14. Sriraam, N.; Raghu, S. Classification of Focal and Non Focal Epileptic Seizures Using Multi-Features and SVM Classifier. *J. Med Syst.* **2017**, *41*, 160. [CrossRef]
15. Bhattacharyya, A.; Pachori, R.; Acharya, U. Tunable-Q wavelet transform based multivariate sub-band fuzzy entropy with application to focal EEG signal analysis. *Entropy* **2017**, *19*, 99. [CrossRef]
16. Sharma, R.; Kumar, M.; Pachori, R.B.; Acharya, U.R. Decision support system for focal EEG signals using tunable-Q wavelet transform. *J. Comput. Sci.* **2017**, *20*, 52–60. [CrossRef]
17. Sharma, R.; Pachori, R.; Acharya, U. Application of entropy measures on intrinsic mode functions for the automated identification of focal electroencephalogram signals. *Entropy* **2015**, *17*, 669–691. [CrossRef]
18. Das, A.B.; Bhuiyan, M.I.H. Discrimination and classification of focal and non-focal EEG signals using entropy-based features in the EMD-DWT domain. *Biomed. Signal Process. Control.* **2016**, *29*, 11–21. [CrossRef]
19. Zeng, W.; Li, M.; Yuan, C.; Wang, Q.; Liu, F.; Wang, Y. Classification of focal and non focal EEG signals using empirical mode decomposition (EMD), phase space reconstruction (PSR) and neural networks. *Artif. Intell. Rev.* **2019**, *52*, 625–647. [CrossRef]
20. Bhattacharyya, A.; Sharma, M.; Pachori, R.B.; Sircar, P.; Acharya, U.R. A novel approach for automated detection of focal EEG signals using empirical wavelet transform. *Neural Comput. Appl.* **2018**, *29*, 47–57. [CrossRef]
21. Chatterjee, S.; Pratiher, S.; Bose, R. Multifractal detrended fluctuation analysis based novel feature extraction technique for automated detection of focal and non-focal electroencephalogram signals. *IET Sci. Meas. Technol.* **2017**, *11*, 1014–1021. [CrossRef]
22. Singh, P.; Pachori, R.B. Classification of focal and nonfocal EEG signals using features derived from Fourier-based rhythms. *J. Mech. Med. Biol.* **2017**, *17*, 1740002. [CrossRef]
23. Taran, S.; Bajaj, V. Clustering variational mode decomposition for identification of focal EEG signals. *IEEE Sens. Lett.* **2018**, *2*, 1–4. [CrossRef]
24. Deivasigamani, S.; Senthilpari, C.; Yong, W.H. Computer Aided Automatic Detection and Classification of EEG Signals for Screening Epilepsy Disorder. *J. Inf. Sci. Eng.* **2018**, *34*, 687–700.
25. San-Segundo, R.; Gil-Martín, M.; D'Haro-Enríquez, L.F.; Pardo, J.M. Classification of epileptic EEG recordings using signal transforms and convolutional neural networks. *Comput. Biol. Med.* **2019**, *109*, 148–158. [CrossRef]
26. Daoud, H.; Bayoumi, M. Deep Learning Approach for Epileptic Focus Localization. *IEEE Trans. Biomed. Circuits Syst.* **2019**. [CrossRef]

27. Frances-Villora, J.; Rosado-Muñoz, A.; Bataller-Mompean, M.; Barrios-Aviles, J.; Guerrero-Martinez, J. Moving Learning Machine towards Fast Real-Time Applications: A High-Speed FPGA-Based Implementation of the OS-ELM Training Algorithm. *Electronics* **2018**, *7*. [CrossRef]
28. Abbaszadeh, A.; Iakymchuk, T.; Bataller-Mompeán, M.; Francés-Villora, J.V.; Rosado-Muñoz, A. AnScalable Matrix Computing Unit Architecture for FPGA, and SCUMO User Design Interface. *Electronics* **2019**, *8*. [CrossRef]
29. Fasil, O.K.; Rajesh, R.; Thasleema, T.M. Influence of differential features in focal and non-focal EEG signal classification. In Proceedings of the 2017 IEEE Region 10 Humanitarian Technology Conference (R10-HTC), Dhaka, Bangladesh, 21–23 December 2017; IEEE: Dhaka, Bangladesh, 2017; pp. 646–649. [CrossRef]
30. Raghu, S.; Sriraam, N.; Kumar, G.P. Classification of epileptic seizures using wavelet packet log energy and norm entropies with recurrent Elman neural network classifier. *Cogn. Neurodyn.* **2017**, *11*, 51–66. [CrossRef]
31. Aarabi, A.; Wallois, F.; Grebe, R. Automated neonatal seizure detection: A multistage classification system through feature selection based on relevance and redundancy analysis. *Clin. Neurophysiol.* **2006**, *117*, 328–340. [CrossRef]
32. Bedeeuzzaman, M.; Farooq, O.; Khan, Y.U. Automatic seizure detection using higher order moments. In Proceedings of the 2010 International Conference on Recent Trends in Information, Telecommunication and Computing, Kerala, India, 12–13 March 2010; pp. 159–163.
33. Greene, B.; Faul, S.; Marnane, W.; Lightbody, G.; Korotchikova, I.; Boylan, G. A comparison of quantitative EEG features for neonatal seizure detection. *Clin. Neurophysiol.* **2008**, *119*, 1248–1261. [CrossRef]
34. Aarabi, A.; Grebe, R.; Wallois, F. A multistage knowledge-based system for EEG seizure detection in newborn infants. *Clin. Neurophysiol.* **2007**, *118*, 2781–2797. [CrossRef]
35. Hernández, D.; Trujillo, L.; Z-Flores, E.; Villanueva, O.; Romo-Fewell, O. Detecting Epilepsy in EEG Signals Using Time, Frequency and Time-Frequency Domain Features. In *Computer Science and Engineering—Theory and Applications*; Springer: Berlin/Heidelberg, Germany, 2018; pp. 167–182.
36. Rosado Muñoz, D.A. Desarrollo de Técnicas de Detección de Fibrilación Ventricular Basadas en Algoritmo Tiempo–Frecuencia. Ph.D. Thesis, Universidad de Valencia, Valencia, Spain, 2011.
37. Sharma, N.; Gedeon, T. Objective measures, sensors and computational techniques for stress recognition and classification: A survey. *Comput. Methods Programs Biomed.* **2012**, *108*, 1287–1301. [CrossRef]
38. Petrantonakis, P.C.; Hadjileontiadis, L.J. Emotion recognition from EEG using higher order crossings. *IEEE Trans. Inf. Technol. Biomed.* **2009**, *14*, 186–197. [CrossRef]
39. Urbanowicz, R.J.; Meeker, M.; La Cava, W.; Olson, R.S.; Moore, J.H. Relief-based feature selection: Introduction and review. *J. Biomed. Inform.* **2018**, *85*, 189–203. [CrossRef]
40. Aydın, S.; Saraoğlu, H.M.; Kara, S. Log energy entropy-based EEG classification with multilayer neural networks in seizure. *Ann. Biomed. Eng.* **2009**, *37*, 2626. [CrossRef]
41. Şen, B.; Peker, M.; Çavuşoğlu, A.; Çelebi, F.V. A comparative study on classification of sleep stage based on EEG signals using feature selection and classification algorithms. *J. Med. Syst.* **2014**, *38*, 18. [CrossRef]
42. Xilinx' Vivado Design Suite. High-Level Synthesis User Guide, UG902 (v2016.1) 6 April. Xilinx. 2011.
43. ARM' AMBA AXI and ACE Protocol Specification: AXI3, AXI4, and AXI4-Lite, ARM IHI 0022E (ID022613). Xilinx. 2011.
44. Xilinx. Vivado Design Suite. AXI Reference Guide, UG1037 v3.0, 24 June. Xilinx. 2015.
45. ARM. AMBA AXI4-Stream Protocol, ARM IHI 0051A (ID030510). Xilinx. 2010.
46. Andrzejak, R.G.; Lehnertz, K.; Mormann, F.; Rieke, C.; David, P.; Elger, C.E. Indications of nonlinear deterministic and finite-dimensional structures in time series of brain electrical activity: Dependence on recording region and brain state. *Phys. Rev.* **2001**, *64*, 061907. [CrossRef]
47. Rahman, M.M.; Bhuiyan, M.I.H.; Das, A.B. Classification of focal and non-focal EEG signals in VMD-DWT domain using ensemble stacking. *Biomed. Signal Process. Control.* **2019**, *50*, 72–82. [CrossRef]
48. Sharma, R.; Sircar, P.; Pachori, R.B. A New Technique for Classification of Focal and Nonfocal EEG Signals Using Higher-Order Spectra. *J. Mech. Med. Biol.* **2019**, *19*, 1940010. [CrossRef]

applied sciences

MDPI

Article

# Optimization of Physical Activity Recognition for Real-Time Wearable Systems: Effect of Window Length, Sampling Frequency and Number of Features

**Ardo Allik [1,*], Kristjan Pilt [1], Deniss Karai [1], Ivo Fridolin [1], Mairo Leier [2] and Gert Jervan [2]**

1 Department of Health Technologies, Tallinn University of Technology, Ehitajate tee 5, 12616 Tallinn, Estonia; kristjan.pilt@taltech.ee (K.P.); deniss.karai@taltech.ee (D.K.); ivo.fridolin@taltech.ee (I.F.)

2 Department of Computer Systems, Tallinn University of Technology, Ehitajate tee 5, 12616 Tallinn, Estonia; mairo.leier@taltech.ee (M.L.); gert.jervan@taltech.ee (G.J.)

* Correspondence: ardo.allik@taltech.ee

Received: 20 October 2019; Accepted: 7 November 2019; Published: 12 November 2019

**Abstract:** The aim of this study was to develop an optimized physical activity classifier for real-time wearable systems with the focus on reducing the requirements on device power consumption and memory buffer. Classification parameters evaluated in this study were the sampling frequency of the acceleration signal, window length of the classification fragment, and the number of classification features, found with different feature selection methods. For parameter evaluation, a decision tree classifier was created based on the acceleration signals recorded during tests, where 25 healthy test subjects performed various physical activities. Overall average F1-score achieved in this study was about 0.90. Similar F1-scores were achieved with the evaluated window lengths of 5 s (0.92 ± 0.02) and 3 s (0.91 ± 0.02), while classification performance with 1 s were lower (0.87 ± 0.02). Tested sampling frequencies of 50 Hz, 25 Hz, and 13 Hz had similar results with most classified activity types, with an exception of outdoor cycling, where differences were significant. Using forward sequential feature selection enabled the decreasing of the number of features from initial 110 features to about 12 features without lowering the classification performance. The results of this study have been used for developing more efficient real-time physical activity classifiers.

**Keywords:** accelerometer; activity classification; activity trackers; machine learning; wearable systems

## 1. Introduction

It is important to propagate active lifestyle, since routine physical activity has been found to have multiple benefits, such as preventing chronic diseases and increasing psychological well-being [1,2], while prolonged inactivity has been shown to lead to an increase of chronic diseases and obesity [1,3]. Advancement of technology has brought a surge of popularity for many activity trackers in the form of mobile phone apps or wearable systems. With these devices, users are able to keep track of their training schedule, exercises and lost calories [4]. Since this makes training more interactive and allows users to have better overview of their progress, then it often motivates the users to have a more active lifestyle and lose weight over sustained periods [5–7].

Wearable systems are used to conveniently measure, collect and analyze the user's psychological data. This requires wearables to be small and unobtrusive, which in turn puts significant demand on reducing power consumption of the system [8]. This is also significant for real-time physical activity classification, which can be used in wearables for online activity recognition by allowing automatic recognition of the activities the user is performing [9,10]. Real-time activity recognition provides valuable information for improving online feedback of the activity trackers or for providing extra safety by monitoring the status of the users working in high-risk environments [11].

Power consumption required for physical activity classification is determined by multiple different components. Some of these components are based on the processing of the acceleration values, such as sampling rate of the signal and filtering [12]. Other elements are based on classification mechanics, such as classification window length, feature calculation, and the used machine learning algorithm. While studies have explored classification mechanics such as training times of different physical activity classification algorithms [13,14], they do not provide valuable information for real-time classification, since classifier training can be done previously on a desktop computer and later implemented into the wearable system. For classification systems working in real time, it is important to focus on processing time of the calculations the system has to do online [13,15].

In an earlier study, our group explored how different accelerometer sampling frequencies, classification window lengths, and the number of correlating features affect the classifier performance [16]. Few studies before have evaluated how different window lengths (commonly chosen between 1.5 s [17] and 5 s [13]) affect physical activity classification performance [15,18], but the lack of gold standard in physical activity classification makes it difficult to compare these results [19]. It has been stated that frequencies above 20 Hz cannot be expected to arise from voluntary movement [20], but comparable performance has been reported while using lower sampling frequencies [12,21]. Various methods have been used for feature selection, such as the ReliefF algorithm [22], principal component analysis [13], or information gain [15], but not in connection with window length and sampling frequency.

The aim of this study was to create an optimized physical activity classifier that would be suitable for implementation on real-time wearable systems. The focus was on testing various sampling frequencies, window lengths and number of features in order to reduce the power consumption, and to decrease the required memory buffer without compromising classification performance. Other classification elements were chosen based on the results of other studies with emphasis on high classification performance and low power consumption.

## 2. Materials and Methods

Physical activity classification often uses machine learning methods, where the classification is usually based on acceleration signals. Overview of the steps taken to create and evaluate the classifier used in this study are shown in Figure 1.

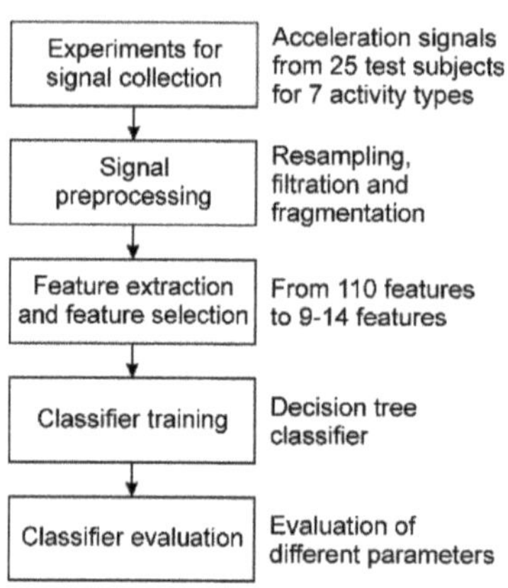


**Figure 1.** A summary of methods used in the study.

### *2.1. Instrumentation*

Acceleration signals were measured with Shimmer3 (from here on Shimmer) sensor platform (Shimmer Research, Dublin, Ireland). While sensor fusion between accelerometers and gyroscopes has shown to increase classification performance in some studies [23], then others have found that gyroscope information does not contribute to activity recognition performance [22]. Due to the emphasis on designing physical activity classifier with low power consumption, gyroscope data were disregarded in this study.

The Shimmer sensor system has two built-in triaxial accelerometers: low noise accelerometer with the dynamic range of ±2 g and a wide range accelerometer with the dynamic range switchable between ±2 g to ±16 g (where 1 g equals to about 9.81 m/s$^2$). Since acceleration values during human motion surpass ±2 g [24], the data from wide range accelerometer was used with the dynamic range set to ±16 g. The wide range accelerometer uses STMicroelectronics LSM303AHTR sensor (Geneva, Switzerland), which has a numeric resolution of 16-bit. Acceleration was measured with a sampling rate of 512 Hz.

### 2.2. Study Group

The study was approved by the Tallinn Medical Research Ethics Committee. The main study group consisted of 25 healthy 21–45 year old test subjects (with an exception of one 57-year-old male), of whom 13 were male and 12 female. Average age was 32.0 ± 8.8 years (median 30.0) for the whole group, 32.8 ± 10.0 years (median 30.0) for males, and 31.0 ± 7.7 years (median 30.0) for females. A separate study group was used to measure the signals of outdoor cycling. This group consisted of 5 males with an average age of 38.4 ± 5.3 years (median 37.0).

### 2.3. Test Overview and Recorded Signals

Test subjects performed various physical activities during which acceleration signals were measured and recorded using the Shimmer sensor system. The sensor was located on the left wrist for feasibility of implementing the results in an activity tracker worn on the wrist. Even though using multiple sensors has been shown to increase the classification performance [25,26], having a wearable system with only one sensor is more comfortable and convenient for the user.

Each test subject conducted activities based on a precise schedule, where each activity was carried out for a fixed amount of time, shown in Table 1. For classification, these activities were grouped into different activity types, shown in Table 2. Indoor activities were divided into three different parts, during which each activity was performed for 3 min, with the exception of lying down, which lasted 4 min. There were short pauses between each activity, which were later discarded from the signals.

**Table 1.** Conducted activities and their duration in minutes.

| Indoor Test 1 | Indoor Test 2 | Indoor Test 3 (% Shows Angle) | Outdoor Test |
|---|---|---|---|
| Walking (3) | Sitting on chair (3) | Walking (3 km/h) (3) | Cycling (14) |
| Running (3) | Lying on bed (4) | Walking (5 km/h) (3) | Cycling uphill (4) |
| Walking upstairs (3) | Typing on computer (3) | Walking (3 km/h, 10%) (3) | Cycling downhill (1) |
| Walking downstairs (3) | Folding clothes (3) | Walking (5 km/h, 10%) (3) | |
| | Cleaning surface (3) | Running (6 km/h) (3) | |
| | | Running (10 km/h) (3) | |
| | | Running (12 km/h) (3) | |
| | | Running (6 km/h, 10%) (3) | |

**Table 2.** Classified activity types.

| Activity Type | Activities Concluded |
|---|---|
| Static | Lying, sitting, standing |
| Low Intensity | Typing on computer |
| Moderate Intensity | Folding clothes |
| Rhythmical Intensity | Cleaning a surface with a towel |
| Walking | Walking in a corridor, walking on a treadmill, walking upstairs, walking downstairs |
| Running | Running in a corridor, running on a treadmill |
| Outdoor Cycling | Cycling outdoors on different terrains |

In the first part, test subjects walked in a corridor, ran in the corridor, walked upstairs, and walked downstairs. Altogether, a total of 12 min of acceleration signals were used from this part.

The second part consisted of sitting on a chair, lying on a bed, typing on a computer while sitting, standing, folding clothes while standing, and cleaning a surface while standing. A total of 19 min of signals were used from the second part.

The third indoor part consisted of walking on a treadmill at different speeds and angles (3 km/h, 5 km/h, 3 km/h with uphill angle 10%, 5 km/h with uphill angle 10%) and running on treadmill at different speeds and angles (6 km/h, 10 km/h, 12 km/h, 6 km/h with 10% uphill angle). A total of 24 min of signals were used from this part.

Outdoor cycling signals were recorded separately with a different study group. These signals consist of 14 min of cycling on a plain road, 4 min of cycling uphill, and 1 min of cycling downhill.

### 2.4. Resampling and Sampling Frequency

As an aim of this study, it was tested how different sampling frequencies affect the classification results. Lowering the sampling frequency, $f_s$, decreases the number of samples in the classification fragment, $s_f$, which is calculated as follows:

$$s_f = f_s \cdot w_f, \quad (1)$$

where $w_f$ is the window length of a fragment given in seconds.

To test different sampling frequencies, the signals that were initially recorded with a sampling frequency of 512 Hz were later resampled using a MATLAB function resample (R2016b, MathWorks, Natick, MA, USA). This function applies interpolation and decimation in order to achieve the desired sampling rate. In case of interpolation, the function inserts points with 0-values between each of the original samples of the signal, after which the signal is low-pass filtered at half of the desired sampling rate. To obtain the final result, decimation is applied by selecting samples from the filtered output [27]. The sampling frequencies of 50 Hz, 25 Hz, and 13 Hz were chosen for evaluating the effects of different sampling frequencies on classifier performance.

### 2.5. Filtering

Following resampling, filtering was applied to separate the recorded acceleration signals into static and dynamic components for physical activity classification. The static component in the acceleration signal is mostly affected by gravity and captures the posture information, while the dynamic component is based on motion and captures the human movement information.

In this study, the static component was found using a third order low-pass Butterworth infinite impulse response (IIR) filter. The passband and stopband edge frequencies and ripples were 0.1 Hz and 0.5 Hz, and 1 dB and 20 dB, respectively. The dynamic component was found by subtracting the static component from the original signal by taking into account the group delay of the low pass filter.

### 2.6. Fragmentation and Window Length

For classifier training, acceleration signals were fragmented into shorter consecutive fragments. Before fragmentation, the short pauses in the signals between different conducted activities were removed and only signals recorded during activities listed in Table 2 were kept. While some studies opt for an overlap between windows to increase the classification performance, in this study, no overlap was used to keep the computational power minimal.

In a system with a physical activity classifier working in real time, the window length determines the delay of the system, since each classification is done after signals have been collected for a whole window. The number of samples in the fragment is determined by both the sampling frequency and the window length according to Equation (1).

To evaluate how different window lengths affect the classifier performance, the window lengths of 5 s, 3 s, and 1 s were chosen, which are near the values usually used for physical activity classification in previous studies [13,17].

### 2.7. Feature Extraction

When using machine learning methods for physical activity classification, the classifier training is done based on features that are extracted from signal fragments. The feature set has to capture specific and diverse information of posture and human motion to allow precise activity classification. The initial set of 110 features used in this study were mostly adopted from previous studies by other researchers: (1) 60 various time-domain features from [28]; (2) 10 body posture related, 6 motion shape related features and 6 motion periodicity related features from [15]; (3) 24 various time-domain features from [22]; and (4) 9 separately added additional features.

Only time-domain features were chosen in this study in order to keep computing power minimal. While activity recognition studies have also used frequency-domain and wavelet transform features, the transforms needed to calculate these features would require extra resources. Additionally, it has been found that time-domain features give comparable results to other feature types [29].

### 2.8. Feature Selection

Another major aim of this study was to analyze how different number of features affects physical activity classification and what is the minimal number of features to use without compromising classification performance. For that, two different feature selection schemes were used to optimize the feature set.

One scheme was based on various methods that were used successively (Figure 2). This scheme used the features extracted with sampling frequency of 50 Hz and window length of 3 s and the achieved optimized feature set was later used with other frequency and window length combinations.

First, correlating features were removed based on a large correlation matrix that showed each feature's correlation coefficient with other features. From feature pairs or groups with a very high correlation (correlation coefficient larger than 0.9 or lower than −0.9), only the simpler features in terms of computational power requirements and complexity were kept. By using this method, 67 features were removed from the initial set, and a new subset of 43 features was formed. This method and the results have also been described in the previous study done by the authors [28].

Further feature optimization was done with one-way analysis of variance (ANOVA). The purpose of one-way ANOVA is to determine whether data from several groups of a factor have a common mean. ANOVA was used in this work to find out which features did not differentiate between any of the activities and thus did not provide any useful information for activity classification. Based on ANOVA results, 15 features were removed that were found not to affect classifier performance, and a new subset of 28 features was formed.

Finally, a sequential backward selection (SBS) procedure was repeated, where each feature was again removed one-by-one (those calculated similarly over all axes were removed together), and the feature that decreased the classifier performance the least was removed. After removing features this way, the classifier performance was still persistent with 13 features used. Further removal of features showed a decrease in activity classification sensitivities.

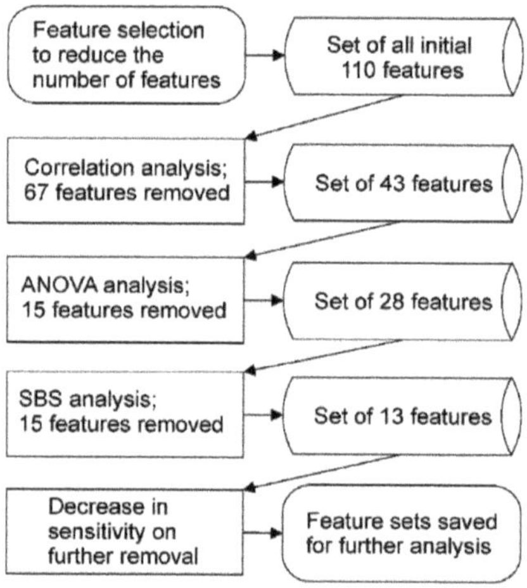


**Figure 2.** First feature selection scheme using correlation analysis, ANOVA, and backwards sequential feature selection with the number of features removed in each step.

The second feature selection scheme used in this study was a sequential forward selection (SFS) method similar to the last steps used in the first scheme (Figure 3). In this method, features were added one-by-one by conducting physical activity classification with each feature and, for every iteration, the best feature was kept. Features were added until the overall average classification sensitivity did not improve by more than 0.001. This method was completed for every sampling frequency and window length combination, and was used to compare the results of the first method.

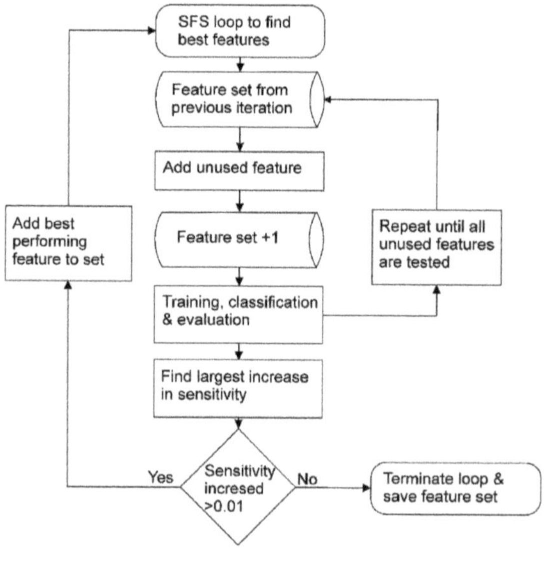


**Figure 3.** Forward sequential feature selection (SFS) method used in the second feature selection scheme.

### 2.9. Classifier Training

A machine learning based decision tree classification algorithm was chosen, which has been previously used in real-time physical activity classification and proposed as the most suitable in terms of performance and computational power needed for real-time classification [15,30]. The classifier was trained based on training data using MATLAB's function fitctree, which returns a fitted binary classification decision tree based on the input variables.

### 2.10. Classifier Evaluation

The classifier performance was evaluated using a leave-one-out cross-validation scheme where each test subject's signals were classified with a classifier that was trained using the signals from all the other test subjects. This method has been previously used in other physical activity classification studies to reduce overfitting errors [29,31].

Sensitivity (also called recall or true positive rate) was chosen as a statistical measure to evaluate classification performance during feature selection. Sensitivity shows the proportion of true positives classified (True_positives) in relation to correct or real ones (Real_positives), i.e., true positives that are correctly identified [32], and it is calculated as follows:

$$\begin{gathered}\text{Sensitivity} = \text{True\_positives}/\text{Real\_positives} = \\ \text{True\_positives}/(\text{True\_positives} + \text{False\_Negatives}).\end{gathered} \tag{2}$$

Classification results were evaluated using F1-score (also called F-score or F-measure), which is calculated as harmonic mean of precision and sensitivity [27], using the following formulas:

$$\begin{gathered}\text{Precision} = \text{True\_positives}/\text{Predicted\_positives} = \\ \text{True\_positives}/(\text{True\_positives} + \text{False\_positives}),\end{gathered} \tag{3}$$

$$\text{F1-score} = (2 \cdot \text{Sensitivity} \cdot \text{Precision})/(\text{Sensitivity} + \text{Precision}). \tag{4}$$

While evaluating the results with different window lengths, sampling frequencies and number of features, F1-scores were calculated separately for each activity type. Additionally, an average F1-score for different parameter combinations was found as a means of the activity type F1-scores.

A paired *t*-test ($p < 0.05$) was used to find statistical differences between the classification F1-scores of different activity types and averages while using different window lengths and sampling frequencies.

## 3. Results

### 3.1. Classifier Performance with Different Window Lengths

An overall average classification F1-score of about 0.90 was achieved for the physical activity classifier in this study, depending on the used window length, sampling frequency, feature set, and classified activity type. To evaluate how each of these parameters affected the classifier individually, classifier F1-scores were averaged over other parameters.

Figure 4 shows the classification F1-score of activity types for the different window lengths when averaged over different sampling frequencies (50 Hz, 25 Hz, 13 Hz) and feature sets (110 features, 43 features, 28 features, 13 features, and SFS feature set). The classifier had better performance with the average F1-score over 0.9 classifying static, walking and running activity types. Window lengths of 5 s and 3 s had similar results with the average F1-scores of 0.92 ± 0.02 and 0.91 ± 0.02, while the result with 1 s was 0.87 ± 0.02.

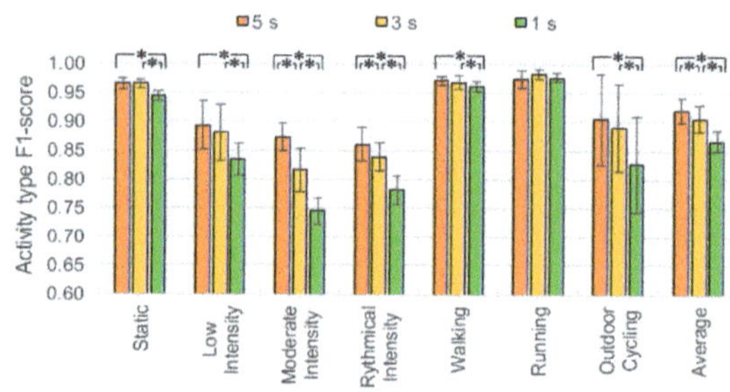


**Figure 4.** F1-scores of different activity types (mean ± SD (Standard deviation)) averaged over sampling frequencies and feature sets using different window lengths. Asterisks show significant statistical difference between different values of the window length ($p < 0.05$).

Statistically significant differences (marked with an asterisk in Figure 4) were found in moderate intensity and rhythmical intensity activity types between window lengths of 5 s and 3 s. Window length of 1 s had a statistical difference classifying every activity type other than running compared to both 5 s and 3 s window length.

### 3.2. *Classifier Performance with Different Sampling Frequencies*

To compare the results with different sampling frequencies, F1-scores were averaged over different window lengths and feature sets (Figure 5). Overall, the classifier had similar average F1-score with 50 Hz (0.92 ± 0.02) and 25 Hz (0.91 ± 0.02), while the average F1-score with 13 Hz was lower (0.87 ± 0.02).

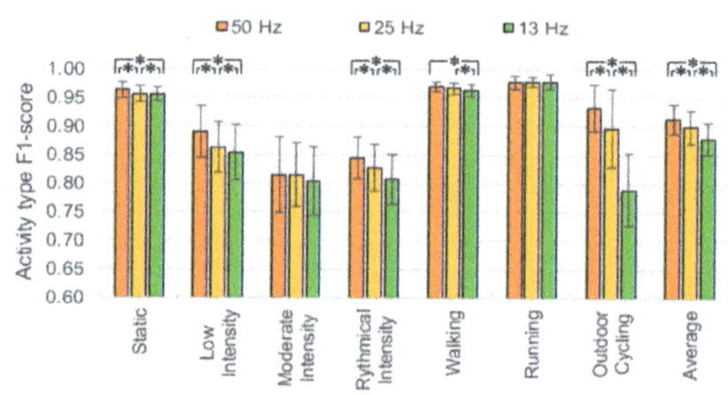


**Figure 5.** F1-score of different activity types (mean ± SD) averaged over window lengths and feature sets using different sampling frequencies. Asterisks show a significant statistical difference between different values of the sampling frequency ($p < 0.05$).

Statistically significant differences between different sampling frequencies (marked with an asterisk in Figure 5) were found for most activity types with the exceptions of moderate intensity and running.

Very large differences in classification performance were noted while classifying outdoor cycling, where the F1-score was 0.93 ± 0.04 with 50 Hz, 0.90 ± 0.07 with 25 Hz and 0.79 ± 0.06 with 13 Hz.

### 3.3. *Classifier Performance with Different Feature Sets*

To evaluate how the feature selection methods and the number of features used for classification affect the classifier performance, the results were averaged over different sampling frequencies and window lengths while using different feature sets (Figure 6). The feature sets of 110 features, 43 features, 28 features and 13 features, achieved with the first feature selection scheme, had similar average F1-scores between 0.89 and 0.90. The SFS feature set had a slightly higher average F1-score of 0.92 ± 0.03.

The SFS feature set had a major increase in performance compared to other feature sets classifying outdoor cycling (0.94 ± 0.04 compared to an average of 0.86 ± 0.09 with other sets) and a slight increase in classifying low intensity activity type (0.90 ± 0.04 compared to an average of 0.86 ± 0.04).

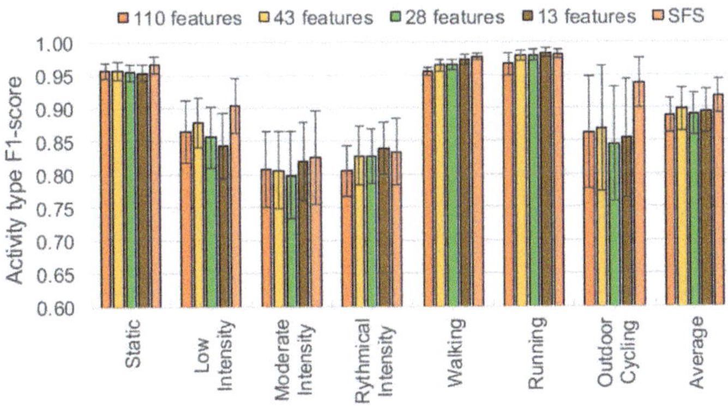


**Figure 6.** F1-scores of different activity types (mean ± SD) averaged over window lengths and sampling frequencies using different feature sets.

Since both classification window length and sampling frequency of the acceleration signal affect the number of samples in classification fragments, it is important to evaluate their combined effect on classification performance. Figure 7 shows the average classification F1-scores with different feature sets using different combinations of sampling frequencies and window lengths. The SD values were large, since the results were averaged over different activity types with different F1-scores.

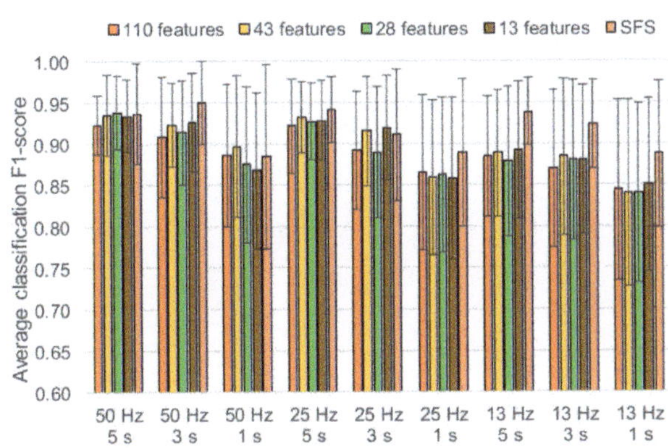


**Figure 7.** F1-scores (mean ± SD) averaged over all activities using different feature sets, window lengths and sampling frequencies.

The average F1-scores of all the combinations of sampling frequencies and window lengths were similar to all of the feature sets of the first feature selection scheme. The classification performance was better with combinations that had more samples per classification fragment, with the highest average of 0.93 ± 0.05 achieved with the combination of 50 Hz and 5 s. The results with the combinations that had either 1 s window length or sampling frequency of 13 Hz were lower compared to other combinations with most feature sets.

Compared to the feature sets of the first feature selection scheme, the SFS method used in the second scheme had higher performance with most window length and sampling frequency combinations. This difference was very noticeable with 13 Hz sampling frequency. The number of features used in

SFS feature sets was between 9 and 14 (Table 3), being remarkably lower than the number of features in most of the feature sets achieved with the first feature selection scheme.

**Table 3.** Number of features in sequential feature selection (SFS) feature sets with different sampling frequencies and window lengths.

| Sampling Frequency and Window Length Combination | Number of Features in the SFS Feature Set |
|---|---|
| 50 Hz, 5 s | 12 |
| 50 Hz, 3 s | 12 |
| 50 Hz, 1 s | 9 |
| 25 Hz, 5 s | 11 |
| 25 Hz, 3 s | 12 |
| 25 Hz, 1 s | 12 |
| 13 Hz, 5 s | 11 |
| 13 Hz, 3 s | 11 |
| 13 Hz, 1 s | 14 |

### *3.4. Best Parameter Combination for Different Activity Types*

While the results of this study generalized the effect of different sampling frequencies, window lengths, and number of features over various activity types, then it might also be useful to know the best combination for each activity type separately. Table 4 shows the parameter combination the highest F1 score for each classified activity type. The values are shown separately for both feature reduction schemes in order to compare the differences.

**Table 4.** Parameter combination with highest F1-score for different activity types and the average for both feature reduction schemes.

| Activity Type | Window Length (s) | Sampling Frequency (Hz) | Number of Features | F1 Score |
|---|---|---|---|---|
| Static | 5 | 25 | 110 | 0.97 |
| | 3 | 50 | 12 (SFS) | 0.98 |
| Low Intensity | 5 | 13 | 110 | 0.93 |
| | 3 | 50 | 12 (SFS) | 0.97 |
| Moderate Intensity | 5 | 50 | 110 | 0.90 |
| | 5 | 13 | 11 (SFS) | 0.91 |
| Rhythmical intensity | 5 | 50 | 13 | 0.90 |
| | 5 | 25 | 12 (SFS) | 0.89 |
| Walking | 3 | 50 | 43 | 0.98 |
| | 3 | 50 | 12 (SFS) | 0.98 |
| Running | 3 | 25 | 13 | 0.99 |
| | 3 | 50 | 12 (SFS) | 0.99 |
| Outdoor Cycling | 5 | 50 | 43 | 0.97 |
| | 3 | 50 | 12 (SFS) | 0.98 |
| Average | 5 | 50 | 28 | 0.94 |
| | 3 | 50 | 12 (SFS) | 0.95 |

## 4. Discussion

In this study it was analyzed for the first time how different window length, sampling frequency, and feature set combinations affect the performance of physical recognition based on decision tree classifiers in order to optimize the classifier for real-time wearable systems. The results of this study have been implemented into a smart work-wear prototype [11]. The main findings were: (1) classification F1-scores with window lengths of 5 s and 3 s were similar, while results with 1 s were lower; (2) all sampling frequencies performed similarly for most activity types, with an exception of outdoor cycling; (3) similar or better results were achieved with the feature sets with 9 to 14 features, achieved with either feature reduction scheme, compared to the initial full feature set of 110 features.

The window lengths of 5 s, 3 s and 1 s were used in this study to analyze how different window lengths affect the performance of physical activity classifier. F1-scores of walking, running and low intensity activity types were similar to all window lengths, while the differences with moderate intensity, rhythmical intensity, and outdoor cycling were larger. Even though window lengths between 3 s and 1 s have been found to be suitable for other studies (2.56 s in [22], 2 s in [26], 1.5 s in [17], 1 s in [18]), in this study, the classifier performance had a larger drop when decreasing the classifier window down to 1 s, while window lengths of 5 s and 3 s had similar results. The window length of 1 s had statistically significant differences with both 3 s and 5 s window lengths while classifying static, moderate intensity rhythmical intensity and outdoor cycling activity types. This could be caused by 1 s window length not being long enough to capture the movement of the body during activities where one period of movement exceeds the window length.

Different sampling frequencies of 50 Hz, 25 Hz, and 13 Hz were used to investigate how sampling frequency affects classification performance. For most classified activity types, no statistical differences were found between tested sampling frequencies, but there were large differences while classifying outdoor cycling. Previously, it had been found that frequencies above 20 Hz cannot be expected to arise from voluntary human movement, where the accelerometer is not in contact with vibrating external sources [20]. It is likely that the 13 Hz sampling frequency was not high enough to capture the vibration during outdoor cycling.

A total of 110 features were extracted from acceleration signals for physical activity classification. To reduce and optimize the number of features, two different feature selection schemes were used in this study. While the first scheme used different consecutive methods to reduce the number of features, the second scheme used forward SFS where features were added one-by-one. The first feature selection scheme enabled the reduction of the feature set from 110 features to 13 features without decreasing the classifier performance. It is possible that the feature set with 13 features was overfit for the conditions used in this study and would perform worse in other conditions.

Compared to the feature sets of the first feature selection scheme, the SFS method used in the second scheme had higher performance with most window length and sampling frequency combinations. This difference was very noticeable when using the sampling frequency of 13 Hz. The number of features used in SFS feature sets were between 9 and 14 (Table 3). The large differences in average F1-scores shown in Figure 7 between SFS feature set and other feature sets while using sampling rates of 25 Hz and 13 Hz were mostly affected by outdoor cycling. Unlike other feature sets, the SFS feature set had a high F1-score while classifying outdoor cycling with all sampling frequency and window length combinations. The highest average classification F1 score was achieved with a parameter combination with SFS feature set (3 s window length, 50 Hz sampling frequency, 12 features), which also had the best performance while classifying static, low intensity, walking and outdoor cycling activity types (Table 4).

It was predictable that the SFS method would provide better results, since the SFS method chose the best features to maximize the classification sensitivity separately for each window length and sampling frequency combination, while, with the first scheme, features were selected based on one sampling frequency and window length combination. The SFS method proved to be a simple comparison method for more comprehensive feature selection and showed that the effect of features depends on different classifier parameters, of which sampling frequency and window length were tested in this study.

Despite the recent advances in deep learning based activity recognition, which reduces the dependency on hand-crafted feature sets and thus could outperform more traditional machine learning methods, it is still far from being used in online mobile systems due to excessive computational power it requires [33]. Thus, the methods and results of this study provide useful information to other researchers for designing and implementing state-of-the-art physical activity recognition for real-time wearable systems.

## 5. Conclusions

This study evaluates the effects of sampling frequency of the acceleration signal, window length of the classification fragment, and number of features on classifier performance. The methods were chosen in order to reduce the requirements on computational power and available memory and are suitable for implementing physical activity classification in real-time systems.

We acknowledge some limitations in our approach that could be improved on in the future studies. First, sampling frequency and window length values evaluated in this study were chosen as a representative of the values used in other studies (low value, mid-range value, high value), but the optimum value could be somewhere between or even out of the explored range. It would be possible to classify larger numbers of different activity types and the acceleration signals should be measured under normal daily living conditions, which would allow for better physical activity classification during everyday life. The results could be evaluated with other machine learning algorithms that are used for physical activity classification, such as support-vector machines, Bayesian networks, and k-nearest neighbor algorithms, in order to see if there are any differences in the effects of the explored parameters.

**Author Contributions:** Conceptualization, A.A., K.P., and I.F.; formal analysis, A.A. and K.P.; methodology, A.A., K.P., and I.F.; investigation, A.A., K.P., D.K., and M.L.; data curation, D.K. and M.L.; writing—original draft preparation, A.A.; writing—review and editing, K.P., I.F., and G.J.; visualization, A.A.; validation, M.L.; supervision, I.F. and G.J.

**Funding:** The research was funded partly by the Estonian Ministry of Education and Research under institutional research financing IUTs 19-1 and 19-2, and by Estonian Centre of Excellence in IT (EXCITE) funded by European Regional Development Fund.

**Acknowledgments:** The authors wish to thank Siiri Mägi and Karl Erlenheim for assistance during the experiments, medical doctors Mari Meren and Ave Nagelmann from the Department of Pulmonology, North Estonia Medical Centre Foundations, Tallinn, Estonia for providing the environment for the study, and also those subjects who so kindly participated in the experiments.

**Conflicts of Interest:** The authors declare no conflict of interest. The funders had no role in the design of the study; in the collection, analyzes, or interpretation of data; in the writing of the manuscript, or in the decision to publish the results.

## References

1. Dunstan, D.W.; Howard, B.; Healy, G.N.; Owen, N. Too much sitting—A health hazard. *Diabetes Res. Clin. Pract.* **2012**, *97*, 368–376. [CrossRef] [PubMed]
2. World Health Organization (WHO). *Obesity: Preventing and Managing the Global Epidemic*; WHO Technical Report Series 894; World Health Organization (WHO): Geneva, Switzerland, 2000.
3. Warburton, D.E.R.; Nicol, C.W.; Bredin, S.S.D. Health benefits of physical activity: The evidence. *Can. Med. Assoc. J.* **2014**, *174*, 801–809. [CrossRef] [PubMed]
4. Evenson, K.R.; Goto, M.M.; Furberg, R.D. Systematic review of the validity and reliability of consumer-wearable activity trackers. *Int. J. Behav. Nutr. Phys. Act.* **2015**, *12*, 1–22. [CrossRef] [PubMed]
5. Coughlin, S.; Stewart, J. Use of Consumer Wearable Devices to Promote Physical Activity: A Review of Health Intervention Studies. *J. Environ. Health Sci.* **2016**, *2*, 1–6. [CrossRef] [PubMed]
6. Maher, C.; Ryan, J.; Ambrosi, C.; Edney, S. Users' experiences of wearable activity trackers: A cross-sectional study. *BMC Public Health* **2017**, *17*, 880. [CrossRef] [PubMed]
7. Middelweerd, A.; Mollee, J.S.; Wal, C.N.; Brug, J.; Velde, S.J. Apps to promote physical activity among adults: A review and content analysis. *Int. J. Behav. Nutr. Phys. Act.* **2014**, *11*, 1–9. [CrossRef] [PubMed]
8. Senevirante, S.; Hu, Y.; Nguyen, T.; Lan, G.; Khalifa, S.; Thilakarathna, K.; Hassan, M.; Senevirante, A. A Survey of Wearable Devices and Challenges. *IEEE Commun. Surv. Tut.* **2017**, *4*, 2573–2620. [CrossRef]
9. Lee, K.; Kwan, M.P. Physical activity classification in free-living conditions using smartphone accelerometer data and exploration of predicted results. *Comput. Environ. Urban Syst.* **2018**, *67*, 124–131. [CrossRef]

10. Wannenburg, J.; Malekian, R. Physical Activity Recognition from Smartphone Accelerometer Data for User Context Awareness Sensing. *IEEE Trans. Syst. Man. Cybern. Syst.* **2017**, *47*, 3142–3149. [CrossRef]
11. Leier, M.; Pilt, K.; Allik, A.; Karai, D.; Jervan, G.; Fridolin, I. Fall detection and activity recognition system for usage in smart work-wear. In Proceedings of the 16th Biennial Baltic Electronics Conference, Tallinn, Estonia, 8–10 October 2018.
12. Yan, Z.; Subbaraju, V.; Chakraborty, D.; Misra, A.; Aberer, K. Energy-Efficient Continuous Activity Recognition on Mobile Phones: An Activity-Adaptive Approach. In Proceedings of the 16th International Symposium on Wearable Computers, Newcastle, UK, 18–22 June 2012; pp. 17–24.
13. Altun, K.; Barshan, B.; Tuncel, O. Comparative study on classifying human activities with miniature inertial and magnetic sensors. *Pattern Recognit.* **2010**, *43*, 3605–3620. [CrossRef]
14. Feng, Z.; Mo, L.; Li, M. A Random Forest-based ensemble method for activity recognition. In Proceedings of the 37th Annual International Conference of the IEEE-EMBC, Milan, Italy, 25–29 August 2015; pp. 5074–5077.
15. Tapia, E.M. Using Machine Learning for Real-time Activity Recognition and Estimation of Energy Expenditure. Ph.D. Thesis, Massachusetts Institute of Technology, Cambridge, MA, USA, June 2008.
16. Allik, A.; Pilt, K.; Karai, D.; Fridolin, I.; Leier, M.; Jervan, G. Activity classification for real-time wearable systems: Effect of window length, sampling frequency and number of features on classifier performance. In Proceedings of the IEEE EMBS Conference on Biomedical Engineering and Sciences (IECBES), Kuala Lumpur, Malaysia, 4–8 December 2016; pp. 460–464.
17. Aktaruzzaman, M.; Scarabottolo, N.; Sassi, R. Parametric estimation of sample entropy for physical activity recognition. In Proceedings of the 37th Annual International Conference of the IEEE-EMBC, Milan, Italy, 25–29 August 2015; pp. 470–473.
18. Bulling, A.; Blanke, U.; Schiele, B. A Tutorial on Human Activity Recognition Using Body-Worn Inertial Sensors. *ACM Comput. Surv.* **2014**, *46*, 33. [CrossRef]
19. Awais, M.; Mellone, S.; Chiari, L. Physical activity classification meets daily life: Review on existing methodologies and open challenges. In Proceedings of the 37th Annual International Conference of the IEEE-EMBC, Milan, Italy, 25–29 August 2015; pp. 5050–5053.
20. Bouten, C.V.C.; Koekkoek, K.; Verduin, M.; Kodde, R.; Janssen, J.D. A triaxial accelerometer and portable data processing unit for the assessment of daily physical activity. *IEEE Trans. Biomed. Eng.* **1997**, *44*, 136–147. [CrossRef] [PubMed]
21. Lee, J.; Kim, J. Energy-Efficient Real-Time Human Activity Recognition on Smart Mobile Devices. *Mob. Inf. Syst.* **2016**, *2016*, 2316757. [CrossRef]
22. Moncada-Torres, A.; Leuenberger, K.; Gonzenbach, R.; Luft, A.; Gassert, R. Activity classification based on inertial and barometric pressure sensors at different anatomical locations. *Physiol. Meas.* **2014**, *35*, 1245–1263. [CrossRef] [PubMed]
23. Wang, Z.; Yang, Z.; Dong, T. A Review of Wearable Technologies for Elderly Care that Can Accurately Track Indoor Position, Recognize Physical Activities and Monitor Vital Signs in Real Time. *Sensors* **2017**, *17*, 341. [CrossRef] [PubMed]
24. Chuang, F.C.; Yang, Y.T.C.; Wang, J.S. Accelerometer-based Energy Expenditure Estimation Methods and Performance Comparison. In Proceedings of the 2nd International Conference on Advances in Computer Science and Engineering (CSE 2013), Los Angeles, CA, USA, 1–2 July 2013; pp. 99–103.
25. Chowdhury, A.K.; Tjondronegoro, D.; Chandran, V.; Trost, S.G. Physical Activity Recognition Using Posterior-Adapted Class-Based Fusion of Multiaccelerometer Data. *IEEE J. Biomed. Health Inform.* **2018**, *22*, 678–685. [CrossRef] [PubMed]
26. Loh, D.; Lee, T.J.; Zihajehzadeh, S.; Hoskinson, R.; Park, E.J. Fitness activity classification by using multiclass support vector machines on head-worn sensors. In Proceedings of the 37th Annual International Conference of the IEEE-EMBC, Milan, Italy, 25–29 August 2015; pp. 502–505.
27. Rajamani, K.; Lai, Y.-S.; Furrow, C.W. An efficient algorithm for sample rate conversion from CD to DAT. *IEEE Signal Process. Lett.* **2000**, *7*, 288–290. [CrossRef]
28. Liu, S.; Gao, R.X.; Freedson, P.S. Computational methods for estimating energy expenditure in human physical activities. *Med. Sci. Sports Exerc.* **2012**, *44*, 2138–2146. [CrossRef] [PubMed]
29. Preece, S.J.; Goulermas, J.Y.; Kenney, L.P.J.; Howard, D. A comparison of feature extraction methods for the classification of dynamic activities from accelerometer data. *IEEE Trans. Biomed. Eng.* **2009**, *56*, 871–879. [CrossRef] [PubMed]

30. Altini, M.; Penders, J.; Amft, O. Energy Expenditure Estimation Using Wearable Sensors: A New Methodology for Activity-Specific Models. In Proceedings of the Wireless Health, San Diego, CA, USA, 23–25 October 2012.
31. Bao, L.; Intille, S.S. Activity Recognition from User-Annotated Acceleration Data. *Pervasive Comput.* **2004**, *3001*, 1–17.
32. Powers, D.M.W. Evaluation: From precision, recall and F-factor to ROC, informedness, markedness and correlation. *J. Mach. Learn. Tech.* **2011**, *2*, 37–63.
33. Wang, J.; Chen, Y.; Hao, S.; Peng, X.; Hu, L. Deep learning for sensor-based activity recognition: A survey. *Pattern Recognit. Lett.* **2019**, *119*, 3–11. [CrossRef]

*applied sciences*

Article

# Analogy Study of Center-Of-Pressure and Acceleration Measurement for Evaluating Human Body Balance via Segmentalized Principal Component Analysis

**Tian-Yau Wu [1,*] and Ching-Ting Liou [2]**

1 Department of Mechanical Engineering, National Chung Hsing University National, Taichung City 402, Taiwan

2 Teco Electric and Machinery Corporation Limited, Taoyuan City 320, Taiwan; aloveasure@gmail.com

* Correspondence: tianyauwu@dragon.nchu.edu.tw

Received: 6 October 2019; Accepted: 5 November 2019; Published: 8 November 2019

**Abstract:** The purpose of this research is to investigate the feasibility of evaluating the human's balancing ability by means of the human body's swaying acceleration measurements instead of the traditional center-of-pressure (COP) measurement. The COP measurement has been used broadly for assessing the balance ability of patients in hospitals. However, the force plate system which is employed to measure the COP signals of the human body is generally restrictive due to the very high cost as well as the bulky portability. In this study, the balancing ability of the human body was evaluated through the measurements of a capacitive accelerometer. The segmentalized principal components analysis (sPCA) was employed to reduce the influence of the gravity component in acceleration measurement projected onto the horizontal components while the accelerometer inevitably tilts. The signal relationship between the COP and the acceleration was derived, so that the swaying acceleration measurements of human body can be utilized to evaluate the human body's balancing ability.

**Keywords:** balance; center of pressure (COP); segmentalized principal component analysis (sPCA); equilibrium score (EQs); empirical mode decomposition (EMD); linear regression; decision support

## 1. Introduction

The aging trend of the social population has been a problem in all the developed countries, and hence, healthcare for the elderly has become an essential issue. The decline of balancing ability is one of the important indicators of the aging process of the human body. In addition to being an important indicator of human aging, the balancing ability is also an index for assessing a person's physical health in medical fields.

The balance of the human body is a sophisticated mechanism. The system that receives the balance-related information consists of three parts: the vision, somatosensory and vestibular system. The vision part provides us with the identification of spatial position, the role of somatosensory is the perception of stimuli on the limbs for the postures and positions and the vestibular system allows us to feel the existence of acceleration in movements. When the human body lacks the balancing ability, the most serious problem is the fall. Therefore, a simple system that can evaluate the balance condition of the human body is definitely beneficial to the balancing ability training and treatment (preventing step) as well as the falling detection (post-mortem remedy step).

The center-of-pressure (COP) has been broadly utilized to evaluate the balancing ability of patients in the fields of clinical medicine and biomedical engineering. The measurements of COP are normally implemented by using the force plate. However, the force-plate-based COP measurement systems are generally highly restrictive due to the very high cost (normally more than 10,000 US dollars) as well as the bulky portability. Therefore, a cheap carry-on device that can be used to estimate the balancing ability of the human body is definitely a merit for the healthcare of the elderly. It is apparent that the swaying dynamics of the human body is related to the balancing ability of the human body. The human body's sway can be detected in terms of the acceleration, which is able to be measured by the accelerometers. As compared with the force-plate-based COP measurement systems, the accelerometers usually have the advantage of much lower price and convenient portability. The price of a capacitive micro-electro-mechanical system (MEMS) accelerometer may be as low as ten US dollars, and thus the MEMS accelerometers can be integrated inside the cellphones as portable devices.

The studies of utilizing the accelerometers for the swaying measurement have been conducted to assess the balancing ability of the human body. Moe-Nilssen [1] applied the accelerometer to estimate the human body's balance and extracted the statistical features for discriminating the difference among balance conditions. Furthermore, different experimental cases were set up for the acceleration measurements of balance control [2]. In addition to the accelerometers, the audio-biofeedback (ABF) approach was also utilized to evaluate the sonic influence of different bandwidth for the human body's balance [3]. Ghasemzadeh et al. [4] investigated the identification of the balance situation through the integration of signals of accelerometer and electromyography (EMG). Their study validated the high correlation between the EMG signal and balance condition of the human body. The acceleration signals were processed by using the wavelet analysis as well as the principal component analysis (PCA) for evaluating the balancing ability of human bodies among the frail, pre-frail and healthy groups [5]. The sample entropy values were utilized to quantify the regularity of COP fluctuations [6]. Their study indicated that the COP fluctuations are more regular for standing than sitting, representing different balance conditions. Huang et al. [7] developed a center-of-pressure and complexity monitoring system (CPCMS) to assess the improvement of human body balance. Their study demonstrated that the CPCMS can achieve similar results to the commercial product. Halicka et al. [8] proposed to examine the effectiveness of visual biofeedback (VBF) signals and accelerometer sensors for improving human balance. They showed that the location of VBF signals had a significant effect on each postural parameter of COP and trunk segments. The approximate entropy values were computed to reflect the amount of irregularity hiding in the COP [9]. The data analysis demonstrated that this method enables us to quantify the postural stability. A point of application (POA) approach was used to determine the accuracy, precision and reliability of COP measurement in a low-cost force plate, called the balance tracking system, and showed an excellent agreement between the POAs and measured COP [10]. In 2018, Adamova et al. [11] used the three-axis accelerometer to quantify the postural stability of patients with cerebellar disorder. Their study demonstrated that the pathological balance control can be identified through the three-dimensional (3D) postural analysis.

Although the COP-based quantification of balancing ability has been employed by medical doctors for decades and is still the major means for evaluating the balance condition of patients, the studies of replacing the COP measurements for balance evaluation have been explored and the feasibility has also been verified. Based on the state-of-art of balance measurement of the human body, a solid and consistent transformation between the COP and acceleration measurements is the crucial step for accurately estimating the balancing ability of the human body. On the other hand, the swaying frequencies of the human body mainly concentrate around the very low-frequency band, as compared to the frequency range of mechanical structural vibration. As for considering the volume, weight, cost and measured bandwidth of the accelerometer, the capacitive accelerometer would be an appropriate one to be bound with the human body and then utilized to measure the swaying acceleration at extremely low frequencies. However, the capacitive accelerometer inherently contains the gravity component on the vertical axis and the gravitation may be projected onto the measurements on the three axes with a time-varying manner, while the bind-in accelerometer tilts along with the human body's swaying. It is definitely an obstacle and there is difficulty deriving the correlation between the COP signal and acceleration measurement.

Based on the problem statements, the objective of this research is to investigate the feasibility of deriving the transformation from the acceleration measurement to the COP signal, so that the COP-based evaluation of the human body's balancing ability can be estimated in terms of the swaying acceleration measurements. The empirical mode decomposition (EMD) method [12] was employed to separate the non-stationary acceleration signals and then extract the swaying-related components at the very low-frequency range. With the EMD process, the traditional filtering process in which the central frequency and bandwidth must be first decided would not be needed. The segmentalized PCA (sPCA) was proposed to alleviate the influence of the time-varying gravitation projection onto the acceleration measurements on the three axes. The equilibrium score (EQs) was also estimated through the measured acceleration signals in this research. The results show that high correlation coefficients of more than 0.7 can be obtained between the processed acceleration signals and the COP measurements. The estimated EQs values have mean average percentage error (MAPE) of 4.89 with respect to the EQs values that were calculated by the commercial computer-aid balance testing apparatus.

## 2. Experiment Design for Relationship Derivation between Acceleration and the Center-of-Pressure (COP)

The sensory organization test (SOT) has been broadly employed to evaluate the balancing ability of the human body in the fields of clinical medicine and biomedical engineering. Among all the external factors that influence the balancing ability of the human body, the majority contains the visual conditions. Based on the SOT, the experimental design in this research contained different visual conditions on a fixed referenced support surface for investigating the effects of the human body's static balancing ability. The examinees were asked to stand on the force plate associated with five visual conditions, including (1) eyes opening, (2) eyes closing, (3) blank reference swaying, (4) static giddy reference and (5) giddy reference swaying, as tabulated in Table 1. The visual conditions of C1 and C2 (eyes opening and closing) mainly assess the balancing ability of the human body with and without the reference. In order to produce the visual perturbation as well as the brain fatigue, three conditions (C3–C5) were conducted in this experiment to investigate the influence of different references upon the balancing ability of the human body. As shown in Figure 1, the referenced wall in the experiment can sway accordingly to simulate the visual conditions C3 to C5. Moreover, the dazzling graph, as shown in Figure 2, was stuck on the referenced wall to produce the factors of visual conditions C4 and C5.

**Table 1.** Visual conditions in the experiment design.

| Visual Condition | Expression |
|---|---|
| C1 | Eyes open (no reference) |
| C2 | Eyes closed |
| C3 | Blank reference sway |
| C4 | Static reference with giddy graph |
| C5 | Swayed reference with giddy graph |

**Figure 1.** The facilities in the experiment design: (**1**) force plate, (**2**) referenced wall.

**Figure 2.** The dazzling graph of the referenced wall (ref: http://richrock.com/illusion.html).

The force plate that the examinee stood on was used to measure the COP locus of the examinee. The force plate measurement system consists of four identical load cells (LDB-30, Jihsense Industril LTD.) and a signal acquisition device (eStrain 4B4V, Chief SI Company). In this experiment, the COP signals were recorded with the sampling rate of 100 Hz. Simultaneously, the capacitive accelerometer (CXL04GP3-R-AL, MEMSIC Inc.) was bound on the waist of the examinee (around the mass center of the human body) to measure the acceleration of human body swaying in the three directions. The acceleration signals were recorded by the data acquisition device (NI 9234) with the same sampling rate as the COP signals. Both the COP loci and acceleration signals of each examinee were measured synchronously for all designated visual conditions with data length of twenty seconds. With the synchronized measurements, the analogy between the COP loci and the acceleration signals of the human body was derived.

## 3. Processing and Analysis of Measurements

### 3.1. Signal Separation and Spectrum Analysis

The major difference between the COP and acceleration measurements is the dimension representation. The COP signals, which are recorded through the force plate, consist of the two independent components (in X and Y directions) that are perpendicular to the gravitational direction. According to the SOT, the COP signal in X-direction is defined as the medial-lateral (ML) component and the Y-component is defined as the anterior-posterior (AP) direction. The acceleration signals that are measured by the tri-axial accelerometer consist of the three components in the typical X-Y-Z directions of the Cartesian coordinate system. Since the accelerometer is bound on the human body's waist and may tilt with human body swaying, the X-Y-Z directions of the tri-axial accelerometer are not fixed in the measurement process.

Through observing the ways that the examinees adjust their posture for balance purposes, both the COP and the acceleration signals have obvious non-stationarity characteristics. The EMD approach that was proposed by Huang et al. [12] is an adaptive data analysis method and can be utilized to separate the non-stationary signals. The EMD process can be simply expressed as:

$$x(t) = \sum_{k=1}^{m} c_k(t) + r_m(t) \tag{1}$$

where, $x(t)$ represents any complicated non-stationary signal, $c_k(t)$ is the $k$-th intrinsic mode function (IMF) of the signal $x(t)$ and $r_m(t)$ is the final residue which can be a constant or the signal mean trend. Based on the concept of EMD, each IMF component satisfies the following conditions [12]: (1) The number of extrema and the number of zero-crossings must be either equal or differ at most by one in the whole data set, and (2) At any point, the mean value of the envelope defined by the local maxima and the envelope defined by the local minima is zeros.

The measurements of COP and acceleration were first decomposed into the IMFs by the EMD method. With the EMD method, the non-stationary COP and acceleration signals can be separated into the independent signal components of different frequency bands. In the experimental process, on the other hand, the external noises or disturbances that are mixed in the measurements may inevitably interfere with the analysis of the COP and the acceleration signals; therefore, it is crucial to remove the uncorrelated signal components from the measurements of COP and acceleration, and reserve the signal components that contain the information correlated to the balancing ability of the human body.

By taking the Hilbert transform of the signal components (IMFs), the analytical signal of $c_k(t)$ can be formed as:

$$z_k(t) = c_k(t) + jH\ \{c_k(t)\} = c_k(t) + j\hat{c}_k(t) = A_k(t)e^{j\phi_k(t)} \tag{2}$$

where, $H\{c_k(t)\}$ represents the Hilbert transform of $c_k(t)$. The time-dependent amplitude, $A_k(t)$, time-dependent phase, $\phi_k(t)$, and instantaneous frequency, $\omega_k(t)$, of $c_k(t)$ can be formulated as:

$$\begin{aligned} A_k(t) &= \sqrt{c_k(t)^2 + \hat{c}_k(t)^2} \\ \phi_k(t) &= \tan^{-1}\tfrac{\hat{c}_k(t)}{c_k(t)} \\ \omega_k(t) &== \tfrac{d\phi_k(t)}{dt} \end{aligned} \tag{3}$$

Therefore, the time-frequency-amplitude distributions of the non-stationary COP and acceleration signals can be expressed as:

$$H(\omega,\ t) = \sum A_k(t)\cos(\int \omega_k(t)dt) \tag{4}$$

In order to clearly observe the energy distributions of measurements in the frequency domain, the marginal spectrum of the COP and acceleration signals can be formulated as:

$$S(\omega) = \sum_k \int_0^T H_k(\omega,\ t)\ dt \tag{5}$$

where, $H_k(\omega,t)$ represents the time-frequency distribution of the $k$-th IMF. The signal compositions and characteristics can then be observed and analyzed within the different frequency scales.

Figures 3 and 4 show the marginal spectra of the COP and acceleration resultants with different experimental visual conditions (as tabulated in Table 1). The marginal spectra of COP and acceleration resultants apparently show that the signal energy concentrates at the low-frequency range. It is intuitive and coincident with the observation that the signal information correlated to the human body's swaying for balance is found in the signal components of low frequencies. Based on the inference, the signal analysis of COP and acceleration in this research was focused on the IMFs whose bandwidths are within 0–5 Hz.

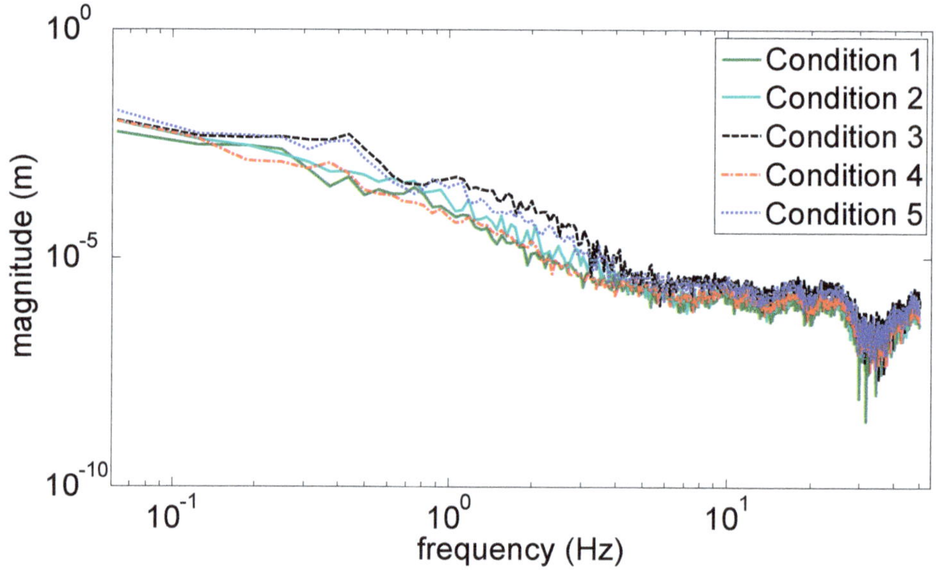


**Figure 3.** Marginal spectrum of center-of-pressure (COP) resultants.

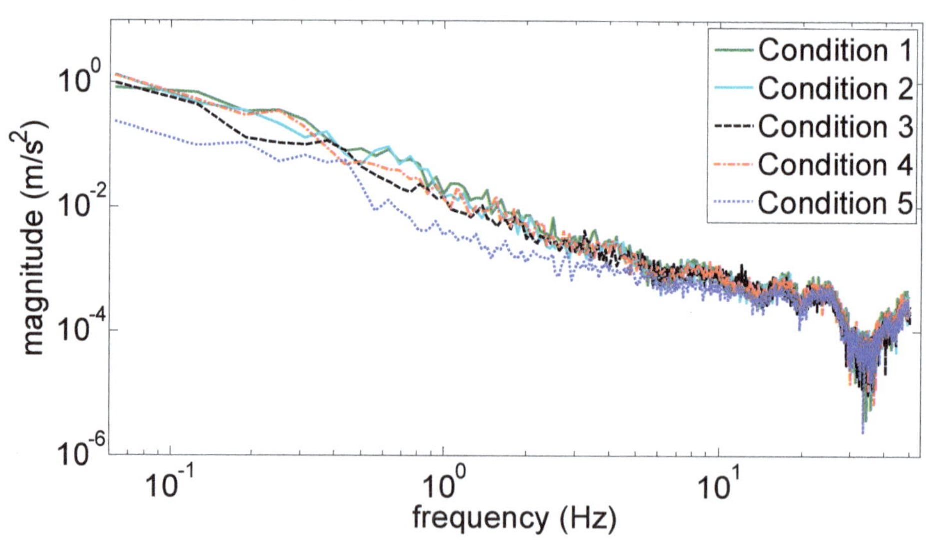


**Figure 4.** Marginal spectrum of acceleration resultants.

### 3.2. *Segmentalized Principal Component Analysis (sPCA) for Acceleration Signal*

As mentioned before, the frequency characteristics of the human body's sway mainly focuses on the very low-frequency range, the capacitive accelerometer has satisfactory measurement performance within the bandwidth of very low frequencies, and thus, was utilized to detect the acceleration of the examinee's body sway in the experiment. However, the bind-in accelerometer inevitably tilted along with the examinee's body sway and hence, the X-Y-Z directions of the acceleration measurements were time-varying with respect to the fixed reference frame. Furthermore, the measurement of the capacitive accelerometer also includes the constant gravitation ($g = 9.81$ m/s$^2$) in the vertical direction and thus, the acceleration measurements in the three directions may contain the gravitation projection components while the accelerometer was tilted along with the examinee's body sway. To the contrary, the COP signal components in ML and AP directions were obtained in the horizontal plane and were independent of the gravitation in the vertical direction. Therefore, it is definitely beneficial for deriving the correlation between the COP and acceleration measurements if the gravitation projection onto the three directions of acceleration signals can be removed.

The concept of PCA that was proposed by Pearson [13] is to convert the measurements in which the variables may be correlated with each other into another sets of variables which are uncorrelated. The sets of variables in the other space are called principal components. The mathematical procedure of PCA is to transform the data sets of measurement into another coordinate system, where the principal components are located through an orthogonal matrix. In this coordinate system, the first principal component has the highest variance, and the subsequent components in turn have as high variance as possible. Therefore, the principal components in the new space are uncorrelated with each other or even orthogonal to each other.

In order to resolve the stated problem of the axial directions variation as well as the gravitation projection issue, and to thereafter derive the relationship between the COP measurement and the acceleration signal, the PCA was employed to process the acceleration signals in this research. The acceleration measurements in X-, Y- and Z-directions were first transformed into another coordinate system where the principal components were located. The principal components in the new coordinate system are theoretically orthogonal to each other, and thus the operation of PCA can reflect the transformed signals in the way of most possible variability in the data [14,15]. In this research, therefore, the variability of acceleration measurements along the three directions was first analyzed.

Simply consider that the accelerometer is bound at the point A of the examinee's waist and sways in the AP direction, as shown in Figure 5. As the examinee sways to maintain the body's balance status, the accelerometer moves from A to A', as shown in Figure 5. The acceleration measurements consist of the components perpendicular to the Z-direction as well as the component parallel to the Z-direction. Since the two acceleration components in Y- and Z-direction are proportional to the distance $a$ and $d$ respectively, it can be briefly proven that the acceleration has larger variance in Y-direction than in Z-direction if $a$ is always greater than $d$. First, assume $d > a$ contradictorily. The distance $a$ in Y-direction can be determined from the geometric relationship, $a = \sqrt{h^2 - (h-d)^2}$, and hence, $d > \sqrt{h^2 - (h-d)^2}$. It is easy to derive the contradiction that the condition for $d > a$ implies $d > h$. Similarly, the same inference can be applied in the X-direction instead of the Y-direction. Therefore, it is reasonable to infer that the acceleration variance in Z-direction is smaller than those in X- and Y-directions.

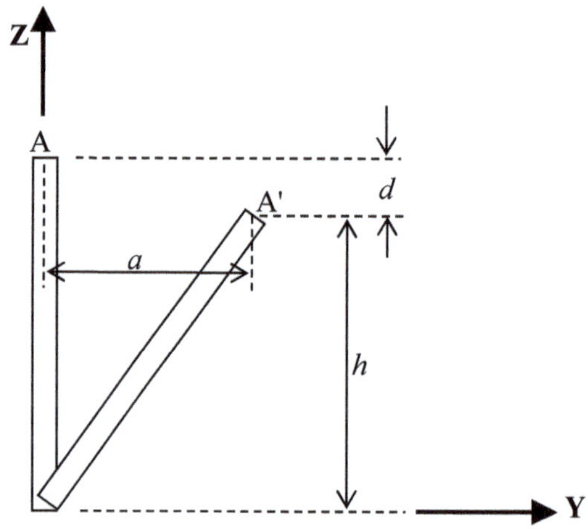


**Figure 5.** Schematic plot of acceleration variance in Y- and Z-directions.

Figures 6 and 7 illustrate the COP measurements of the five visual conditions (Conditions 1 to 5, as shown in Table 1) in the SOT of this research. It is found in the figures that the examinee generally has more sway variance in the AP direction than in the ML direction. A similar phenomenon can also be observed in most of the other COP measurements. The inference is reasonable because the structure of legs and ankles has more capability against the perturbation in the ML direction than in the AP direction while the examinee stands in the normal posture. Therefore, it is assumed in this study that the acceleration measurement in AP direction dominates the balancing characteristics as the examinee sways for most tests.

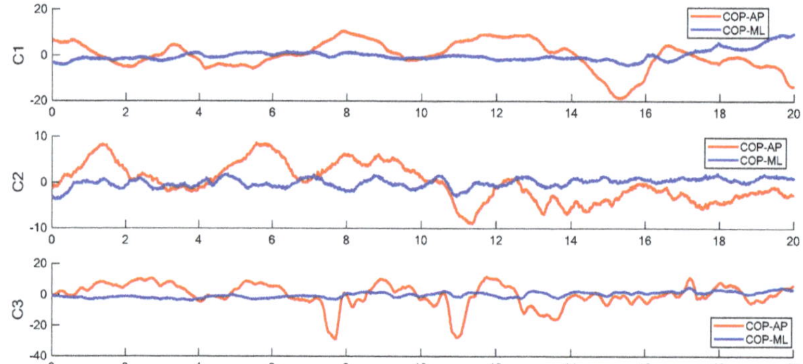


**Figure 6.** COP measurements in medial-lateral (ML) and anterior-posterior (AP) direction under Conditions 1–3.

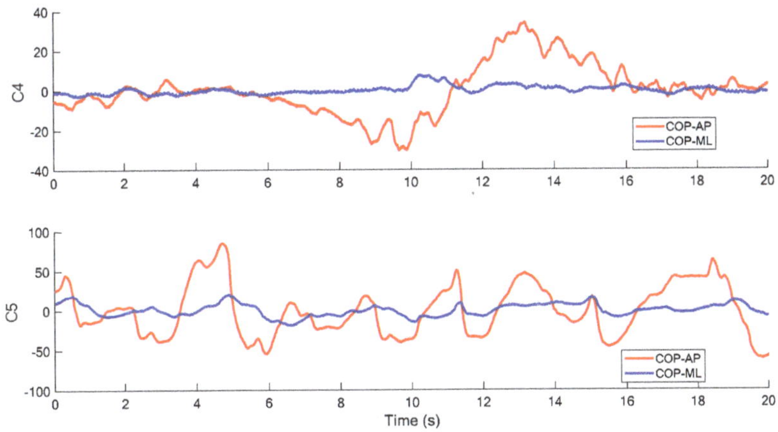


**Figure 7.** COP measurements in ML and AP direction under Conditions 4–5.

Based on the concept of PCA, the first principal component corresponds to the COP signal component in the AP direction. The second principal component corresponds to the COP signal component in the ML direction. The third principal component is attributed to the gravitation of the vertical component. This inference was drawn from the fact that the acceleration measurement had less variance in the direction parallel to the gravitation than the directions perpendicular to the gravitation.

Although the PCA is capable of transforming the acceleration signals into the three principal components corresponding to the AP, ML and vertical directions conceptually, it is, however, apparently unable to reflect the instantaneous variation in the projection of gravitation onto the three measured components in case of the time-varying tilting accelerometer if the PCA is employed to process the signal throughout the whole data length (20 s). In order to address this concern and to deal with the time-varying problem of directivity variation on the oblique accelerometer, the segmentalized PCA (sPCA) was proposed in this research to accurately extract the time-varying components corresponding to the acceleration components that are perpendicular to the gravity at the different instants during the human body's sway.

As illustrated in Figure 8a, the acceleration measurements in X-, Y- and Z-directions were divided into four segments and then the PCA was applied for each segment to determine the three principal components (PC-1, PC-2 and PC-3), as shown in Figure 8b. With the connected principal components of each segment, Figure 9a shows the comparison between the COP measurement in the Y-axis (AP) and the first principal component (PC-1) of acceleration signals. It is noted that the PC-1 of the acceleration measurements has similar variation trend with the COP measurement in the AP direction and the correlation coefficient between the two sets of series was calculated as 0.841. As compared with the PC-2 and PC-3 of the acceleration signals, their variation trends are very different from the COP measurement in the AP direction (as shown in Figure 9b), and thus, they have correlation coefficients of −0.512 and −0.089, respectively. The high correlation coefficient between the COP measurement in the AP direction and the PC-1 of the acceleration measurements demonstrates that it is sufficient to accurately predict the COP measurement in the AP direction. Therefore, it is feasible to utilize the processed acceleration measurement to represent the COP signal for evaluating the human body's balancing ability.

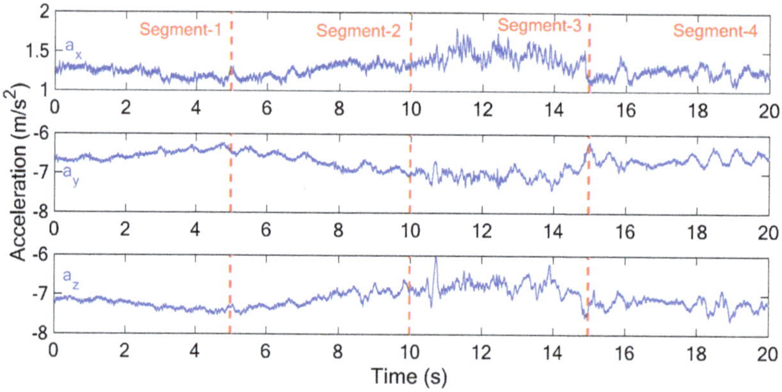


(a)

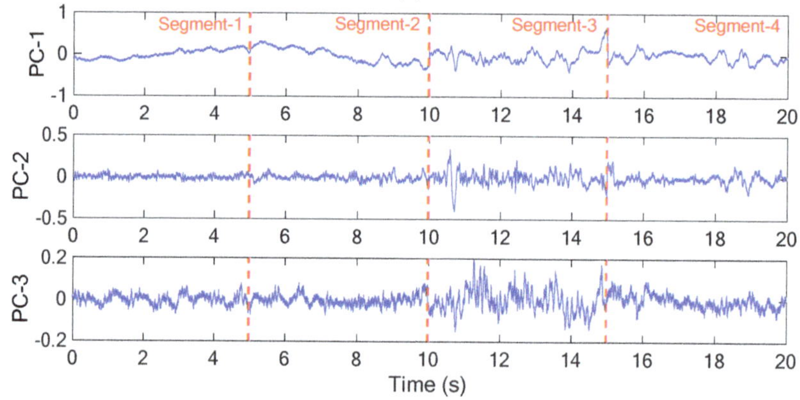


(b)

**Figure 8.** (**a**) Segments of acceleration measurements, (**b**) Principal components in each segmentalized part.

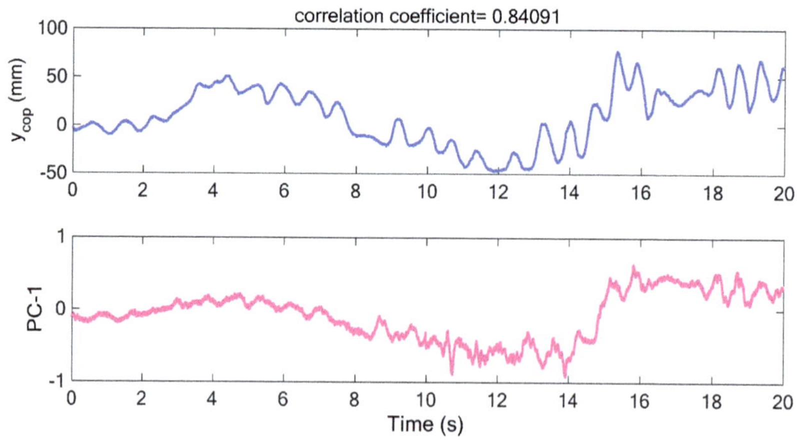


(a)

**Figure 9.** *Cont.*

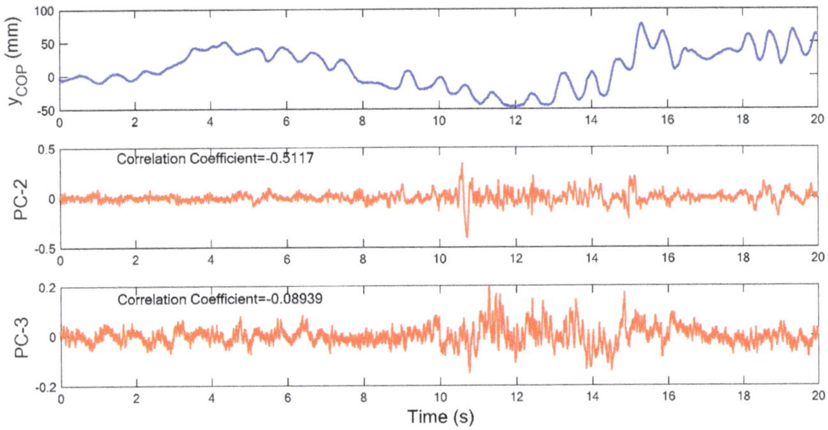


(b)

**Figure 9.** (**a**) Comparison between COP in the Y-axis and first principal component of acceleration signals, (**b**) Comparison between COP in the Y-axis, second and third principal components of acceleration signals.

## 4. Equilibrium Score (EQs) Estimation

A high correlation between the COP in the Y-axis and the first principal component of acceleration signals has been validated in the previous section, and thus it is feasible to use the acceleration measurements representing the COP-based evaluation for human balancing ability. Furthermore, in the field of clinical medicine, the balancing ability of patients is generally assessed through the commercial computer-aid balance testing apparatus in SOT and is broadly quantified in terms of the equilibrium score (EQs). The EQs of patients' balancing ability is mainly computed through the measurement of COP. The EQs is defined from zero to one hundred. The EQs value of one hundred means an ideal stability while the EQs value of zero represents a tumble. In order to verify the accuracy of analogy between the acceleration measurements and the COP-based EQs evaluation at the balance testing apparatus, the datasets that include the synchronous measurements of body swaying acceleration and COP as well as the EQs evaluation were collected from the Department of Physical Therapy and Assistive Technology in National Yang-Ming University, in which the commercial computer-aid balance testing apparatus was utilized to assess the balancing ability of patients in the five visual conditions. The collected data consisted of 85 sets of acceleration signals and COP measurements that were recorded in the SOT of different examinees. The examinees that had normal balancing abilities in a majority were evaluated through some of C1 to C5 of SOT. All the data were available as the format of digits in text which can be processed and analyzed in MATLAB software.

Before the EQs estimation was investigated, all the collected data was first processed through the correlation analysis to obtain the statistical result of correlation coefficients between the COP measurements and the acceleration signals. Figure 10 shows the results of correlation coefficients between the COP measurements in the AP direction and the three data sets that are the PC-1 of acceleration measurements through the sPCA (with segment length of 1.0 second), the PC-1 of acceleration measurements by using the PCA for the whole signal length, and the original acceleration measurements in the Y-direction, respectively. As shown in the figure, the COP signals in the AP direction have higher overall correlation with the PC-1 of the acceleration measurements with the sPCA process (mean correlation coefficient around 0.78) than the PC-1 through the pure PCA for the whole signal length (mean correlation coefficient around 0.41), as well as the original acceleration measurements in the Y-direction (mean correlation coefficient around 0.40). The correlation analysis apparently shows that the acceleration measurements can be utilized to accurately predict the COP signal in the AP direction through the sPCA, and hence, it is feasible to use the cheap bind-in accelerometer for estimating the human being's balancing ability instead of the expensive bulky COP measurement system.

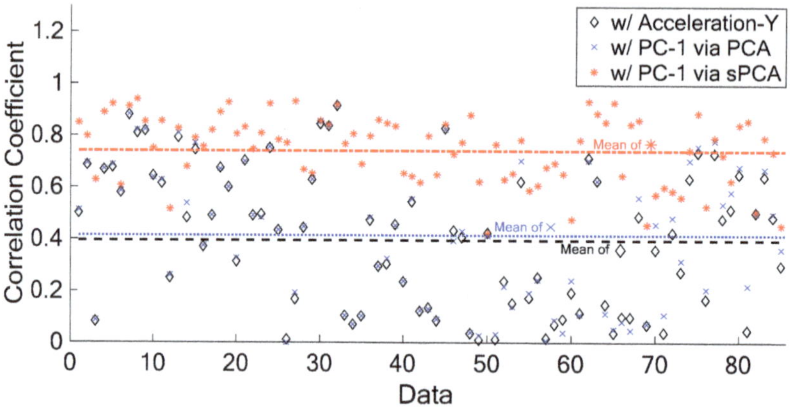


**Figure 10.** Correlation coefficient: COP-AP signal versus acceleration signal in the Y-axis (black ◊); COP-AP signal versus PC-1 of acceleration signals with principal component analysis (PCA) for the whole data length (blue ×); COP-AP signal versus PC-1 of acceleration signals via segmentalized PCA (sPCA) (red *); mean value of ◊ (black —); mean value of × (blue ...); mean value of * (red -.-.-).

Since the transformation from the COP measurement to the EQs is unavailable in this study, the statistical computation can be used to simply derive the relationship between the COP signals and the EQs. Define the swaying level to be (100 EQs). The root-mean-square (RMS) values of COP measurement resultants and the corresponding (100 EQs) values are shown in Figure 11. It is apparent that the transformation between the RMS values of COP measurement resultants and corresponding (100 EQs) values can be derived through the regression analysis. Similar results were also obtained by linear regression to derive the relationship between the RMS values of COP measurements in the AP direction and the corresponding swaying levels, as shown in Figure 12. As shown in these two figures, it is noted that the EQs values can be accurately derived through the linear regression process only using COP measurements in the AP direction.

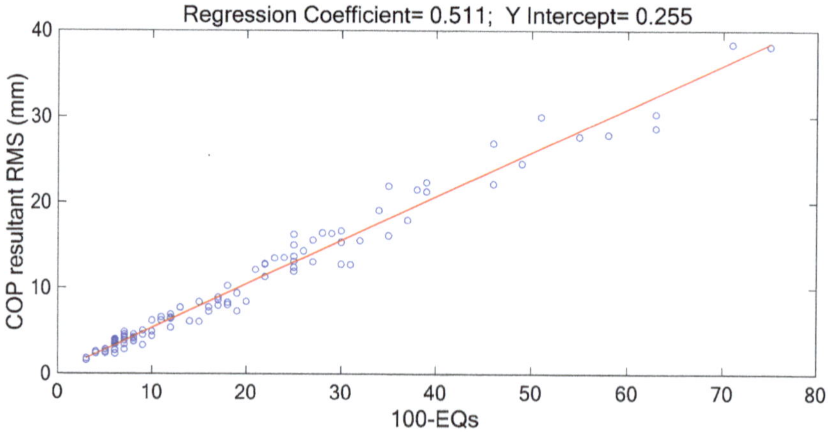


**Figure 11.** Regression analysis for root-mean-square values of COP measurement resultants and the swaying level (100 equilibrium score (EQs)).

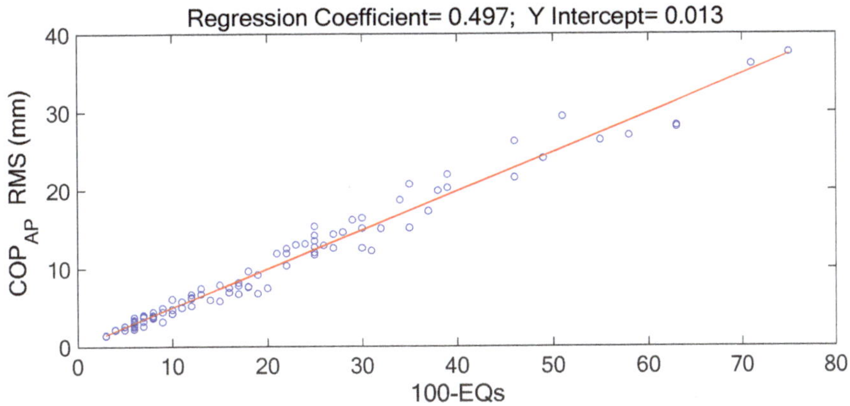


**Figure 12.** Regression analysis for RMS values of COP measurement in the AP direction and the swaying level (100 EQs).

As mentioned previously, the analysis shows that the COP measurements in the AP direction have high correlation with the first principal component of the acceleration measurements. Therefore, it is feasible to estimate the EQs values through the analysis of the swaying acceleration signals that are measured by the tri-axis capacitive accelerometer. Figure 13 shows the flow chart of signal processing in estimating the EQs values in terms of the acceleration measurements. The human body swaying accelerations were measured by the bind-in tri-axis capacitive accelerometer that is capable of effectively capturing the signals at very low frequencies of human body swaying. The acceleration signals were then processed by using the proposed sPCA to extract the first principal component (PC-1). The signal of PC-1 was separated by the EMD method and the IMFs within the frequency range of 0–5 Hz were synthesized for the further procedure. The RMS value of the synthesized filtered PC-1 signal was calculated and then transformed to estimate the RMS value of COP signal in AP-direction by means of the a priori correlation analysis. Through the a priori linear regression analysis between (100 EQs) and the COP-AP RMS, the EQs values can be estimated. It is noted that the correlation analysis and the linear regression method were utilized as the decision support tools for data transformations among the acceleration signals, COP measurements and (100 EQs) values. The EQs values that were obtained by the commercial computer-aid balance testing apparatus were also utilized to compare with the ones that were estimated from the acceleration measurements. Figures 14–18 show the EQs values which were estimated through the acceleration measurements and the EQs values which were obtained by the commercial computer-aid balance testing apparatus in C1 to C5 of SOT. These figures demonstrate that most of the estimated EQs values are close to the EQs values that were obtained by the balance testing apparatus. Table 2 shows the mean absolute percentage error (MAPE) of the estimated EQs values in each visual condition of SOT compared with the EQs values that were displayed in the balance testing apparatus, which is defined as:

$$MAPE = \frac{|EQs_{est} - EQs_{app}|}{EQs_{app}} \times 100\% \tag{6}$$

where, $EQs_{est}$ represents the estimated EQs value and $EQs_{app}$ represents the EQs value displayed in the balance testing apparatus. As shown in Table 2, high accurate EQs estimation can be obtained through the swaying acceleration measurement of the human body as well as the proposed signal processing steps in most of the SOT cases except for some of the cases in C5. The overall accuracy of the estimated EQS values was 95.11%.

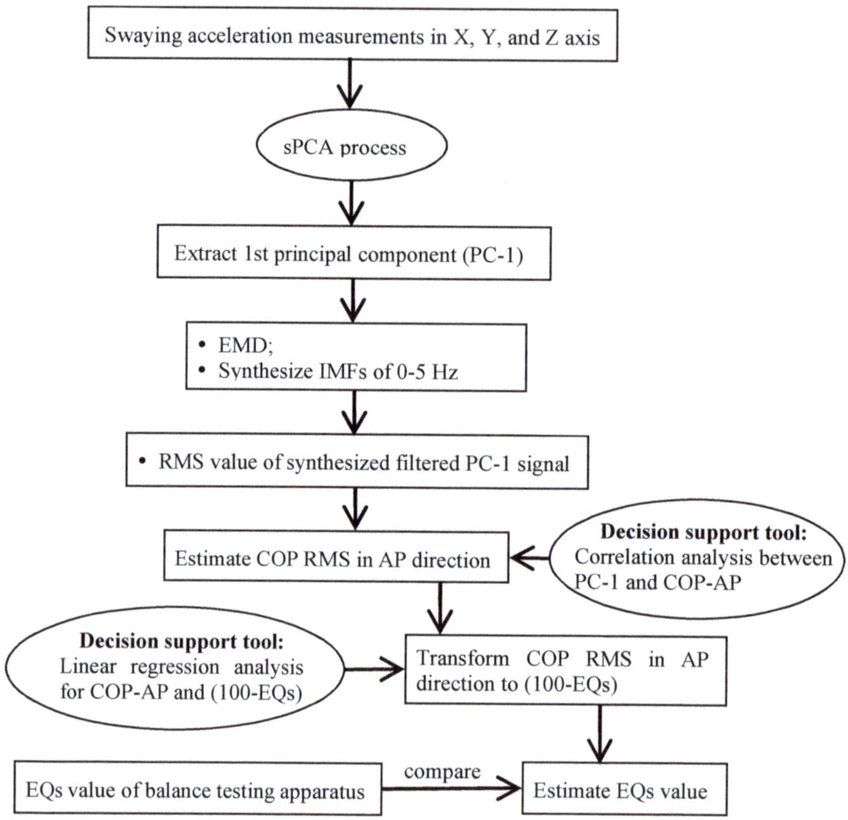


**Figure 13.** Flow chart of acceleration signal processing in estimating the EQs values.

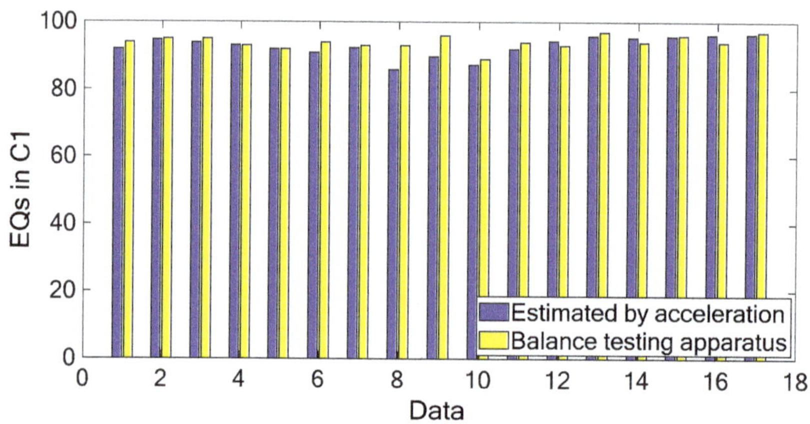


**Figure 14.** Estimated EQs values compared with EQs values of commercial computer-aid balance testing apparatus in C1.

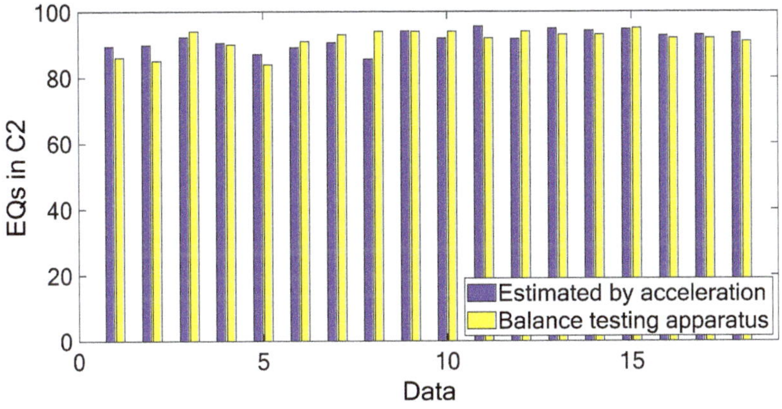


**Figure 15.** Estimated EQs values compared with EQs values of commercial computer-aid balance testing apparatus in C2.

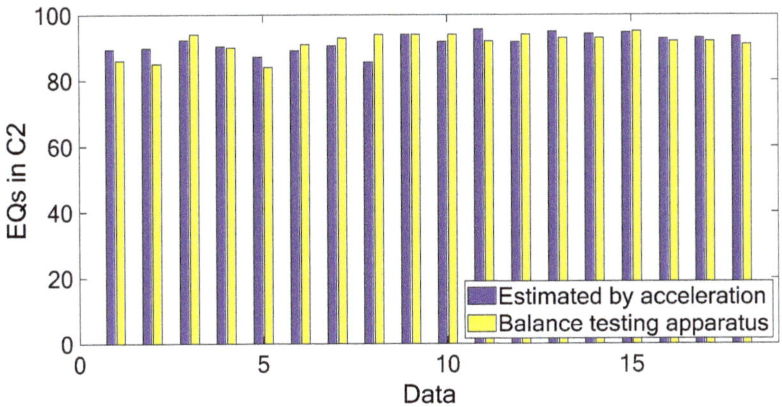


**Figure 16.** Estimated EQs values compared with EQs values of commercial computer-aid balance testing apparatus in C3.

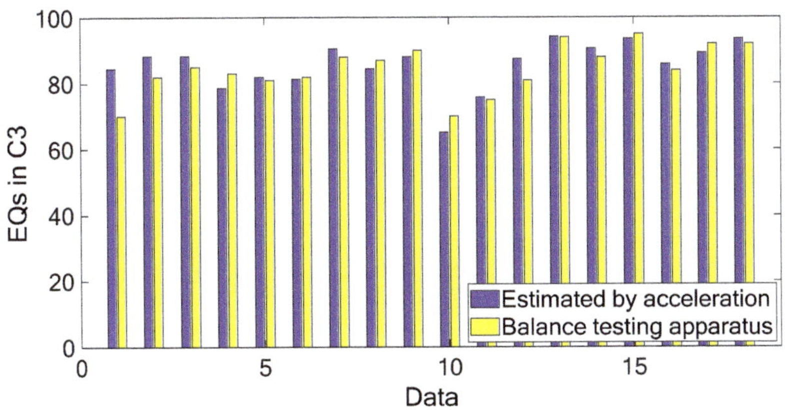


**Figure 17.** Estimated EQs values compared with EQs values of commercial computer-aid balance testing apparatus in C4.

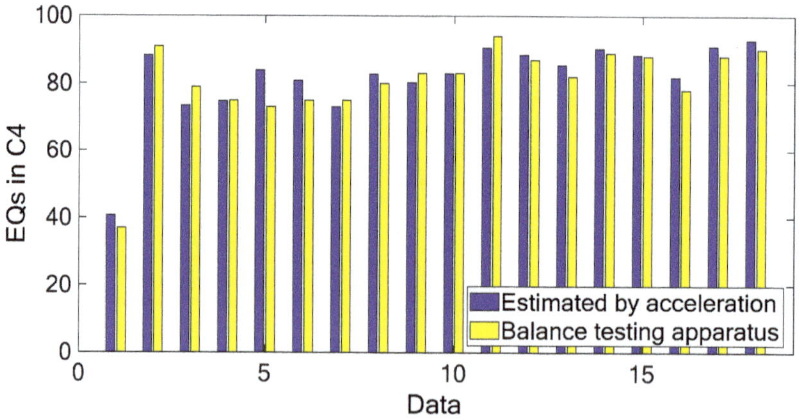


**Figure 18.** Estimated EQs values compared with EQs values of commercial computer-aid balance testing apparatus in C5.

**Table 2.** Mean absolute percentage error (MAPE) of the estimated EQs values in each visual condition of the sensory organization test (SOT).

| | C1 | C2 | C3 | C4 | C5 |
|---|---|---|---|---|---|
| MAPE | 1.95% | 2.56% | 4.16% | 4.17% | 11.63% |

## 5. Conclusions

In this research, the proposed sPCA method was employed to reduce the influence of the measured gravitation component projected onto the horizontal components while the bind-in capacitive accelerometer was utilized to measure the human body's sway and tilt inevitably. A high correlation between the acceleration measurements and the COP signals of the human body's sway can thus be derived. Therefore, the acceleration measurements of the human body's sway can be utilized to represent the human balancing ability with lower hardware expense. Furthermore, the RMS values of COP can be estimated to quantify the EQs values of the human body's balancing ability in the SOT. The analysis results show that the estimation MAPE of 4.89% can be obtained through the collected SOT data.

**Author Contributions:** T.-Y.W. was in charge of the supervision in this research, including initiating the research project, conducting the experiment setup, advising the data analysis methods and final manuscript writing. C.-T.L. was in charge of the experiment implementation, swaying signal acquisition and signal analysis. All the authors have approved the research contents that were written in the final manuscript.

**Funding:** This research was funded by the Ministry of Science and Technology in Taiwan, Republic of China, grant number MOST 105-2221-E-005-025-MY3, MOST 108-2634-F-005-001 and MOST 108-2221-E-005-052.

**Conflicts of Interest:** The authors declare no conflict of interest.

## References

1. Moe-Nilssen, R. Test-retest reliability of trunk accelerometry during standing and walking. *Arch. Phys. Med. Rehabil.* **1998**, *79*, 1377–1385. [CrossRef]
2. Moe-Nilssen, R.; Helbostad, J.L. Trunk accelerometry as a measure of balance control during quiet. *Gait Posture* **2002**, *16*, 60–68. [CrossRef]
3. Chiari, L.; Dozza, M.; Cappello, A.; Horak, F.B.; Macellari, V.; Giansanti, D. Audio-Biofeedback for Balance Improvement: An Accelerometry-Based System. *IEEE Trans. Biomed. Eng.* **2005**, *52*, 2108–2111. [CrossRef] [PubMed]

4. Ghasemzadeh, H.; Jafari, R.; Prabharan, B. A body sensor network with electromyogram and inertial sensor: multimodal interpretation of muscular activities. *IEEE Trans. Inf. Technol. Biomed.* **2010**, *14*, 198–206. [CrossRef] [PubMed]
5. Martinez-Ranmirez, A.; Lecumberri, P.; Gomez, M.; Rodriguez-Manas, L.; Garcia, F.J.; Izquierdo, M. Frailty assessment based on wavelet analysis during quiet standing balance test. *J. Biomech.* **2011**, *44*, 2213–2220. [CrossRef] [PubMed]
6. Roerdink, M.; Hlavackova, P.; Vuillerme, N. Center-of-pressure regularity as a marker for attentional investment in postural control: A comparison between sitting and standing postural. *Hum. Mov. Sci.* **2011**, *30*, 203–212. [CrossRef] [PubMed]
7. Huang, C.W.; Sue, P.D.; Abbod, M.F.; Jiang, B.C.; Shieh, J.S. Measuring center of pressure signals to quantify human balance using multivariate multiscale entropy by designing a force platform. *Sensors* **2013**, *13*, 10151–10166. [CrossRef] [PubMed]
8. Halicka, Z.; Lobotkova, J.; Buckova, K.; Hlavacka, F. Effectiveness of different visual biofeedback signals for human balance improvement. *Gait Posture* **2014**, *39*, 410–414. [CrossRef] [PubMed]
9. Zhu, Y. Design and validation of a low-cost portable device to quantify postural stability. *Sensors* **2017**, *17*, 3. [CrossRef] [PubMed]
10. Goble, D.J.; Khan, E.; Baweja, H.S.; O'Connor, S.M. A point of application study to determine the accuracy, precision and reliability of a low-cost balance plate for center of pressure measurement. *J. Biomech.* **2018**, *71*, 277–280. [CrossRef] [PubMed]
11. Adamova, B.; Kutilek, P.; Cakrt, O.; Svoboda, Z.; Viteckova, S.; Smrcka, P. Quantifying postural stability of patients with cerebellar disorder during quiet stance using three-axis accelerometer. *Biomed. Signal Process. Control* **2018**, *40*, 378–384. [CrossRef]
12. Huang, N.E.; Shen, Z.; Long, S.R.; Wu, M.C.; Shih, H.H.; Zheng, Q.; Yen, N.-C.; Tung, C.C.; Liu, H.H. The empirical mode decomposition and the Hilbert spectrum for nonlinear and non-stationary time series analysis. *Proc. R. Soc. A-Math. Phys. Eng. Sci.* **1998**, *454*, 903–995. [CrossRef]
13. Pearson, K. On lines and planes of closest fit to systems of points in space. *Philos. Mag.* **1901**, *2*, 559–572. [CrossRef]
14. Jackson, J.E. *A User's Guide to Principal Components*; John Wiley and Sons: New York, NY, USA, 1991.
15. Yang, C.Y.; Wu, T.Y. Diagnostics of gear deterioration using EEMD approach and PCA process. *Measurement* **2015**, *61*, 75–87. [CrossRef]

*applied sciences*

MDPI

*Article*

# Dynamic Handwriting Analysis for Neurodegenerative Disease Assessment: A Literary Review

**Gennaro Vessio**

Computer Science Department, University of Bari, 70125 Bary, Italy; gennaro.vessio@uniba.it

Received: 18 September 2019; Accepted: 25 October 2019; Published: 1 November 2019

**Abstract:** Studying the effects of neurodegeneration on handwriting has emerged as an interdisciplinary research topic and has attracted considerable interest from psychologists to neuroscientists and from physicians to computer scientists. The complexity of handwriting, in fact, appears to be sensitive to age-related impairments in cognitive functioning; thus, analyzing handwriting in elderly people may facilitate the diagnosis and monitoring of these impairments. A large body of knowledge has been collected in the last thirty years thanks to the advent of new technologies which allow researchers to investigate not only the static characteristics of handwriting but also especially the dynamic aspects of the handwriting process. The present paper aims at providing an overview of the most relevant literature investigating the application of dynamic handwriting analysis in neurodegenerative disease assessment. The focus, in particular, is on Parkinon's disease (PD) and Alzheimer's disease (AD), as the two most widespread neurodegenerative disorders. More specifically, the studies taken into account are grouped in accordance with three main research questions: disease insight, disease monitoring, and disease diagnosis. The net result is that dynamic handwriting analysis is a powerful, noninvasive, and low-cost tool for real-time diagnosis and follow-up of PD and AD. In conclusion of the paper, open issues still demanding further research are highlighted.

**Keywords:** neurodegenerative diseases; handwriting analysis; survey

## 1. Introduction

### *1.1. Motivations and Purposes*

Neurodegenerative diseases, such as Parkinson's disease (PD) and Alzheimer's disease (AD), affect the structure and functions of brain regions resulting in a progressive cognitive, functional, and behavioural decline. PD is caused by the degeneration of the dopaminergic nigrostriatal neurons of the basal ganglia, resulting primarily in motor deficits: akinesia, bradykinesia, rigidity, and tremor are typically observed [1]. AD, on the other hand, is characterized by short-term memory loss in its early stages, followed by a progressive decline in other cognitive and behavioral functions as the disease advances: therefore, the dominant feature of AD is mainly of cognitive nature [2]. Unfortunately, in the case of signs of brain degeneration, there is no cure and the gradual decline of the patient can only be somehow managed during disease progression. However, an early diagnosis of neurodegeneration would be crucial in the perspective of proper medical treatment to be administered and for improving the quality of life of the patient. In addition, the assessment of signs and manifestations of a specific disease is useful for its diagnostic differentiation with respect to similar disorders and for monitoring and tracking its progression as the disease advances. To this end, a special attention is devoted to mild cognitive impairment (MCI) signs, as an individual with MCI is at a high risk of developing dementia, especially of the Alzheimer's type [3].

The evaluation of the patient's clinical status and their responsiveness to medication is typically achieved via a clinical workup including a thorough medical history, a neuropsychological test battery, and rating scales. Mini Mental State Examination (MMSE) [4], for example, is used extensively to assess cognitive impairment. However, there is still no one certain test to determine if someone is affected by a neurodegenerative disorder and a precise diagnosis is possible only *postmortem*. Getting a reliable diagnosis can require months, and symptoms need to be constantly monitored. In addition, the traditional evaluations depend to some extent on the experience of the clinician performing the assessment, and this makes the determination of the exact type of disease as well as its degree of severity difficult. For these reasons, identifying accurate *biomarkers* for early and differential diagnosis, prognosis, and response to therapy is a primary goal of the research on neurodegenerative disorders today (e.g., References [5,6]).

Changes in the brain caused by neurodegeneration—brain atrophy, neuronal loss, synaptic dysfunction, etc.—particularly result in a dysfunction of the motor system as well as in impairments of the performance of previously learned motor skills. Therefore, a key role in the context of neurodegenerative diseases assessment can be assumed by handwriting. Handwriting, in fact, is a complex activity entailing motor as well as cognitive components [7], of which the changes are promising as a *biomarker* for disease assessment. First, handwriting exercises are already part of neuropsichological test batteries. For instance, the Clock Drawing Test (CDT), which is part of the Mini-Cog test, requires the patient to draw a clock from memory and to put the hands at a given time: the goal is to evaluate executive functions [8]. Second, it is worth noting that, in several studies, researchers examined handwriting difficulties by using writing tests: their results showed that these difficulties are well correlated to the disease severity as well as the concomitant cognitive impairment. For example, in the seminal paper by McLennan et al. [9], it was pointed out how micrographia, which is an abnormally small writing typically associated with PD, can be easily detected by simple pen-and-paper exercises. Other studies, e.g., References [10,11], used analogous tasks and found that agraphia, which encompasses a progressive disorganization of the various components of handwriting, is an early symptom of AD.

Although several advancements have been so far obtained through the analysis of static characteristics of handwriting, i.e., the ones that can be analyzed after the writing process has already occurred, with the advent of new technologies, novel, dynamic features of handwriting have been available to the research community. These features concern the *dynamic* characteristics of handwriting that can be acquired while the writing process still occurs. Typical acquisition tools are inexpensive commercially available digitizing tablets and/or electronic pens. Through these devices, one can measure not only temporal and spatial variables of handwriting but also the pressure exerted over the writing surface and measures of pen inclination and pen orientation. Moreover, these devices can capture pen movement not only while the pen is in contact with the writing surface but also when the pen is in close proximity of the surface, i.e., "in-air".

In the context of neurodegenerative diseases assessment, dynamic handwriting analysis has been employed for studying several issues and has attracted considerable research interest from psychologists to neuroscientists and from physicians to computer scientists. A large part of the literature on this topic investigated fine motor control in healthy and unhealthy people. Examining changes in the handwriting of impaired patients, in fact, facilitates the understanding of the brain–body functional relationships and can lead to identifiable patterns of the sensorimotor dysfunction associated with PD or AD. Several other studies focused on the effects of medication on handwriting: these changes can provide a useful tool for monitoring and tracking disease progression. More recently, an increasing research effort has been made towards the development of an automatic tool for the discrimination between impaired subjects and healthy controls on the basis of dynamic handwriting features. The goal is to provide a complementary approach to the pathology evaluation performed by expert clinicians that is quantitative, noninvasive, and very low-cost.

This paper aims at providing an overview of the most relevant literature investigating the application of dynamic handwriting analysis to the assessment of neurodegenerative disorders. In particular, PD and AD, as the two most widespread and most extensively investigated disorders, are taken into account.

### 1.2. Related Surveys

It is worth remarking that surveys on this topic have already been provided in References [12–15]. Neils-Strunjas et al. [12] discussed papers focusing only on the static characteristics of handwriting in AD. Letanneux et al. [13] considered papers that focused both on static and dynamic features for PD assessment. In particular, the authors proposed to extend the concept of "dysgraphia" also to PD, as it encompasses all deficit characteristics of Parkinsonian handwriting. De Stefano et al. [14] and Impedovo and Pirlo [15] recently proposed surveys focusing both on PD and AD. In Reference [14], De Stefano et al. made a categorization of works into statistical and classification studies, based on the methodological approach followed by the reviewed experiments. However, while the paper extensively reviews works on AD and MCI as well, at the time of writing, less research was done on classification of PD, thus forcing the authors to exclude several findings that are currently available. In particular, only one classification study on PD is reviewed in the paper. Conversely, in Reference [15], the authors considered the problem at hand only from a pattern recognition perspective. For this reason, they did not consider a body of previous literature not using the established machine learning experimental workflow. This paper is intended to provide a more comprehensive overview of the topic, providing the reader with a broad and organized view covering a wider spectrum of methodological approaches and analyses. In particular, the present survey aims at covering papers using either statistical or classification approaches, starting from the earlier papers, which reports the first attempts to investigating dynamic handwriting analysis for neurodegenerative diseases assessment, to the very recent works. The topic received an exploding attention in the last few years; thus, this papers aims at covering also the very recent advancements achieved.

### 1.3. Structure of the Survey

The present survey is intended to provide the reader not only with a historic, state-of-the-art, and future perspective on the topic but also with some guidelines. These guidelines may be useful to the reader to enter this line of research or to easily compare their findings with the existing literature. For this reason, the literary review provided in this paper is divided in two parts. The first part, which is reported in the next section, describes the experimental design typically adopted: the process of dynamic handwriting analysis is sketched, and the main issues arising from its application to health care are pointed out. Almost all surveyed studies, in fact, share a common experimental design including data acquisition, feature extraction, and data analysis. In particular, different studies reported the results of the application of different techniques, depending on the research question to be investigated. The second part, which is reported in Section 3, discusses the main research questions that have been addressed. As previously mentioned, the literature on this topic mainly followed three research directions: providing insights into the motor control mechanisms of handwriting; monitoring and tracking disease progression and the responsiveness of patients to therapies; and providing novel instruments for the (possibly early) real-time disease diagnosis. The last section concludes the paper and provides some considerations about directions for further research on the topic.

## 2. Typical Experimental Design

The studies investigating the application of dynamic handwriting analysis in neurodegenerative diseases assessment typically follow a common experimental setup including data acquisition, feature extraction, and data analysis (Figure 1). These issues are discussed separately in the following subsections.

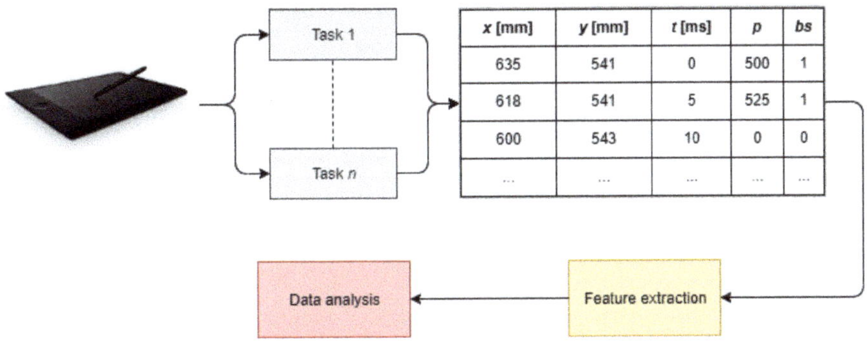


**Figure 1.** Typical workflow of dynamic handwriting analysis (*t* stands for timestamp, *p* stands for pressure, and *bs* stands for button status).

## 2.1. Data Acquisition

At this step, issues arising concern participant recruiting, apparatus choice, and acquisition protocol definition. The currently available datasets are also described.

### 2.1.1. Participant Recruiting

In recruiting study participants, three aspects should be taken care of. The first important issue is to have the groups under study balanced under some criteria. Besides their cardinality, the study groups should be balanced at least in terms of age; otherwise, handwriting changes may be attributed to age differences instead of underlying pathological conditions. A balanced education (typically expressed in years) should also be considered, as there is evidence that education can influence the dynamics of handwriting despite the presence of cognitive decline [16].

The second aspect, instead, is whether the patient is on/off medication. For example, some studies on PD dealt with patients under treatment of antiparkinson medication (e.g., References [17,18]). These studies showed how handwriting significantly changes depending on the level of medical treatment administered.

Finally, the third aspect concerns the disease severity, in accordance with some standard clinical score. The unified Parkinson's disease rating scale—UPDRS (part V) score, corresponding to the Modified Hoehn and Yahr Scale [19], is a commonly used rating scale for describing how PD symptoms evolve during time. Conversely, standard assessments of probable AD include cognitive and functional tests such as the already mentioned MMSE or the Trail Making Test [20]. MMSE, in particular, is a 30-point questionnaire which includes questions and problems in many areas: from orientation to time and place, and attention and calculation, etc. Having data of patients at different degrees of disease severity can better support the early disease diagnosis or the multi-class classification problem.

It is also important to pay attention to individuals who suffered injuries that could have significantly affected their handwriting: these participants should be excluded.

### 2.1.2. Apparatus

Current technology makes available a multitude of devices for data acquisition, some of them providing immediate visual feedback to the writer. The dynamic handwriting data are generally acquired by using digitizing tablets and/or electronic pens. The dominant attributes acquired are the $x$- and $y$-coordinates of the pen position and their time stamps. Moreover, pen tablets usually capture more information than the pen trajectory, namely pen orientation (azimuth and altitude) and pen pressure. In addition, pen tablets also detect the pen trajectory while the tip is not in contact with the pad surface, allowing trajectory acquisition pen-ups. One measure, in fact, is the so-called button status, which is a binary variable evaluating 0 for pen-ups (in-air movement) and 1 for pen-downs (on-surface movement).

It is worth remarking that elderly people may be unfamiliar with technological tools: to make writing conditions as close as possible to the usual ones, writing with an inking pen on a sheet of paper fixed to the tablet is an effective option (e.g., Reference [21]).

Electronic pens (also called "smart pens") have been also adopted in alternative to tablets. For example, in References [22,23] a Biometric Smart Pen (BiSP) biometric smart pen was used. The BiSP pen is a multi-sensor pen system which is capable of capturing position, acceleration, and tilt angle of the pen, as well as the pressure and vibration generated in the refill during writing and the grip pressure of the fingers holding the pen.

Contrary to other diagnostic methods, such as medical imaging, data acquisition through these devices can be carried out even in the patient's home; moreover, the task performance is quite simple and natural and does not require timing or exhaustive repetitions.

#### 2.1.3. Acquisition Protocol

A crucial step in designing a computerized tool based on handwriting concerns the choice of the most appropriate handwriting tasks to be administered for data acquisition. Some tasks, in fact, may be redundant with other ones; others may even introduce noise in the data. Some recent works [24–26], in fact, employed ensembles of classifiers, each built on the feature space of every single task, emphasizing how a performance-driven selection of a subset of tasks can improve classification performance against the use of all tasks simultaneously. Generally speaking, handwriting tasks can be classified into simple drawing, simple writing, and complex tasks: they are described in the following paragraphs. It is worth noting that, in order to allow participants to familiarize with the equipment, some preliminary trials are typically required before the effective experimental session is carried out.

##### Drawing Tasks

Spirals, as well as meanders and circles, have been frequently used for the evaluation of motor performance. Spiral drawing on a digitizing tablet, in particular, was pioneered by Pullman [27] for assessing tremor. In fact, it is particularly suited to study motor control deficits in PD patients. The task is very easy to perform and is usually well tolerated. In general, simple drawings have been used for trajectory, tremor, dimension, and velocity evaluations, e.g., References [23,28,29]. Fine motor control problems may be caused by a reduced capability to coordinate the fingers and wrist and by a reduced control of wrist flexion. In Reference [28], for example, MCI and AD patients produced less automated, accurate, and regular movement compared to controls when drawing a spiral with the dominant hand. Differentiation between MCI and controls increased when subjects were requested to press a device, with the nondominant hand, while drawing the spiral. In Reference [30], excellent sensitivity in discriminating AD and MCI patients from controls with in-air movement was observed: the task consisted in copying a simple 3-D house with two windows, a door, and a chimney. Drawing a simple figure is very easy to perform and is usually well tolerated by all subjects. Complexity increases in the intersecting pentagon copying task, which is part of the MMSE test. Patients, in fact, typically exhibit constructional apraxia: drawing may contain fewer angles, spatial alterations, lack of perspective, and simplifications [31]. Patients can be unable to perform the task even if they understand what they should do. In particular, they typically show different drawing strategies: some trace the contours of the figure to be copied, others put points first and then connect them with segments, and so on. These issues can be reflected in the dynamic features of handwriting [31,32].

##### Writing Tasks

No-sense words composed by one or more character repetitions, for example *lll* and *lele*, can be used, e.g., References [22,33–36]. These characters are easy to write in a recursive and continuous fashion. One of the most typical evidence of PD is rigidity and tremor; thus, in contrast to controls, which show an automated handwriting, PD may produce slower and more irregular movements. In addition, PD patients may write letters in a more segmented fashion and show micrographia over time within the task [35]. The difficulty to anticipate the upcoming letter, in particular, may be the expression of a general difficulty in producing simultaneous actions. For this reason, this task can be discriminant also in the case of AD [34]. Writing words/sentences is suited to assess agraphia. A sentence requires a high degree of simultaneous processing and may have a higher neuromotor programming load than a sequence of the same characters, since it also involves linguistic skills, attention, and memory (for example, in the case where the sentence must be remembered). It also provides the possibility to evaluate the motor-planning activity between a word and the following one (in general, a hesitation between two words could highlight the necessity to replan the writing activity, while fluid writing can reveal the presence of an anticipated motor plan). A sentence allows one to capture a large number of in-air movements between words; by contrast, a word could be written without leaving the pen from the tablet surface [37,38]. Some works, e.g., Reference [39], also consider handwritten signatures. A signature represents an automatic gesture rather than a programmed one, as it is repeated very frequently during the lifespan. Since it requires only a minimal consciousness, a signature can remain preserved even when the subject is no longer able to write. Therefore, signatures may be weak predictors of cognitive impairment. Nevertheless, a signature carries a huge amount of information about the person who signs. Indeed, features of handwritten signatures emphasizing subtle deterioration of signature apposition have been successfully used to differentiate among groups [39].

Complex Tasks

Finally, the handwriting task can be part of a more complex task also involving cognitive and functional issues. For example, handwriting has been examined together with a simultaneous hearing and tone counting or has been part of a functional task (e.g., copying a bank cheque [40]). In Reference [40], participants with MCI and AD showed a significantly longer in-air time than controls. Moreover, they exerted more pressure on the surface: mean pressure, indeed, provided the best information for classification. The well-known Clock Drawing Test involves not only executive functions but also numerical knowledge, visual memory, planning, reconstruction, and visuospatial abilities. When drawing the clock, people with better cognitive-functional level generally divide the circle into different quadrants, placing the numbers 12, 3, 6, and 9 first and then the others. Conversely, patients with dementia start writing from 1 or 12 (sometimes from 11), filling the whole space with the following numbers; often, the clock is filled leaving out either the first or last number. The hands are indicated with a simple segment, and this is not a sign of cognitive decline. Instead, missing the position of the hands is a typical sign of cognitive or neurological deficit. In Reference [30], excellent sensitivity and good specificity in discriminating MCI patients from controls were obtained with in-air time. Finally, Trail Making Test and Attentional Matrices explore cognitive abilities and executive functions, in particular attentional skills, visuomotor planning and problem solving. The examiner is interested in evaluating the time of completion and the number of errors. In the Trail Making Test, the test taker is asked to connect a sequence of numeric or alphanumeric targets. Recently, Reference [41] showed that features related to timing (including times between and inside circles and rates between and inside circles) and features related to mobility (including pauses, lifts, pressure, and size) provide additional information not captured by the traditional paper-based Trail Making Test. The Attentional Matrice test, instead, is a cancellation test in which the subject is asked to mark target digits assigned among several distractors. In Reference [42], it was shown how the perceptual decision while scanning, easily captured by in-air movement analysis, is impaired in cognitively deteriorated subjects.

2.1.4. Datasets

Unfortunately, very few datasets are currently available to the research community. A schematic description of each of them is provided in Table 1.

**Table 1.** Datasets (PD = Parkinson's disease; AD = Alzheimer's disease; EC = elderly controls).

| Dataset | Groups | Apparatus | Tasks | Reference |
|---|---|---|---|---|
| PaHaW | 37 PD, 38 EC | Wacom Intuos 4M | Spiral drawing, repetition of characters, words, and sentence writing | Drotár et al. [38] |
| NewHandPD | 31 PD, 35 EC | Biometric Smart Pen | Spiral and meander drawing | Pereira et al. [23] |
| ParkinsonHW | 62 PD, 15 EC | Wacom Cintiq 12WX | Spiral drawing and stability test | Isenkul et al. [43] |
| ISUNIBA | 29 AD, 12 EC | Wacom Intuos Touch 5 | Repetition of a single word | Impedovo et al. [44] |

The Parkinson's Disease Handwriting Database (PaHaW) consists of multiple handwriting samples from 37 Parkinsonian patients and 38 age- and gender-matched controls. Subjects were requested to complete eight handwriting tasks in accordance with a prefilled template: drawing an Archimedes spiral; writing in cursive the letter *l*, the bigram *le*, and the trigram *les*; writing in cursive the word *lektorka* ("female teacher" in Czech), *porovnat* ("to compare"), and *nepopadnout* ("to not catch"); and writing in cursive the sentence *Tramvaj dnes už nepojede* ("The tram won't go today").

The original HandPD dataset comprises handwritten exams from healthy and PD people; thus, it was primarily intended for static analysis. However, the dataset was further extended for dynamic analysis purposes, comprising data from 66 individuals (35 healthy controls and 31 PD patients). The extended version is called NewHandPD. Each individual was asked to draw 12 exams, with 4 of them related to spirals, 4 related to meanders, 2 circled movements (one circle in the air and another on the paper), and left- and right-handed diadochokinesis. During the exam, the handwritten dynamics was captured by using the BiSP smart pen.

The ParkinsonHW database collects 62 PD patients and 15 healthy individuals. From all subjects, three types of handwriting recordings, namely Static Spiral Test (SST), Dynamic Spiral Test (DST), and Stability Test on Certain Point (STCP), were considered. The images of the spirals drawn by patients are also provided. In the SST test, three Archimedes spirals appeared on the graphic tablet and patients were asked to retrace them. Unlike SST, in the DST test, the Archimedes spiral just appeared and disappeared at certain time stamps. This forced the patient to keep the pattern in mind and to continue to draw. In the STCP test, there was a certain red point in the screen and the subjects were asked to hold the digital pen on that point without touching the surface. The purpose of this test was to determine the patient's hand stability or hand tremor level.

Finally, the ISUNIBA dataset collected the data of 29 probable AD patients and 12 healthy controls, who were requested to write the word *mamma* ("mother" in Italian) over different writing sessions. This is one of the first words learned and one of the last words used before dying.

At the time of writing, Castrillón et al. [45] are developing a large set of Parkinsonian handwritten patterns, including samples from adult and young healthy individuals. Concerning AD, the Handwriting Analysis against Neuromuscular Disease (HAND) project, among its goals, intended to release a large dataset of a battery of handwriting tasks performed by elderly controls and by people suffering from MCI and neurodegenerative dementia [32,46].

### *2.2. Feature Extraction*

The horizontal and vertical components of handwriting, as recorded by the tablet, are typically segmented into on-surface and in-air strokes in accordance with the button status. A *stroke* corresponds to a single trait of the handwritten pattern which is connected and continuous, i.e., between two consecutive pen-lifts. By using the Cartesian coordinates of the sampled points and their time stamps, several features can then be calculated for both on-surface and in-air strokes.

Kinematic features include number of strokes; tangential, horizontal, and vertical displacement, velocity, acceleration, and jerk; number of changes of velocity/acceleration (NCV/NCA); and NCA and NCV relative to writing duration. Displacement corresponds to the straight-line distance between two consecutive sampled points: it provides a good approximation of the pen trajectory. From displacement, velocity, acceleration, and jerk can be straightforwardly calculated as the first, second, and third derivatives of displacement, respectively. Analogously, displacement, velocity, acceleration, and jerk can be calculated with respect to both the horizontal and vertical directions. NCV and NCA are the mean number of local extrema of tangential velocity and acceleration, respectively.

Spatiotemporal features include stroke size and duration; speed and stroke speed; stroke height and width; on-surface and in-air time; total time; normalized on-surface and in-air time; and in-air/on-surface ratio.

In order to make use of the pressure signal, the following dynamic handwriting measures are also typically calculated: mean pressure; number of changes of pressure (NCP); and relative NCP. NCP was proposed in Reference [38], and its meaning is analogous to the concept of NCV/NCA, explained above.

In order to capture the randomness and irregularity of fine movements, which are difficult to analyze using only the abovementioned features, the following features can also be computed for both the on-surface and in-air horizontal and vertical components of handwriting [37]: Shannon and Rényi entropy; signal-to-noise ratio (SNR); and empirical mode decomposition (EMD). EMD iteratively decomposes the signal into so-called intrinsic mode functions (IMFs), which are functions that satisfy two requirements: (1) the number of extrema and the number of zero crossings are either equal or differ at most by one, and (2) the mean of their upper and lower envelopes equals zero.

It is worth noting that, to obtain complete statistical representations of the available features, statistical functions of the feature vector are also computed. They include means, percentiles, moments, and other statistical functions (range, median, mode, standard deviation, etc.). In addition, note that features are generally normalized before classification so as to have zero mean and unit variance.

An alternative approach to modeling the handwritten patterns is to use the Kinematic Theory of Rapid Human Movements [47,48] and, in particular, the so-called sigma-lognormal ($\Sigma\Lambda$) model [49]. This model has been used with successful results in many practical applications, for example, for developing an online signature verification system [50] and for analyzing graphomotor performance in kindergarten children [51]. The main advantage of this approach is that it is based on a physiological model of the human movement production which can lead to an improved characterization of the hidden specificity of the writers.

Finally, due to their increasing popularity, a robust alternative to more classic "hand-crafted" features is to use features automatically learned by deep learning models. Some works, in particular, used (possibly pretrained) convolutional neural networks for automatically extracting features from static images obtained by exploiting dynamic information of the handwriting, e.g., Reference [52].

A schematic overview of the features most commonly used in the different studies is provided in Table 2. Some features provide different perspectives on the same aspect of handwriting, e.g., kinematic and spatiotemporal features are able to capture the fluency and (ir)regularities of handwriting movements, leading to similar results. Some others, in particular those automatically learned by deep learning models, are difficult to correlate with the other ones; however, they may provide novel and nonoverlapping information. In general, almost all features, either directly measured by the digitizing tablet or derived from them, have been used with promising results in every single study. Only the pen angle information is typically discarded: its applicability appears to be not useful, even if a very recent work applied it and reported encouraging results [53].

**Table 2.** Most commonly used features: they are typically intended both on-surface and in-air.

| Feature | Description | Observation |
|---|---|---|
| *Direct* | | |
| Position | $(x, y)$-coordinates of the sampled points | They are used to derive the geometrical pattern of handwriting |
| Time stamp | Temporal information of the sampled points | It is used to derive the temporal duration of the hand movement |
| Pressure | Pressure exerted over the writing surface | In patients, pressure takes on erratic values due to cognitive and muscular difficulties |
| Tilt-$x$ | Angle between the pen and the surface plane | Their use is typically ignored and their usefulness is controversial |
| Tilt-$y$ | Angle between the pen and the plane vertical to the surface | |
| Button status | Boolean variable of whether the pen is on-surface or in-air | It enables the separation between on-surface and in-air movements. It has been shown how the two handwriting modalities carry on nonredundant information |
| *Kinematic* | | |
| Displacement | Trajectory during handwriting | It is generally used to derive other kinematic features |
| Velocity | Rate of change of displacement with respect to time | Patients suffering from neurodegenerative diseases do not write with the same constancy of healthy subjects but are affected by a lower writing speed, with continuous acceleration peaks |
| Acceleration | Rate of change of velocity with respect to time | |
| Jerk | Rate of change of acceleration with respect to time | |
| NCV/NCA | Number of local extrema of velocity/acceleration | These measures are tailored to capture the fluency of the handwriting movement. Highly automated movements are characterized by bell-shaped and smooth velocity and acceleration profiles |
| *Spatio-temporal* | | |
| Stroke size | Strokes' path length | Patients suffering from PD can exhibit micrographia |
| Stroke height/width | Height/width of strokes | |
| Stroke duration | Movement time per stroke | The average writing duration of a patient affected by a neurodegenerative disease is typically longer than in a healthy subject |
| Time | Time spent on-surface/in-air during writing | |
| *Entropy and energy* | | |
| Entropy | Entropy-based features | These measures are tailored to capture the randomness and irregularities of fine movements |
| SNR | Signal-to-noise ratio | |
| EMD | Empirical mode decomposition | |
| *Model-based* | | |
| $\Sigma\Lambda$-based | Parameters of the $\Sigma\Lambda$ reconstruction of the handwritten pattern | The $\Sigma\Lambda$ model can help investigate the dynamics of handwriting during the generation of the action plan |
| *Automatically learned* | | |
| Deep-learning based | Features automatically learned by deep learning models trained on static representations of the handwriting dynamics | Their meaning is typically hard to interpret: they can express subtle between-group differences not captured by traditional hand-crafted features |

### 2.3. Data Analysis

The goal of this final step is to uncover useful patterns able to support decision making. Mostly, the literature investigating handwriting changes due to aging and relies on statistical analyses to perform this step. For example, the classic analysis of variance (ANOVA) is typically used to test group differences across different measures of handwriting.

In the last years, the studies focusing on the development of computer aided diagnosis systems have made use of machine learning and statistical pattern recognition strategies to discriminate between unhealthy and healthy subjects [54]. In a series of experiments, for instance, Drotár et al. found that support vector machines, fed with kinematic and spatiotemporal features, provide better prediction accuracy than other classic approaches, such as Naïve Bayes, e.g., Reference [37].

More recently, due to their increasing popularity in a plethora of recognition tasks, some works investigated the usefulness of deep learning approaches [25,55]. The features automatically extracted by a convolutional neural network can be used to feed a fully connected layer stacked on top of the convolutional base or a more classic statistical classifier.

It is worth noting that, since the data at disposal are typically small, several resampling methods are usually adopted to achieve more reliable evaluations of the classification performance, such as cross-validation and leave-one-out [56].

A simple analysis, based on the visual inspection of the performed task and the velocity and pressure profiles of handwriting is sketched in Figure 2.

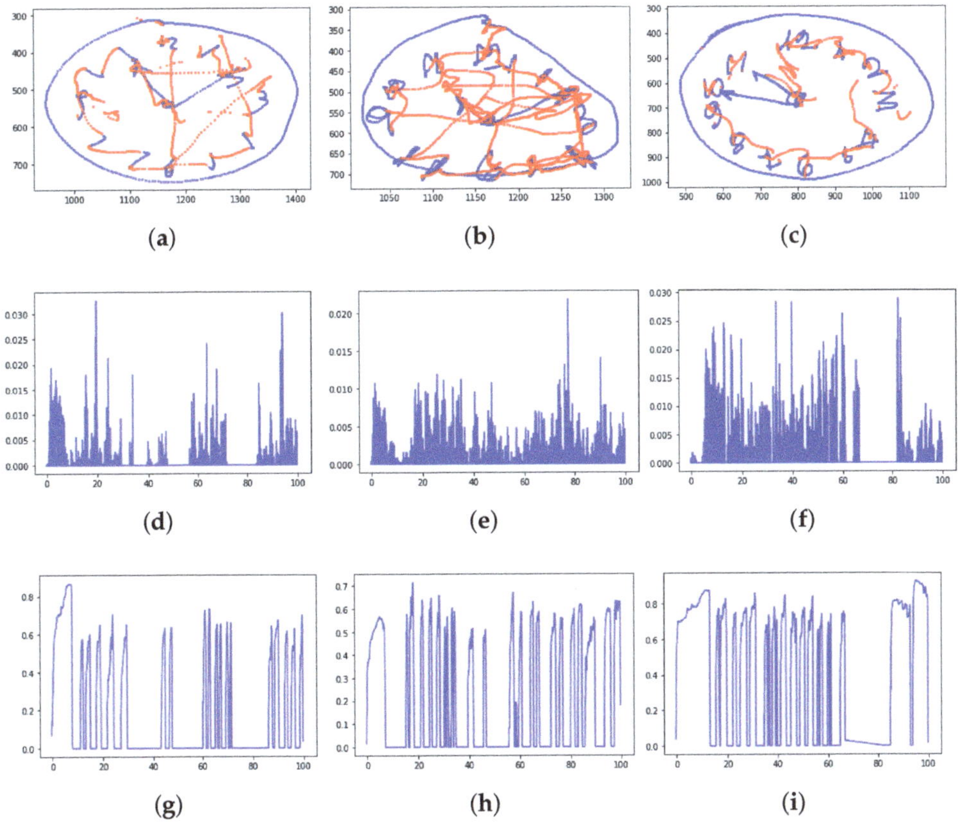


**Figure 2.** Clock drawing test (CDT) performed by a healthy control, a PD patient, and an mild cognitive impairment (MCI) subject: From top to bottom are the rendered task (the on-surface movement is in blue color; the in-air movement is in red), the velocity profile, and the pressure profile. It is recognizable how the PD patient tends to alter the figure dimensions, while the MCI subject missed the correct time. The velocity and pressure profile show more peaks, highlighting a movement which is characterized by less fluency and more changes of direction than the healthy handwriting. (Data have been acquired within the Handwriting Analysis against Neuromuscular Disease project—http://hand-project.di.uniba.it/). (**a**) Healthy CDT, (**b**) PD CDT, (**c**) MCI CDT, (**d**) healthy velocity, (**e**) PD velocity, (**f**) MCI velocity, (**g**) healthy pressure, (**h**) PD pressure, and (**i**) MCI pressure.

## 3. Research Directions

The studies involving dynamic handwriting analysis for neurodegenerative diseases assessment can be broadly classified in accordance with the disease taken into account: Parkinson's and Alzheimer's disease. For each of them, different lines of investigation can be identified: they are discussed in the following.

### 3.1. Parkinson's Disease

The papers focusing on Parkinsonian handwriting can be further classified depending on three main research directions:

- *Disease insight*: the first category (including the oldest papers) have been devoted to providing an insight into the fine motor control of handwriting and its relationship with the concomitant impairment. The main goal is to better understand the involved mechanisms underlying PD;
- *Disease monitoring*: other papers studied the effects of medication on handwriting with the aim to evaluate the effectiveness of handwriting analysis on monitoring disease progression;
- *Disease diagnosis*: the third category (including the most recent works) investigated the use of handwriting as an inexpensive objective tool for automatic disease diagnosis.

This section is structured in accordance with this classification. A schematic overview is provided in Table 3.

**Table 3.** Summary of studies on PD (EC = elderly controls; YC = young controls; and SZ = schizophrenia patients).

| Reference | Groups | Tasks | Main Features | Main Findings |
|---|---|---|---|---|
| *Disease insight* | | | | |
| Phillips et al., 1991 | 7 PD, 7 EC, 7 YC | Meander writing | Kinematic and spatiotemporal | Dynamic analysis can be useful to characterize PD handwriting |
| Teulings and Stelmach, 1991 | 6 PD, 6 EC | No-sense word and a sentence | Kinematic and spatiotemporal | PD patients experience problems with the production and regulation of force amplitude |
| Fucetola and Smith, 1997 | 20 PD, 20 EC | Figure drawing | Kinematic and spatiotemporal | PD patients have difficulties in adjusting the size of their drawing to compensate for distortions in visual feedback |
| Oliveira et al., 1997 | 11 PD, 14 EC | No-sense word | Kinematic and spatiotemporal | PD patients write with a normal amplitude when given external cues |
| Teulings et al., 1997 | 17 PD, 12 EC | Lines, circles, and no-sense words | Kinematic | PD fine motor control problems may be caused by a reduced capability to coordinate the fingers and wrist and by a reduced control of wrist flexion |
| Van Gemmert et al., 1998 | 9 PD, 9 EC, 9 YC | Sentence writing | Kinematic and spatiotemporal | PD patients are vulnerable to a moderate level of secondary task load |
| Van Gemmert et al., 1999 | 13 PD, 15 EC | Lines, circles, and no-sense words | Spatiotemporal | PD patients may have trouble in maintaining a constant force amplitude |
| Swinnen et al., 2000 | 13 PD, 13 EC | Triangle drawing | Kinematic | PD patients benefit from practice to alleviate their basic motor deficits |
| Van Gemmert et al., 2001 | 7 PD, 7 EC | No-sense word and sentence writing | Kinematic and spatiotemporal | PD patients show micrographia when cuncurrent processing load increases |
| Teulings et al., 2002 | 11 PD, 16 EC, 10 YC | No-sense word | Spatiotemporal | PD patients rely on previous or ongoing feedback to program subsequent strokes |
| Van Gemmert et al., 2003 | 13 PD, 13 EC | No-sense words | Spatiotemporal | PD patients produce inadequate stroke sizes when the size equals or exceeds 1.5 cm |
| Caligiuri et al., 2006 | 13 PD, 10 SZ, 12 EC | "hello hello" | Kinematic | PD patients exhibit impaired movement velocity and velocity scaling |
| Ponsen et al., 2008 | 13 PD, 13 EC | Sentence writing | Kinematic and spatiotemporal | Impairments in performing tasks involving complex uni-manual upper limb movements are an early sign of PD |
| Broderick et al., 2009 | 16 PD, 16 EC, 16 YC | Shape drawing | Kinematic | PD patients show smaller-than-required movement amplitude |
| Dounskaia et al., 2009 | 9 PD, 9 EC | Line and circle drawing | Kinematic | PD causes deficits in coordination patterns between wrist and finger motions |
| Gangadhar et al., 2009 | 34 PD, 25 EC | No-sense word | Spatiotemporal | PD handwriting exhibits smaller size and larger velocity fluctuations than normal handwriting |
| Bidet-Ildei et al., 2011 | 7 PD, 7 EC | No-sense words | Kinematic and spatiotemporal | PD handwriting does not show signs of motor anticipation |
| Ma et al., 2013 | 15 PD, 15 EC | A Chinese character | Spatiotemporal | Micrographia is not evident in vertical writing |

**Table 3.** *Cont.*

| Reference | Groups | Tasks | Main Features | Main Findings |
|---|---|---|---|---|
| Broeder et al., 2014 | 18 PD, 11 EC | Loops while counting | Kinematic and spatiotemporal | PD patients exhibit deficits in handwriting under dual-task conditions |
| Smits et al., 2014 | 10 PD, 10 EC | Circle, star, and spiral drawing, a no-sense word and a sentence | Kinematic and spatiotemporal | Handwriting can provide objective measures for bradykinesia, micrographia, and tremor |
| Senatore and Marcelli, 2019 | 30 PD, 30 EC, 30 YC | Loop shapes | Kinematic | Parkinsonian handwriting is similar to that produced by beginner writers |
| *Disease monitoring* | | | | |
| Eichhorn et al., 1996 | 29 PD, 40 EC | Circle drawing | Kinematic | Dynamic handwriting analysis is useful for quantifying dopamimetic effects |
| Contreras-Vidal et al. and Poluha et al., 1998 | 10 PD | No-sense words | Spatiotemporal | Handwriting measures show significant trends across the levedopa cycle |
| Siebner et al., 1999 | 12 PD | Lines, circles, and a no-sense word | Spatiotemporal | subthalamic nucleus stimulation improves handwriting performance in PD patients |
| Cobbah and Fairhurst, 2000 | 6 PD | No-sense words | Kinematic | Dynamic handwriting measures are indicative of positive response to dopamine |
| Boylan et al., 2001 | 10 PD | Spiral drawing | Temporal | repetitive transcranial magnetic stimulation provides beneficial effects on PD movements |
| Lange et al., 2006 | 12 PD, 12 EC | Sentence writing | Kinematic | Participants with an altered dopaminergic neurotransmission shifted from an automatic to a controlled processing of movement execution |
| Tucha et al., 2006 | 27 PD, 27 EC | Sentence writing | Kinematic | Dopaminergic medication results in improved dynamics of movement execution |
| Randhawa et al., 2013 | 10 PD | No-sense word | Kinematic and spatiotemporal | Fine motor performance benefit from rTMS stimulation |
| Smits et al., 2015 | 14 PD | Graphical tasks | Spatiotemporal | Graphical tasks are useful to assess upper limb functioning |
| Danna et al., 2018 | 20 PD, 20 EC | Spiral drawing | Kinematic and spatiotemporal | PD handwriting performance may not be impacted by handedness |
| *Disease diagnosis* | | | | |
| Ünlü et al., 2006 | 28 PD, 28 EC | Meanders, sentence, words, and no-sense word writing | Pressure | Pressure features can distinguish between PD patients and controls |
| Drotár et al., 2013a | PaHaW | | Kinematic and spatiotemporal | Good accuracy is obtained by using only on-surface features |
| Drotár et al., 2013b; 2014 | | | Kinematic and spatiotemporal | Accuracy improved with in-air features |
| Drotár et al., 2015a; 2015b | | | Kinematic, spatiotemporal, entropy, and energy | Accuracy improved using entropy and energy-based features |
| Drotár et al., 2016 | | | Kinematic, spatiotemporal, and pressure | Pressure features can be profitably used |
| Rosenblum et al., 2013 | 20 PD, 20 EC | Functional tasks | Kinematic and spatiotemporal | Patients write smaller letters, applying less pressure and requiring more performance time than controls |
| Pereira et al., 2016; 2018; Afonso et al., 2018 | NewHandPD | | Deep learning-based | The deep learning approach is promising for the recognition problem at hand |
| San Luciano et al., 2016 | 138 PD, 150 EC | Spiral drawing | Kinematic and spatiotemporal | Spiral analysis is a promising quantitative biomarker for the early diagnosis |
| Kotsavasiloglou et al., 2017 | 24 PD, 20 EC | Line drawing | Kinematic | High predictive accuracy can be obtained even using very simple tasks |
| Zham et al., 2017 | 27 PD, 28 EC | Spiral drawing | Kinematic and pressure | Spiral drawing can be partly used to differentiate among degrees of disease severity |
| Impedovo et al., 2018 | PaHaW | | Kinematic, spatiotemporal, entropy, energy, and pressure | Accuracy decreased in distinguishing between controls and patients only at the early stage of disease |
| Gallicchio et al., 2018 | ParkinsonHW | | Deep learning-based | Recurrent neural networks are profitably used |
| Mucha et al., 2018 | PaHaW | | Fractional derivative-based | Promising results are obtained with fractional derivative-based features |
| Impedovo, 2019 | PaHaW | | Velocity-based | New velocity-based features are proposed |
| Jerkovic et al., 2019 | 33 PD, 10 EC | Sentence writing | Kinematic | The best results are obtained when combining in-air to on-surface features |
| Loconsole et al., 2019 | 4 PD, 7 EC | Word and no-sense word writing | Kinematic | Gyroscope-based features are used with promising results |
| Rios-Urrego et al., 2019 | 39 PD, 39 EC, 40 YC | Spiral drawing and sentence writing | Kinematic, geometrical, and nonlinear features | Irregularities of handwriting increase with aging |
| Diaz et al., 2019 | PaHaW | | Deep learning-based | A new dynamically enhanced static representation of handwriting is proposed |
| Ribeiro et al., 2019 | NewHandPD | | Deep learning-based | Recurrent neural networks can be fruitfully used to capture tremor in time-dependent handwriting signals |
| Ammour et al., 2020 | 28 PD, 28 EC | Text copying | Kinematic, pressure, and pen inclination | A new semi-supervised approach is proposed to discriminate among groups |

### 3.1.1. Disease Insight

Phillips et al. [57] and Teulings and Stelmach [58] were among the first to use digitizing tablets to assess Parkinsonian handwriting. The work of Phillips et al. was the first showing how dynamic handwriting features (in particular in the velocity domain) can successfully differentiate between patients and healthy controls. Teulings and Stelmach (1991), instead, asked participants to alter their usual handwriting in an attempt to study the extent to which the patients' motor system can adjust size, force, and speed parameters. Results showed that Parkinsonians, as well as controls, were generally able to modify stroke size, peak accelerations, and stroke duration as they wrote the required patterns. However, a signal-to-noise analysis suggested that the movement deficits were primarily due to an impaired force-amplitude component rather than an impaired stroke-duration component.

Contreras-Vidal and Stelmach [59] were the first to integrate previous experimental data on the anatomy of the basal ganglia to the motor impairments in PD; the aim was to develop a neural model of the basal ganglia useful to explain normal and Parkinsonian movements. The model consists of a model of basal ganglia, in which each nucleus is represented by a single unit, combined with a model capable of learning and generating simple handwriting movements. This model was able to reproduce many aspects of the normal and PD movement control including hypometria, bradykinesia, akinesia, impairments in the coordination of multiple joints, micrographia, effects of levodopa on movement size and speed, and pallidotomy. The simulation data of this model, in fact, were confirmed by the experimental data obtained in some other studies, e.g., References [33,60].

Since proprioceptive, kinaesthetic, and visual feedback are essential for the completion of many movements, impaired utilization of sensory feedback may retard the effective learning of motor programs. Based on this hypothesis, Fucetola and Smith [61] investigated the effects of a distorted visual feedback on the drawing performance of Parkinsonian patients. They observed that patients were less able than controls to adjust the size of their drawing to compensate for distortions in visual feedback. The effect was particularly pronounced when patients were required to draw smaller than normal. Nevertheless, with practice, PD patients showed a similar degree of improvement in size as controls, although they did not match the control group's level of performance.

Oliveira et al. [62] investigated whether micrographia in individuals with PD is lessened either by giving visual targets or by continually reminding them that they should write in a normal way. In a first trial of free writing, patients showed micrographia, as they reduced their letter size over time within the trial. However, the letter size increased significantly when they were given either visual targets or constant auditory reminders. This improvement persisted when, shortly afterwards, the patients were requested to write freely without external cues.

Teulings et al. [33] investigated whether Parkinsonism reduces coordination of fingers, wrist, and arm in fine motor control. These movement problems contribute to an increase in jerk levels, as jerk represents the rate of change of acceleration over time. In the PD group, back-and-forth strokes involving coordination of fingers and wrist showed larger normalized jerk than strokes performed using either the wrist or the fingers alone. Moreover, wrist flexion showed greater normalized jerk in comparison to wrist extension. The elderly control subjects showed no such effects as a function of coordination complexity. Thus, the authors hypothesized that fine motor control problems in PD patients may be caused by a reduced capability to coordinate the fingers and wrist and by a reduced control of wrist flexion.

Van Gemmert et al. [63] tested the hypothesis that PD patients are more vulnerable to a moderate level of secondary task load than elderly or young controls due to a heightened variability in the motor system. Patients and the two control groups were requested to write a sentence under four load conditions: start writing after they heard the recorded word "start"; ignore auditory presented digits while writing; repeat orally the presented digits; and subtract the number 2 from each presented digit and pronounce the outcome aloud. The results obtained showed that, in contrast to young and elderly controls, PD patients tended to increase movement time and normalized jerk when the secondary task consisted primarily of motor load. Furthermore, it was shown that PD patients did not reduce writing size because of a high level of mental load: this suggested that writing in an automated fashion does not cause micrographia.

Van Gemmert et al. [60] investigated whether PD patients can have difficulty in increasing stroke size, decreasing stroke duration, or both during the execution of noncomplex handwriting tasks. To this end, they designed an experiment comprising simple writing patterns, such as straight lines and circles, requesting participants to vary writing size and speed. Although the different handwriting patterns affected movement time and writing size significantly, patients did not show increasing difficulty in maintaining writing size and/or stroke duration because of a decrease of pattern complexity: thus, it was argued that the complexity of a pattern is likely to not be a relevant factor in handwriting.

In Reference [64], Swinnen et al. addressed the problem of determining whether practice modifies the temporal and spatial features of handwriting in PD patients. The findings obtained showed that PD patients can change their performance thanks to practice, suggesting that practice may help them to partly overcome bradykinesia. Nevertheless, they never reached the performance level obtained by the elderly control group.

Van Gemmert et al. [65] hypothesized that the cause of micrographia in PD patients can be associated with the concurrent processing demands that result from the coordination and control of fingers, wrist, and arm during writing and processing of future words. In their experiments, patients and controls were requested to write four different phrases of various word counts. All phrases started with an *llll* pattern, and this pattern was repeated later in the phrase. PD patients reduced stroke size when the number of words increased in the phrase, i.e., when the processing demand increased. This finding suggested that the motor system of PD patients anticipates increased processing demands by reducing stroke size rather than increasing stroke duration.

Teulings et al. [66] compared PD patients to elderly people and young adults with respect to their ability to use visual feedback to control handwriting size. Participants wrote sequences of cursive *l*-shaped loops on a digitizer display, which enabled the authors to distort the visual feedback without the participant's knowledge by altering the vertical dimension of handwriting. The results showed that controls gradually corrected loop size by enlarging (or reducing) the size of the entire loop sequence. Conversely, PD patients showed an entirely different response: instead of correcting for the distortions, they progressively amplified its effect. This suggested that PD patients do not adapt their visuomotor map in response to the distorted visual feedback of handwriting. Instead, they seem to rely constantly on the visible trace feedback during the ongoing movement. The authors thus hypothesized that they either plan their writing based on the visual feedback of their previous strokes or that they attempt to track the ongoing, distorted handwritten trace.

Van Gemmert et al. [67] evaluated the ability of PD patients to increase stroke size independently of stroke duration for different sizes. Patients and controls were requested to write cursive patterns at different sizes (1, 1.5, 2, 3, and 5 cm). Each target pattern was displayed at its required size on the tablet but disappeared as soon as the pen touched the surface of the screen. In contrast to controls, patients with PD undershot the target size of 2 cm and, when required to write as fast as possible, they even undershot the 1.5 cm target size. These findings support the hypothesis that the range in which stroke size can be manipulated without significant changes in stroke duration is smaller in Parkinsonian handwriting than in the healthy handwriting.

Caligiuri et al. [68] examined the handwriting dynamics of patients with idiopathic PD, schizophrenia, and drug-induced Parkinsonism (SZ) and of healthy control. Participants were instructed to write the word *hello* twice at three vertical height scales. The (in)ability to scale movement velocity with increasing movement distance was quantified. Four observations were drawn: (1) both SZ patients with drug-induced Parkinsonism and PD patients exhibited impaired movement velocities and velocity scaling; (2) performance on the velocity scaling measure can distinguish drug-induced Parkinsonism from controls with 90% accuracy; (3) SZ, but not PD, participants displayed abnormalities in movement smoothness; and (4) there was a positive correlation between age and magnitude of the velocity scaling deficit in PD participants.

Ponsen et al. [69] were among the first to analyze handwriting in newly diagnosed, untreated PD patients. The results of the study showed that newly diagnosed patients are impaired in performing complex uni-manual upper limb motor tasks in comparison to healthy subjects. They appeared to be particularly impaired in their handwriting, exhibiting reduced sentence length and writing velocity and a decrease in letter height during writing. Therefore, the authors concluded that impairments in performing tasks involving complex uni-manual upper limb movements are an early characteristic of PD; thus, they could be used for the early disease diagnosis.

Broderick et al. [70] considered a drawing task instead of handwriting ones to address the hypothesis that PD patients exhibit deficits in controlling acceleration when the task involves an increase in inertial load, specifically under the requirement to increase movement amplitude and/or speed, and in the weight of the pen. Patients showed significantly lower mean velocity, lower acceleration, higher constant error of stroke length, and higher normalized jerk scores than controls. Nevertheless, these effects were not worsened by adding weight to the pen. The observed smaller-than-required movement amplitude suggested a relationship between hypometria and bradykinesia in drawing and/or handwriting.

In Reference [71], Dounskaia et al. tested the hypothesis that PD affects differently handwriting movements depending on the coordination pattern of wrist and finger motions. To investigate this hypothesis, the groups under study were requested to perform three types of cyclic wrist and finger movements: drawing two lines and a circle. Although both groups deformed the circle during fast movements, the deformation was more pronounced in patients than in controls. A possible reason for this is that PD patients may be unable to properly regulate the influence of biomechanical factors on wrist and finger motion.

In the model of PD handwriting proposed in Reference [59], basal ganglia nuclei is modeled as lumped units, with activity levels represented by rate codes. Basal ganglia dynamics is described in terms of fixed-point behavior; thus, only magnitude-related aspects of handwriting—faster/slower, larger/smaller, etc.—can be captured. Gangadhar et al. [72] presented an alternative model of Parkinsonian handwriting, which produces a stable rhythm in a network of oscillators and resolves the stroke output in a Fourier-style. In the paper, the model predictions were compared to handwriting data obtained by patients and controls. PD handwriting statistically exhibited smaller size and larger velocity fluctuation compared to normal handwriting. These findings were reflected in both experimental data and network predictions.

Bidet-Ildei et al. [35] hypothesized that, if it is true that PD patients produce sequential movements in a more segmented fashion, then they should have difficulties in anticipating the forthcoming letter. Their experimental findings revealed that handwriting in PD patients did not exhibit any sign of motor anticipation: although they could write three letters without pauses, PD patients tended to produce each letter in a more independent manner. In order to explain this, the authors suggested that the difficulty in anticipating the upcoming letters may be the expression of a general difficulty in producing simultaneous actions.

Ma et al. [73] noticed that all the published studies investigating micrographia in PD examined handwriting only in the horizontal direction, as the handwritten samples analyzed were primarily in Western languages. However, several other languages, as those from Eastern Asia, can be written not only horizontally but also vertically, from top to bottom. Since different directions require different joint coordination patterns and writing horizontally requires more wrist extension than writing vertically, the micrographia reported in horizontal writing may not be generalized to characters written vertically. To investigate on this problem, the authors asked patients and controls to write Chinese character *Zheng*. The main finding was that the PD group had a linear decrease in overall character size and horizontal strokes along the writing sequence in the horizontal direction but not in the vertical direction. This observation confirms that micrographia in PD may be associated with wrist extension.

Broeder et al. [74] obtained results in line with the abovementioned study by Van Gemmert et al. [63]: they observed that PD patients experience more dual-task interference during writing than controls when performing a cognitive tone-counting task and a writing task simultaneously. Dual-task interference refers to the decreased performance experienced during dual tasking, i.e., when two motor tasks with different goals are combined. More specifically, the secondary task consisted in counting high and low tones during writing. The results obtained showed that dual-task performance was affected in PD patients. In particular, they suggested that the control of writing at small amplitudes requires more compensation brain-processing resources in PD than controls.

Smits et al. [36] investigated handwriting tasks that may be helpful to provide a quantitative method to differentiate between PD patients and healthy controls: circle, star, and spiral drawing; *elel*; and writing a sentence. The drawing and writing tasks were analyzed to evaluate the speed of movement to assess bradykinesia and the size of writing to assess micrographia. In addition, a frequency analysis was carried out to assess rest tremor. The results showed that Parkinson patients tend to be slower than healthy control participants. PD patients also wrote smaller than controls. Furthermore, rest tremor was detected in the group of patients who were clinically assessed as having rest tremor.

In a very recent work [75], Senatore and Marcelli proposed a novel paradigm aimed at emulating the early stage of handwriting learning in proficient writers by asking them to produce a familiar *l*-shape with a novel, unfamiliar motor plan. In other words, participants were asked to produce the sequence of strokes by using a motor plan different from the one an individual is used to. The authors involved young and elderly healthy participants comparing them with the data of the pathological group of the PaHaW dataset. The authors found that Parkinsonian writing during a familiar movement is characterized by lack of fluency, slowness, and abrupt changes of direction, as the handwriting produced by beginner writers. These results support the hypothesis that the fine tuning of the motor plan parameters involved during the production of handwriting is deteriorated by PD.

#### 3.1.2. Disease Monitoring

Papers falling in this category mainly investigated two kinds of treatment: antiparkinson medication and neurostimulation. Concerning the former, in particular, PD treatment often involves the administration of levedopa to reduce the associated rigidity and bradykinesia. During this treatment, a conversion process occurs in the brain so that levedopa becomes dopamine and the reduced level of the body own's dopamine is compensated.

Eichhorn et al. [17] used a computational analysis of open-loop handwriting movements to monitor the effect of levodopa and apomorphine in three groups of Parkinson patients: those with untreated probable Parkinson's disease, those with fluctuating PD, and some other patients with known levedopa unresponsive Parkinsonism. Subjects were instructed to draw fluently concentric circles. After apomorphine injection, the group with untreated probable PD and the group with long-standing PD showed significant improvement of kinematic features. The patients with levedopa unresponsive Parkinsonism did not change significantly in any of the parameters under study. In conclusion of the paper, the authors observed that the improvement of handwriting kinematics by dopamimetic stimulation may be helpful to predict responsiveness to levodopa treatment in Parkinsonian syndromes.

Levedopa levels decay over several hours; thus, every few hours, another dose of levedopa should be taken. In light of this, Contreras-Vidal et al. [76] and Poluha et al. [77] hypothesized that Parkinsonian handwriting would change across the levedopa cycle. The most remarkable finding of these studies was that handwriting up-stroke duration varied significantly across the medication cycle.

In Reference [78], Siebner et al. investigated the effect on handwriting of high-frequency stimulation of the subthalamic nucleus (STN), which is a therapeutic approach in patients with severely disabling PD. During high-frequency STN stimulation, handwriting movements became faster and smoother, indicating a partial restoration of the open-loop automatic performance. In addition, a stimulation-related reduction in micrographia was observed.

Cobbah and Fairhurst [79] investigated the dynamic changes evident in ordinary handwriting under strict dopamimetic challenge test conditions. Patients with Parkinsonism were requested to write handwriting patterns before medication and once again at peak motor performance, after doses of apomorphine or levodopa were administered. The results obtained suggested that a differentiation between on and off states in dopamimetic tests is possible by using ordinary handwritten samples. The effects reflected on kinematic features of handwriting, in fact, suggested improvements in movement efficiency in the on state.

Boylan et al. [80] studied the therapeutic potential of repetitive transcranial magnetic stimulation (rTMS) for PD by delivering stimulation at high intensity and frequency over time. rTMS is a noninvasive technique that allows the cortical excitability to be altered; thus, it can induce a dopamine release in the stratum of people with PD. Among some other tests, assessment included spiral drawing as handwriting task. The major finding of the study was the worsening of motor performance on spiral drawing with active rTMS to the supplementary motor area (SMA) of patients.

Lange et al. [81] carried out a battery of experiments to study the role of dopamine in movement execution during handwriting. The findings of the experiments showed that alterations of the dopamine system adversely affect movement execution during handwriting. All experiments showed that the number of inversions of the direction of velocity is increased in participants with an altered dopaminergic neurotransmission.

A study analogous to the one reported in Reference [81], with the same apparatus and procedure, allowed the authors to reach some other conclusions on the dopaminergic effects on handwriting movements [18]. The main finding was that dopamine medication results in a partial restoration of automatic movement execution: although dopaminergic treatment in PD patients resulted in marked improvements in the handwriting dynamics, patients never reached an undisturbed level of performance.

Analogously to Reference [80], Randhawa et al. [82] investigated whether the delivery of rTMS impacts handwriting performance. The authors found that 5-Hz rTMS over SMA increased the global size of handwriting. Moreover, the stimulation led to a decrease in the amount of pen pressure. These findings suggested that 5-Hz rTMS over SMA can influence key aspects of handwriting including vertical size and axial pressure, at least in the short term.

In Reference [83], Smits et al. evaluated the validity of a battery of graphical tasks useful to assess upper limb functions in individuals with probable PD. The Purdue Pegboard Test (PPT), in which metal pins have to be placed within holes, was used as a reference test. Only PD patients, who were on and off medication, performed the tasks. Moderate correlations between performance on graphical tasks and the PPT test were obtained, suggesting that the set of graphical tasks is a valid tool to assess and monitor upper limb functions in PD. In addition, the study showed that this set can be used to detect subtle changes in performance after medication that are barely visible by only observing the patient.

Considering that handwriting involves linguistic processes that can be influenced by cognitive impairments and sociocultural factors, Danna et al. [84] focused only on drawing tasks, particularly spiral drawing, which have the advantage to involve exclusively motor mechanisms. Different analyses were carried out to evaluate the effectiveness of digitized spiral drawing in distinguishing patients with and without medical treatment. The results obtained confirmed this hypothesis. Surprisingly, the general performance of PD patients was not impacted by handedness, suggesting that the side-dominance of PD symptoms can prevail over handedness.

#### 3.1.3. Disease Diagnosis

In Reference [22], Ünlü et al. focused on approaches for PD diagnosis based on the pressure information provided by the electronic biosensor BiSP pen. It turned out that the most discriminating feature, which achieved an Area Under the ROC Curve (AUC) equal to 0.933, was based on the difference between the controlled writing pressure in the $x$-$y$ direction and the tilt tremor of the pen. It was observed, in fact, that, for PD patients, the tremor control is better achieved during movements (like handwriting) instead of constant pressure (pen tilt).

A remarkable contribution to the application of machine learning algorithms to the automatic discrimination of PD was provided by Drotár et al. All their studies were carried out on a same dataset, i.e., PaHaW, which the authors made freely available. In Reference [85], by comparing the predictive potential of models built on every task individually and models trained merging all tasks, the authors found that the best classification performance was reached by the combination of all tasks. In Reference [86], the authors investigated the extent to which classification performance can be improved considering not only on-surface but also in-air movement, since the two modalities appear to carry on nonredundant information. They found that in-air features outperform on-surface features. These findings were further improved in Reference [87], where different feature selection strategies were employed. In addition to conventional kinematic handwriting measures, Drotár et al. [37] also computed novel measures based on entropy, signal energy, and empirical mode decomposition of the handwriting signals. These features provided more insight and better understanding of the data. It is worth noting that, in this study, only on-surface movement was considered. In Reference [88], instead, the authors employed these novel features also considering in-air movement. In Reference [38], the authors introduced additional features based on the pressure exerted over the writing surface. The fundamental pressure features were the value of pressure as captured by the tablet during the particular task and the rate at which pressure changes with respect to time. Then, they introduced correlation coefficients to capture the relationship between pressure and kinematic features.

It is worth remarking that, in all of the studies by Drotár et al., the spiral task was undertaken with no significant impact on classification. This may have been due to the use of measures only tailored to handwriting; instead, visual features, for example, those provided by deep learning algorithms [25,89], seem to overcome this issue.

Rosenblum et al. [21] assessed whether simple characteristics of handwriting can provide quantitative measures to accurately differentiate between PD patients and controls. Study participants were requested to write their name and to copy an address. Significant group effects were observed: compared to controls, patients wrote smaller letters, applying less pressure and requiring more performance time. A discriminant function was found for the effective group classification of all participants. Furthermore, the authors highlighted the importance to analyze handwriting not only on-paper but also in-air, as significant differences were observed between these two writing conditions. In fact, as the authors wrote, in-air time is a manifestation of "planning the next movement", which can reflect cognitive ability and supply information about the writer.

In Reference [23], Pereira et al. proposed NewHandPD, a dataset of signals extracted from the BiSP smart pen comprising spiral and meander drawings. Each sensor of the device outputs the whole signal acquired during the handwriting tasks; thus, it can be subsequently represented as a time series. The authors used CNNs and meta-heuristic-based optimization techniques to fine-tune the network hyper-parameters due to their ability to learn without human intervention. Hence, the main contribution of the work was the application of a deep learning-oriented approach to aid PD diagnosis as well as the design of a signal-based dataset.

The abovementioned work was extended by the authors in References [52,90]. In Reference [90], CNNs were used to learn features directly from time-series-based images. The main hypothesis was that texture-oriented features are able to encode the tremors during handwriting. In Reference [52], the recurrence plot technique was used to map the pen signals into the image domain; then, these images were used to instruct a CNN on how to learn discriminating features. A recurrence plot enables to visualize repeated events of higher dimensions through projections onto low-dimensional representations.

San Luciano et al. [29] assessed the validity of the digitized Archimedes spiral drawing as a biomarker for the early diagnosis of PD. Spatial and temporal variables of handwriting were, in general, significantly different between PD subjects and controls. A model using all features showed high discriminating validity. Therefore, the authors claimed spiral analysis to be a promising quantitative biomarker for the early disease diagnosis.

Kotsavasiloglou et al. [91] asked patients and healthy subjects to draw a horizontal line on the tablet's surface, keeping the pen's velocity as constant as possible. The choice of this simple task was made with the expectation that one should be able to detect differences between the groups even in very simple tasks, as the impairment manifests independently of the complexity of the task. Indeed, good accuracy performance were obtained with a Bayesian classifier. It is worth noting that, as an additional contribution, the authors introduced a new metric, termed normalized velocity variability, which quantifies the variability of the pen's horizontal speed as the line is drawn.

The majority of the studies focused on the binary discrimination healthy vs unhealthy, independently of the degree of disease severity. In other words, the Parkinsonian group is typically considered as a single cluster in which all subjects share the same degree of disease severity. In Reference [92], Zham et al. addressed this issue by investigating the correlation between the speed and pen pressure while sketching a spiral and the severity of disease symptoms. The strongest correlation was found with a combination of these two parameters, which turned out to be useful for the automatic differentiation between the low and high degree of severity. However, this measure was not able to differentiate between low and middle and between middle and high disease severity. In Reference [93], classification accuracy was refined by focusing on angular features and the count of direction inversion during the sketching of the spiral.

In Reference [24], Impedovo et al. also addressed this problem by performing a classification study on only a subset of the PaHaW dataset, focusing on the earlier and mild degree of disease severity. They found that classification performance significantly drops when considering this subset, instead of taking into account the entire dataset including the more severe cases. In this work, the authors also showed how a multi-expert approach based on ensembling the different tasks at disposal can provide better results than combining the features coming from each task into a unique high dimensional feature vector.

Gallicchio et al. [94] further explored the application of deep learning techniques to aid PD diagnosis through handwriting by exploiting recurrent neural networks. These networks were used to obtain automatically significant features without human intervention from the time series data of the ParkinsonHW dataset [43].

Mucha et al. proposed a new methodology for the kinematic feature analysis of PD handwriting based on fractional derivatives of arbitrary order. Promising results using this techniques have been reported in Reference [95].

In Reference [96], the author improved the results obtained on the PaHaW dataset [37] by combining more classic features to new velocity-based features. The extended set of features include parameters obtained from the application of the sigma-lognormal model, the Maxwell–Boltzmann distribution, and the Discrete Fourier Transform to the velocity profile of handwriting.

Confirming the findings reported in Reference [21,87], Jerkovic et al. [97] found that in-air and on-surface movement on the tablet tend to be statistically independent and to carry on nonredundant information. The highest prediction accuracy in discriminating patients with PD and atypical Parkinsonism from controls, in fact, resulted from the combination of both in-air and on-surface parameters.

In Reference [98], Loconsole et al. were among the first to use features based on the gyroscope signal obtained by the tablet. Unfortunately, their classification study was based on a very small sample of participants.

Rios-Urrego et al. [99], in addition to using kinematic features, proposed to use geometrical and nonlinear dynamic features. The latter, in particular, was meant to capture the distortions and irregularities of handwriting, which are assumed to increase as the disease advances.

Diaz et al. [25] recently proposed a "dynamically enhanced" handwriting representation which consists of synthetically generated images obtained by exploiting simultaneously static and dynamic properties of handwriting. Specifically, they proposed a static representation that embeds dynamic information based on drawing the points of the samples, instead of linking them, so as to retain temporal/velocity information, and that adds pen-ups in the same way. The new handwriting representation was able to outperform the results obtained by using static and dynamic handwriting separately on the PaHaW dataset.

Ribeiro et al. [55] focused on the analysis of tremor, being one of the most distinctive characteristics of PD. In particular, they proposed to learn temporal information from time-dependent signals collected from handwriting exams by exploiting bidirectional gated recurrent units along with an attention mechanism. These units are a gated mechanism in recurrent neural network architectures. In addition, the authors also introduced the concept of "bag of samplings" as a compact representation of the signals. Experimental results on the NewHandPD dataset compared favorably with the previous literature.

In Reference [53], Ammour et al. proposed to use a clustering method to analyze several factors (i.e., age, intellectual level, frequency of writing per week, etc.), which can intervene in the characterization of the groups under study. Then, by using a semi-supervised approach, the authors developed a model for distinguishing the aspects of handwriting pertaining those factors from those related to pathological conditions. A balanced cohort of healthy subjects and PD patients were involved, and they were asked to copy a given Arabic text. Interestingly, among the features used, the authors also considered measures of pen inclination based on the azimuth and altitude information provided by the tablet. During data analysis, three clusters were observed: one where the pathological factor appeared to be the only discriminating element of the corresponding subpopulation; another cluster with mostly healthy people; and one characterized by a mixture of elderly controls (ECs) with medium intellectual level and PD patients with high intellectual level and writing frequency. This finding corroborates the hypothesis that education level may act as a resilience mechanism against the deterioration caused by neurodegeneration [16].

### 3.2. Alzheimer's Disease

Similarly to the classification of studies on PD made in the previous subsection, papers focusing on handwriting in AD can be grouped in accordance with two main research questions:

- *Disease insight*: a group of papers examined changes in handwriting of AD and MCI patients to identify patterns of sensorimotor dysfunction associated with the disease;
- *Disease diagnosis*: another group of works applied dynamic handwriting analysis for the purpose to develop a computer-aided diagnosis system.

A schematic overview is provided in Table 4.

It is worth noting that, in contrast to studies focusing on PD, less research effort has been made towards the investigation of AD; moreover, the literature still lacks studies involving the application of dynamic handwriting analysis to support monitoring of disease progression. This is largely due to the absence of effective cures that slow down disease symptoms. However, as MCI patients are at high risk to develop in AD, handwriting changes found in this condition may be used not only for the early disease diagnosis but also to monitor disease progression.

**Table 4.** Summary of studies on AD (EC = elderly controls; YC = young controls; DEP = depressed patients).

| Reference | Groups | Tasks | Main Features | Main Findings |
|---|---|---|---|---|
| *Disease insight* | | | | |
| Slavin et al., 1999 | 16 AD, 16 EC | No-sense word | Kinematic and spatiotemporal | AD patients exhibit less consistent movement than controls |
| Schröter et al., 2003 | 35 AD, 39 MCI, 39 DEP, 40 EC | Circle drawing | Kinematic | AD and MCI patients differ from healthy subjects in automation parameters and regularity of movement sequence |
| Yan et al., 2008 | 9 AD, 9 MCI, 10 EC | Handwriting patterns requiring the coordination of finger and/or wrist | Kinematic and spatiotemporal | Patients show slow and uncoordinated movements |
| Impedovo et al., 2013 | ISUNIBA | | Kinematic | Handwriting in impaired subjects presents strongly irregular velocity profiles |
| Faundez-Zanuy et al., 2014 | Not specified | House drawing | Kinematic and pressure | Group differences are reflected by cognitive impairments than motor ones |
| Yu and Chang, 2016 | 20 AD, 12 MCI, 16 EC | Line and circle drawing | Kinematic | The degree of motor impairment may help identify those at risk for AD |
| *Disease diagnosis* | | | | |
| Werner et al., 2006 | 22 AD, 31 MCI, 41 EC | Functional tasks | Kinematic and spatiotemporal | Handwriting measures, especially those related to in-air movements, are promising for the automatic discrimination |
| Pirlo et al., 2015 | 29 AD, 30 EC | Signature | Sigma-lognormal-based | Signature deterioration can be a disease predictor |
| Garre-Olmo et al., 2017 | 23 AD, 12 MCI, 17 EC | Sentence copying and writing, figure copying, and CDT | Kinematic and pressure | Higher specificity in distinguishing between normal and impaired condition and higher sensitivity in distinguishing between AD and MCI |
| Kawa et al., 2017 | 37 MCI, 37 EC | Single letter and sentence writing | Spatiotemporal | MCI writing is significantly slower than the normal one |
| Müller et al., 2017a; 2017b | 20 AD, 30 MCI, 20 EC | House drawing and CDT | Spatiotemporal | In-air time is a good predictor for disease diagnosis |
| El-Yacoubi et al., 2018 | Different aging conditions | Word and sentence writing | Kinematic and spatiotemporal | Classification based on temporal representations improves |
| Ghaderyan et al., 2018 | 15 AD, 13 MCI, 15 EC | Spiral drawing and word writing | Kinematic and spatiotemporal | Individual variability of handwriting can be mitigated by noise-robust methods such as singular value decomposition |
| Angelillo et al., 2019 | 36 MCI, 29 EC | Attentional matrices | Kinematic and entropy | Digitized attentional tasks are promising for discriminating cognitively impaired individuals from controls |
| Impedovo et al., 2019 | 71 MCI, 34 EC | Standard and nonstandard writing and drawing tasks | Kinematic | An integrated protocol for disease diagnosis based on handwriting is proposed |
| Ishikawa et al., 2019 | 10 AD, 25 MCI, 36 EC | Standard writing and drawing tasks | Kinematic | The usefulness of digitizing neuropsychological tests on the tablet is supported |

Disease Insight

Slavin et al. [34] were among the first to assess handwriting dynamics in patients with dementia of Alzheimer's type by making use of a digitizing tablet. Irrespective of medication or disease severity, patients wrote strokes of significantly less consistent length than controls and were disproportionately impaired by a reduction of visual feedback. Moreover, patients' strokes had a significantly less consistent duration and a significantly less consistent peak velocity than controls. The authors suggested that the more variable performance of patients indicates a degradation of the base motor program and resembles that of Huntington's disease rather than PD. It may indeed reflect frontal rather than basal ganglia dysfunction; thus, it seems that relative movement duration may be useful to differentiate between subcortical dementias (like PD) and cortical dementias (like AD).

In Reference [28], Schröter et al. adopted dynamic handwriting analysis to quantify differences in fine hand motor function in patients with probable AD and MCI compared to depressed patients and controls. All participants were instructed to perform two tasks. The first one consisted in drawing concentric superimposed circles as fast and fluently as possible with the dominant hand; the second task was identical to the first one but, in addition, participants were requested to simultaneously perform a distraction task (pressing a counting device as often as possible) with the nondominant hand. The results obtained showed that kinematic handwriting parameters were effectively related to cognitive status in elderly patients. Patients with MCI and probable AD exhibited a loss of fine motor performance: especially when compared to control subjects, movements of AD patients were significantly less automated, accurate, and regular.

Yan et al. [100] investigated whether the decline in fine motor control and coordination characterizes sensorimotor deficiencies of cognitively impaired patients with AD or MCI. Their findings supported this hypothesis. Specifically, when performing handwriting tasks, movement slowing was associated to MCI and AD. When performing fine movements, the AD patients also showed more jerky movement than the other groups.

In Reference [44], Impedovo et al. investigated the relationship between the delta-lognormal and the sigma-lognormal models [49] and the early signs and symptoms of AD. The previously mentioned dataset ISUNIBA was collected and used to perform the analysis. By looking at the speed profile along the writing process, it was observed that the maximum speed value was almost always regular in healthy subjects; instead, this regularity was strongly reduced for the patients at the beginning of the disease and completely lost in the patients at advanced stages of the disease.

Faundez-Zanuy et al. [101] compared dynamic characteristics of drawing tasks performed by patients with probable AD and controls. Although some pathological drawings looked "normal" if only considering on-surface movements, in-air patterns and pressure appeared quite entangled. Interestingly, pressure and in-air information were significantly different between the groups even when controls were requested to perform the tasks with the nondominant hand. This suggested that the differences between the groups may not reflect physical problems but cognitive ones.

Yu and Chang [102] explored the motor impairments of individuals with probable AD and amnestic MCI through handwriting analysis. The results showed that slowness and irregularity of movement of AD and MCI patients were not present in all the proposed tasks. For example, impairments were not found when drawing straight lines and cursive-connected loops. Instead, AD and MCI participants had more difficulty than the control group when drawing circles. The study mainly provided evidence that MCI is characterized also by motor dysfunction.

### 3.3. Disease Diagnosis

The study by Werner et al. [40] was aimed at examining kinematically the handwriting process of individuals with MCI compared with those with mild Alzheimer's disease and healthy controls; assessing the importance of the kinematic measures for the differentiation of the groups; and assessing characteristics of the handwriting process across different functional tasks. Participants were requested to perform five functional writing tasks, such as copying a phone number and a grocery list. Two underlying assumptions guided the selection of these tasks: they are functional tasks related to the performance of daily activities; moreover, they reflect an increase in difficulty, as they are long and involve cognitive effort. An ANOVA test was used to test group differences across measures (both on-surface and in-air) for each writing task. Furthermore, a discriminant analysis was carried out to determine which features would be the best predictors for classification. The results of the work showed significant differences between the three groups under study in almost all measures, with the MCI group assuming, as expected by the authors, a position between the other two groups. Temporal measures (especially in-air time) were higher in the more cognitively deteriorated groups, while the mean pressure was lower. The results also showed that kinematic measures of the handwriting process, together with cognitive status measures, provide an efficient way to differentiate between the groups, although the classification of MCI was relatively poor. Finally, the writing characteristics of participants in all groups showed that, although measures of velocity and pressure remained stable across the different tasks, the temporal and spatial measures increased as the difficulty of the task increased.

Although this finding might be obvious, it is interesting that the increase was reflected mainly in the in-air movements.

In Reference [39], Pirlo et al. investigated the extent to which the analysis of dynamic features extracted from handwritten signatures can be fruitfully used for the binary classification healthy vs. AD. A signature, in fact, is well-known to convey a huge amount of information related not only to the representation of the name and surname of the signer but also to the writing system (hand, arm, etc.) as well as the psychophysical state. The feature extraction phase was accomplished in accordance with the sigma-lognormal model of the Kinematic Theory of Rapid Human Movements. The best classification performance was obtained by using a Bagging CART (classification and regression tree) classifier. It is worth noting that some contrasting result was obtained by Renier et al. [103], who found no significant correlation between signature deterioration and level of cognitive decline.

Garre-Olmo et al. [31] compared the dynamic characteristics of handwriting and drawing between patients with probable AD and MCI and healthy controls. Participants were asked to copy one sentence, to write a dictated sentence and an own sentence, to copy two and-three dimensions drawings, and to execute the Clock Drawing Test. By means of discriminant analyses, the authors explored the value of several kinematic features in order to classify participants depending on their degree of cognitive functioning. The degree of correct classification was dependent on the nature of the groups to be classified and the specific task. Classification performance showed higher specificity values when distinguishing between normal and impaired cognition (MCI and AD) and higher sensitivity when distinguishing between impaired cognition levels (MCI and AD). Interestingly, the results obtained showed that, for the same task, the discriminant parameters differed depending on the type of group to be discriminated, suggesting that they are not the dimensional features of the parameters but rather the qualitative combination of these parameters that are relevant for group discrimination.

Kawa et al. [104] evaluated the usefulness of handwriting features obtained with an electronic pen to distinguish MCI patients from controls. Subjects with confirmed MCI needed more time to complete two out of three writing tasks, as their writing was significantly slower. These results were associated with a longer time to complete a single stroke of written text. The written text was also noticeably larger in the MCI group in all three tasks.

Müller et al. [30] investigated movement kinematics between patients with early dementia due to probable AD, patients with amnestic MCI, and cognitively healthy control individuals while copying a three-dimensional house using a digitizing tablet. Receiver operating characteristic (ROC) curves and logistic regression analyses were conducted to explore whether alterations in movement kinematics could be used to discriminate patients with MCI and AD from controls. In-air time differed significantly between the three groups, showing an excellent sensitivity and a moderate specificity to discriminate MCI subjects from normal elderly and an excellent sensitivity and specificity to discriminate patients affected by mild AD from healthy individuals. On-surface time differed only between controls and patients with AD but not between controls and patients with MCI. Furthermore, the total time (i.e., in-air plus on-surface time) did not differ between patients with MCI and early dementia due to AD.

In Reference [105], Müller et al. reported the results of an experiment analogous to the previous one, employing the same apparatus and the same participants. What differed from the previous study was the task the participants were requested to perform, which consisted in a digitized version of the classic Clock Drawing Test. While the traditional CDT revealed only poor sensitivity but excellent specificity in discriminating MCI patients from healthy individuals, excellent sensitivity and a good specificity were obtained in discriminating these groups when considering the digital version of the test. In Reference [106], this research was extended by comparing the digital Clock Drawing Test with the traditional Consortium to Establish a Registry for Alzheimer's Disease (CERAD) neuropsychological test score. Digital Clock Drawing Test (dCDT) provided a slightly better diagnostic accuracy for the discrimination of amnestic MCI from controls than using the CERAD score. Instead, in differentiating patients with mild AD from controls, both tests provided excellent accuracy.

El-Yacoubi et al. [107] proposed a novel paradigm for studying handwriting changes due to cognitive decline (or aging) by addressing the major limitations of the state-of-the-art solutions. The first one is the assumption of a unique behavioural trend for each cognitive profile. This restriction was relaxed by allowing, for each profile, the emergence of a multi-modal behavioural pattern reflecting the diversity of behaviours within a given healthy condition. The authors achieved this by using unsupervised or semi-supervised learning algorithms to uncover homogeneous groups of subjects and byanalyzing how much information these clusters carry on about the cognitive profiles. The second main limitation is the encoding of handwriting spatiotemporal dynamics only by using global or semi-global parameters, assumed implicitly to be discriminant. The proposed method is based on a representation learning approach which is suitable for treating sequential data from which temporal feature representations can be uncovered. A key advantage of this temporal representation learning is that it is fully explainable, as it can be visualized and easily understood. The main finding of the work, from a diagnostic perspective, is that MCI individuals tend to form clusters either with controls or AD patients, revealing that MCI individuals have fine motor skills with characteristics from both the two extreme groups.

An approach similar to the previous one was followed by the authors in Reference [108]. In particular, they modeled the velocity trajectory of loop-like movements through a temporal clustering based on dynamic time warping as a dissimilarity measure. For classification, the authors used a Bayesian framework, which aggregates the contributions of the clusters by combining the discriminating power of each oh them probabilistically.

In Reference [109], Ghaderyan et al. pointed out how shape and writing style differences among individuals are sources of undesired variability in the handwriting signals which may affect the recognition performance. In order to mitigate this effect, the authors proposed a noise-robust method based on the singular value decomposition and a sparse nonnegative least-square classifier to make the handwritten patterns less dependent on small individual variations.

In Reference [42], Angelillo et al. proposed a digitized version of the Attentional Matrices Test for selective attention assessment: it is based on three matrices of increasing difficulty, and the subject is asked to the mark target digits assigned. The authors observed how, although a pathological matrix may look "normal" if considering only the on-surface pattern, the information provided by in-air movements reveal a completely deteriorated search strategy of the targets to be marked among the distractors. An ensemble built over three different classifiers trained on the matrices separately provided the best classification performance.

In Reference [32], Impedovo et al. proposed a handwriting-based protocol for screening and follow-up of dementia based on a digitized version of standardized cognitive and functional tests (such as Mini-Cog and MMSE), together with handwriting and drawing tasks currently under investigation by researchers. The proposed protocol achieved good specificity in distinguishing MCI patients and controls. A similar work has been recently presented in Reference [110], where Ishikawa et al. proposed to use a digitized version of neuropsychological tests developed on a digitizing tablet to learn to distinguish between AD, MCI, and EC subjects.

## 4. Conclusions and Future Directions

The body of evidence on computerized handwriting analysis supports the hypothesis that physical, cognitive, and psychological characteristics of individuals can be captured by handwriting measures. In particular, changes in handwriting seem to be a prominent *biomarker* for the evaluation of neurodegenerative diseases. Several works, in fact, provided evidence that the automatic discrimination between unhealthy and healthy people can be accomplished on the basis of features obtained through simple and easy-to-perform handwriting tasks. In this view, as the number of devices for data capturing and processing is increasing all over the world, the use of handwriting to detect and monitor health conditions is becoming more and more attractive. In particular, the advent of digitizing tablets and electronic pens allow researchers to investigate not only the static characteristics of handwriting available only after the writing process has occurred but also dynamic characteristics collected when the handwriting task is still in progress. Based on dynamic analyses, several works provided evidence that certain aspects of the handwriting process are more vulnerable than others and may therefore present diagnostic signs.

A handwriting-based decision support system has the potential to assist clinicians at the point of care, providing a novel diagnostic tool while reducing the expenditure of public health care. Moreover, it can be used to quantify aspects of the motor system and its disorders in order to better understand the involved underlying mechanisms, e.g., the difficulties in coordinating the components of a motor sequence movement. Finally, it can help study the effects of medication on handwriting with the aim to monitor the responsiveness of the patient to therapy. More in general, handwriting can provide a simple, user-friendly, and easy-to-use instrument to support the daily clinical trials. Artificial intelligence and machine learning, in fact, are changing the way we think of health care from many perspectives, and the use of computer aided tools within medical practice is continuously increasing. The best perspective of this line of research is the integration of new medical tools which can increase the level of diagnostic accuracy. Doctors can be provided with user-friendly tools in their daily practice, even though they are not necessarily skilled in high-level algorithms. In this sense, a handwriting-based tool is attractive since it not only provides the user with an automatic response within few seconds but also allows the doctor to store useful metadata, e.g., the patient's information and diagnosis, for later use. Of course, handwriting-based intelligent systems are not expected to replace standard techniques or even doctors but rather to provide additional evidence to further support the clinical assessment. Intelligent systems technology is proving beneficial in a number of health domains, also including neuroimaging [111] and cardiovascular risk assessment [112].

The present paper has been devoted to provide a comprehensive overview of the literature dealing with the application of dynamic handwriting analysis to the assessment of Parkinson's disease and Alzheimer's disease. Three well-defined research trends have been identified, ranging from studies aimed at understanding the facets of fine motor control related to the disease to works investigating the application of dynamic handwriting features to disease monitoring and diagnosis. Handwriting features in PD and AD patients can overlap; in particular, handwriting is typically slower, less regular, and less consistent in patients if compared to the healthy counterpart. However, the two diseases also show distinctive characteristics. PD patients tend to exhibit rigidity of movements and unwanted muscle contractions, while preserving, in most cases, their cognitive faculties. AD and MCI patients, on the other hand, tend to exhibit a more pronounced alteration of their visuospatial abilities and executive functions, while preserving their fine motor control. These features can be reflected on the production of handwritten patterns, especially in more complex handwriting tasks.

Encouraging results have been so far obtained; however, there still remain open issues demanding further research. First, interoperability is still a problem, since data are typically obtained from different devices and different handwriting tasks. Some works, e.g., References [29,32,87], provided the community with tentative hand-drawing/handwriting protocols for the assessment of PD and AD. An integrated protocol, as the one proposed in Reference [32], may be useful to the research community to collect different handwritten traits; at the same time, it may be of real use for doctors to support their daily activities.

Another important problem is related to the collection of a statistically significant large amount of samples. The research community still lacks a large benchmark dataset so that different tools, algorithms, and techniques can be effectively evaluated and compared. The few datasets currently available are composed by very few subjects and do not always consider other factors such as the stage of disease, the medical treatment, and so on. Unfortunately, collecting a large benchmark database is a time-consuming and expensive process. Furthermore, as neurodegeneration evolves during time, such a dataset should longitudinally follow a same group over several years.

Senatore et al., in Reference [113], recently raised another important issue: the acceptance of artificial intelligence-based diagnosis systems by physicians may be hampered by the black-box approach implemented by most state-of-art systems. To address this problem, the authors proposed an evolutionary approach based on Cartesian Genetic Programming which allows for the automatic detection of the presence of disease and, simultaneously, provides the explicit classification rules used by the system. This approach can allow physicians to derive guidelines that may be used to define novel testing protocols and intervention strategies. Also classic decision trees, which are capable of providing the decision criteria in terms of both the most relevant features and how their values are used to reach the final decision, can be used for the purpose [114]. Explainable AI can be therefore a prominent research direction within this context.

The lessons learned from the mentioned body of evidence can be profitably used for other health and psychological domains. For example, in Reference [115], handwriting was employed with successful results to the problem of recognizing malingering in health care, i.e., the false information given by patients about their health. Preliminary results suggested that a computerized tool based on handwriting can help detect deception. A similar tool was used in Reference [116] for capturing cognitive load implications during complex figure drawing. The work was then extended in Reference [117], where the contribution of handwriting to classifying cognitive mental workload was assessed. Also, handwriting-based measures recently showed promising results in investigating emotional states, such as stress, depression, and anxiety [118].

It is worth noting that tablet technology enables the implementation of a multi-modal interaction system in which not only the input provided by the electronic pen but also tactile, speech, or visual input can be acquired. Promising results on noninvasive methods based on speech and handwriting analysis for neurological disorder assessment have been obtained, e.g., Reference [119]. Moreover, thanks to such a multimodal interface, the development of a mobile conversational agent appears to be feasible. An example of a mobile conversational agent successfully used in the context of AD has been recently reported in Reference [120]. A purely automatic diagnostic tool paws the way of a quick instrument which enables mass screening of the population or even home training for improving cognitive abilities.

Finally, another open issue is a technological one and is related to the realization of an all-in-one solution specifically devised for the automatic differential diagnosis.

Some other aspects are related specifically to AD assessment. The most obvious observation that can be drawn from the present survey is that less research effort has been done for AD; thus, additional effort is necessary for advancing this line of research. Concerning disease insight, investigating the neural process that underlies handwriting may provide further criteria for selecting the most representative features associated to a writer, i.e., those containing more information about the message the handwriting measures represent and the way to examine them profitably. This goal may be achieved through a multidisciplinary approach, which involves both the analysis/comparison of handwritten data provided by healthy subjects and patients affected by AD, and through the analysis of the behavior of a neural network model that emulates the neural mechanisms occurring in the brain areas involved in handwriting generation and learning. Understanding the neural process involved in learning complex motor behaviors could also provide a meaningful help to the development of devices and important insights in developing more effective treatments for the motor deficits affecting AD patients.

In the context of AD diagnosis, the challenge appears to be to separate from healthy people the mild cognitive impairment subjects, which are likely to evolve in AD. MCI is characterized by slight problems with memory loss, language, or other mental functions; thus, finding deterministic patterns useful to discriminate the impairment from a non-pathological condition is very difficult: this is well-known to the research community working on neuroimaging. Some works, for example, References [40,104], clearly indicated that such difficulty is also reflected on dynamic handwriting analysis.

**Funding:** This research received no external funding.

**Acknowledgments:** The author is thankful to the anonymous reviewers for their valuable suggestions.

**Conflicts of Interest:** The author declares no conflict of interest.

## References


1. Cookson, M.R. Parkinson's disease. In *Disease-Modifying Targets in Neurodegenerative Disorders*; Elsevier: Amsterdam, The Netherlands, 2017; pp. 157–174.
2. Alzheimer's Association. 2018 Alzheimer's disease facts and figures. *Alzheimers Dement.* **2018**, *14*, 367–429. [CrossRef]
3. Petersen, R.C. Mild cognitive impairment. *Contin. Lifelong Learn. Neurol.* **2016**, *22*, 404. [CrossRef] [PubMed]
4. Arevalo-Rodriguez, I.; Smailagic, N.; i Figuls, M.R.; Ciapponi, A.; Sanchez-Perez, E.; Giannakou, A.; Pedraza, O.L.; Cosp, X.B.; Cullum, S. Mini-Mental State Examination (MMSE) for the detection of Alzheimer's disease and other dementias in people with mild cognitive impairment (MCI). *Cochrane Database Syst. Rev.* **2015**, *3*, CD010783.
5. Lella, E.; Amoroso, N.; Diacono, D.; Lombardi, A.; Maggipinto, T.; Monaco, A.; Bellotti, R.; Tangaro, S. Communicability Characterization of Structural DWI Subcortical Networks in Alzheimer's Disease. *Entropy* **2019**, *21*, 475. [CrossRef]
6. Margiotta, N.; Avitabile, G.; Coviello, G. A wearable wireless system for gait analysis for early diagnosis of Alzheimer and Parkinson disease. In Proceedings of the 5th International Conference on Electronic Devices, Systems and Applications (ICEDSA), Ras Al Khaimah, UAE, 6–8 December 2016; pp. 1–4.
7. Plamondon, R.; O'reilly, C.; Galbally, J.; Almaksour, A.; Anquetil, É. Recent developments in the study of rapid human movements with the kinematic theory: Applications to handwriting and signature synthesis. *Pattern Recognit. Lett.* **2014**, *35*, 225–235. [CrossRef]
8. Vyhnálek, M.; Rubínová, E.; Marková, H.; Nikolai, T.; Laczó, J.; Andel, R.; Hort, J. Clock drawing test in screening for Alzheimer's dementia and mild cognitive impairment in clinical practice. *Int. J. Geriatr. Psychiatry* **2017**, *32*, 933–939. [CrossRef]
9. McLennan, J.; Nakano, K.; Tyler, H.; Schwab, R. Micrographia in Parkinson's disease. *J. Neurol. Sci.* **1972**, *15*, 141–152. [CrossRef]
10. Platel, H.; Lambert, J.; Eustache, F.; Cadet, B.; Dary, M.; Viader, F.; Lechevalier, B. Characteristics and evolution of writing impairmant in Alzheimer's disease. *Neuropsychologia* **1993**, *31*, 1147–1158. [CrossRef]
11. Onofri, E.; Mercuri, M.; Salesi, M.; Ferrara, S.; Troili, G.M.; Simeone, C.; Ricciardi, M.R.; Ricci, S.; Archer, T. Dysgraphia in relation to cognitive performance in patients with Alzheimer's disease. *J. Intellect. Disabil.-Diagn. Treat.* **2013**, *1*, 113–124.
12. Neils-Strunjas, J.; Groves-Wright, K.; Mashima, P.; Harnish, S. Dysgraphia in Alzheimer's disease: A review for clinical and research purposes. *J. Speech, Lang. Hear. Res.* **2006**, *49*, 1313–1330. [CrossRef]
13. Letanneux, A.; Danna, J.; Velay, J.L.; Viallet, F.; Pinto, S. From micrographia to Parkinson's disease dysgraphia. *Mov. Disord.* **2014**, *29*, 1467–1475. [CrossRef] [PubMed]
14. De Stefano, C.; Fontanella, F.; Impedovo, D.; Pirlo, G.; di Freca, A.S. Handwriting analysis to support neurodegenerative diseases diagnosis: A review. *Pattern Recognit. Lett.* **2019**, *121*, 37–45. [CrossRef]
15. Impedovo, D.; Pirlo, G. Dynamic handwriting analysis for the assessment of neurodegenerative diseases: A pattern recognition perspective. *IEEE Rev. Biomed. Eng.* **2018**, *12*, 209–220. [CrossRef] [PubMed]
16. Angelillo, M.T.; Impedovo, D.; Pirlo, G.; Vessio, G. Handwriting dynamics as an indicator of cognitive reserve: An exploratory study. In Proceedings of the 2019 IEEE International Conference on Systems, Man, and Cybernetics, Bari, Italy, 6–9 October 2019.
17. Eichhorn, T.; Gasser, T.; Mai, N.; Marquardt, C.; Arnold, G.; Schwarz, J.; Oertel, W. Computational analysis of open loop handwriting movements in Parkinson's disease: A rapid method to detect dopamimetic effects. *Mov. Disord. Off. J. Mov. Disord. Soc.* **1996**, *11*, 289–297. [CrossRef]
18. Tucha, O.; Mecklinger, L.; Thome, J.; Reiter, A.; Alders, G.; Sartor, H.; Naumann, M.; Lange, K.W. Kinematic analysis of dopaminergic effects on skilled handwriting movements in Parkinson's disease. *J. Neural Transm.* **2006**, *113*, 609–623. [CrossRef]
19. Martínez-Martín, P.; Rodríguez-Blázquez, C.; Alvarez, M.; Arakaki, T.; Arillo, V.C.; Chaná, P.; Fernández, W.; Garretto, N.; Martínez-Castrillo, J.C.; Rodríguez-Violante, M.; et al. Parkinson's disease severity levels and MDS-Unified Parkinson's Disease Rating Scale. *Park. Relat. Disord.* **2015**, *21*, 50–54. [CrossRef]

20. Reitan, R.M. Validity of the Trail Making Test as an indicator of organic brain damage. *Percept. Mot. Ski.* **1958**, *8*, 271–276. [CrossRef]
21. Rosenblum, S.; Samuel, M.; Zlotnik, S.; Erikh, I.; Schlesinger, I. Handwriting as an objective tool for Parkinson's disease diagnosis. *J. Neurol.* **2013**, *260*, 2357–2361. [CrossRef]
22. Ünlü, A.; Brause, R.; Krakow, K. Handwriting analysis for diagnosis and prognosis of Parkinson's disease. In *Proceedings of the International Symposium on Biological and Medical Data Analysis, Thessaloniki, Greece, 7–8 December 2006*; Springer: Berlin, Germany, 2006; pp. 441–450.
23. Pereira, C.R.; Weber, S.A.; Hook, C.; Rosa, G.H.; Papa, J.P. Deep Learning-Aided Parkinson. In Proceedings of the 29th SIBGRAPI Conference on Graphics, Patterns and Images, Sao Jose, Brazil, 4–7 October 2016; pp. 340–346.
24. Impedovo, D.; Pirlo, G.; Vessio, G. Dynamic handwriting analysis for supporting earlier Parkinson's disease diagnosis. *Information* **2018**, *9*, 247. [CrossRef]
25. Diaz, M.; Ferrer, M.A.; Impedovo, D.; Pirlo, G.; Vessio, G. Dynamically enhanced static handwriting representation for Parkinson's disease detection. *Pattern Recognit. Lett.* **2019**, *128*, 204–210. [CrossRef]
26. Angelillo, M.T.; Impedovo, D.; Pirlo, G.; Vessio, G. Performance-driven Handwriting Task Selection for Parkinson's Disease Classification. In *Proceedings of the 18th International Conference of the Italian Association for Artificial Intelligence, Rende, Italy, 19–22 November 2019*; Springer: Berlin, Germany, 2019.
27. Pullman, S.L. Spiral analysis: A new technique for measuring tremor with a digitizing tablet. *Mov. Disord.* **1998**, *13*, 85–89. [CrossRef] [PubMed]
28. Schröter, A.; Mergl, R.; Bürger, K.; Hampel, H.; Möller, H.J.; Hegerl, U. Kinematic analysis of handwriting movements in patients with Alzheimer's disease, mild cognitive impairment, depression and healthy subjects. *Dement. Geriatr. Cogn. Disord.* **2003**, *15*, 132–142. [CrossRef] [PubMed]
29. San Luciano, M.; Wang, C.; Ortega, R.A.; Yu, Q.; Boschung, S.; Soto-Valencia, J.; Bressman, S.B.; Lipton, R.B.; Pullman, S.; Saunders-Pullman, R. Digitized spiral drawing: A possible biomarker for early Parkinson's disease. *PLoS ONE* **2016**, *11*, e0162799. [CrossRef] [PubMed]
30. Müller, S.; Preische, O.; Heymann, P.; Elbing, U.; Laske, C. Diagnostic value of a tablet-based drawing task for discrimination of patients in the early course of Alzheimer's disease from healthy individuals. *J. Alzheimers Dis.* **2017**, *55*, 1463–1469. [CrossRef] [PubMed]
31. Garre-Olmo, J.; Faúndez-Zanuy, M.; López-de Ipiña, K.; Calvó-Perxas, L.; Turró-Garriga, O. Kinematic and pressure features of handwriting and drawing: preliminary results between patients with mild cognitive impairment, Alzheimer disease and healthy controls. *Curr. Alzheimer Res.* **2017**, *14*, 960–968. [CrossRef] [PubMed]
32. Impedovo, D.; Pirlo, G.; Vessio, G.; Angelillo, M.T. A Handwriting-Based Protocol for Assessing Neurodegenerative Dementia. *Cogn. Comput.* **2019**, *11*, 576–586. [CrossRef]
33. Teulings, H.L.; Contreras-Vidal, J.L.; Stelmach, G.E.; Adler, C.H. Parkinsonism reduces coordination of fingers, wrist, and arm in fine motor control. *Exp. Neurol.* **1997**, *146*, 159–170. [CrossRef]
34. Slavin, M.J.; Phillips, J.G.; Bradshaw, J.L.; Hall, K.A.; Presnell, I. Consistency of handwriting movements in dementia of the Alzheimer's type: A comparison with Huntington's and Parkinson's diseases. *J. Int. Neuropsychol. Soc.* **1999**, *5*, 20–25. [CrossRef]
35. Bidet-Ildei, C.; Pollak, P.; Kandel, S.; Fraix, V.; Orliaguet, J.P. Handwriting in patients with Parkinson disease: Effect of L-dopa and stimulation of the sub-thalamic nucleus on motor anticipation. *Hum. Mov. Sci.* **2011**, *30*, 783–791. [CrossRef]
36. Smits, E.J.; Tolonen, A.J.; Cluitmans, L.; van Gils, M.; Conway, B.A.; Zietsma, R.C.; Leenders, K.L.; Maurits, N.M. Standardized handwriting to assess bradykinesia, micrographia and tremor in Parkinson's disease. *PLoS ONE* **2014**, *9*, e97614. [CrossRef]
37. Drotár, P.; Mekyska, J.; Rektorová, I.; Masarová, L.; Smékal, Z.; Faundez-Zanuy, M. Decision support framework for Parkinson's disease based on novel handwriting markers. *IEEE Trans. Neural Syst. Rehabil. Eng.* **2015**, *23*, 508–516. [CrossRef] [PubMed]
38. Drotár, P.; Mekyska, J.; Rektorová, I.; Masarová, L.; Smékal, Z.; Faundez-Zanuy, M. Evaluation of handwriting kinematics and pressure for differential diagnosis of Parkinson's disease. *Artif. Intell. Med.* **2016**, *67*, 39–46. [CrossRef] [PubMed]

39. Pirlo, G.; Diaz, M.; Ferrer, M.A.; Impedovo, D.; Occhionero, F.; Zurlo, U. Early diagnosis of neurodegenerative diseases by handwritten signature analysis. In *Proceedings of the International Conference on Image Analysis and Processing, Genova, Italy, 7–11 September 2015*; Springer: Berlin, Germany, 2015; pp. 290–297.
40. Werner, P.; Rosenblum, S.; Bar-On, G.; Heinik, J.; Korczyn, A. Handwriting process variables discriminating mild Alzheimer's disease and mild cognitive impairment. *J. Gerontol. Ser. Psychol. Sci. Soc. Sci.* **2006**, *61*, P228–P236. [CrossRef] [PubMed]
41. Dahmen, J.; Cook, D.; Fellows, R.; Schmitter-Edgecombe, M. An analysis of a digital variant of the Trail Making Test using machine learning techniques. *Technol. Health Care* **2017**, *25*, 251–264. [CrossRef]
42. Angelillo, M.T.; Balducci, F.; Impedovo, D.; Pirlo, G.; Vessio, G. Attentional Pattern Classification for Automatic Dementia Detection. *IEEE Access* **2019**, *7*, 57706–57716. [CrossRef]
43. Isenkul, M.; Sakar, B.; Kursun, O. Improved spiral test using digitized graphics tablet for monitoring Parkinson's disease. In Proceedings of the International Conference on e-Health and Telemedicine, South Wales, UK, 10–12 November 2014; pp. 171–175.
44. Impedovo, D.; Pirlo, G.; Mangini, F.M.; Barbuzzi, D.; Rollo, A.; Balestrucci, A.; Impedovo, S.; Sarcinella, L.; O'Reilly, C.; Plamondon, R. Writing generation model for health care neuromuscular system investigation. In *Proceedings of the International Meeting on Computational Intelligence Methods for Bioinformatics and Biostatistics, Nice, France, 17–22 June 2013*; Springer: Berlin, Germany, 2013; pp. 137–148.
45. Castrillon, R.; Acien, A.; Orozco-Arroyave, J.R.; Morales, A.; Vargas, J.; Vera-Rodrıguez, R.; Fiérrez, J.; Ortega-Garcia, J.; Villegas, A. Characterization of the Handwriting Skills as a Biomarker for Parkinson Disease. In Proceedings of the 14th IEEE International Conference on Automatic Face and Gesture Recognition (FG 2019), Lille, France, 14–18 May 2019; pp. 1–5.
46. Cilia, N.D.; De Stefano, C.; Fontanella, F.; Di Freca, A.S. An Experimental Protocol to Support Cognitive Impairment Diagnosis by using Handwriting Analysis. *Procedia Comput. Sci.* **2018**, *141*, 466–471. [CrossRef]
47. Plamondon, R. A kinematic theory of rapid human movements. *Biol. Cybern.* **1995**, *72*, 295–307. [CrossRef]
48. Plamondon, R. A kinematic theory of rapid human movements. Part II. Movement time and control. *Biol. Cybern.* **1995**, *72*, 309–320. [CrossRef]
49. O'Reilly, C.; Plamondon, R. Development of a Sigma–Lognormal representation for on-line signatures. *Pattern Recognit.* **2009**, *42*, 3324–3337. [CrossRef]
50. Fischer, A.; Plamondon, R. A dissimilarity measure for on-line signature verification based on the sigma-lognormal model. In Proceedings of the 17th Biennial Conference of the International Graphonomics Society, Pointe-à-Pitre, France, 21–24 June 2015.
51. Duval, T.; Rémi, C.; Plamondon, R.; Vaillant, J.; O'Reilly, C. Combining sigma-lognormal modeling and classical features for analyzing graphomotor performances in kindergarten children. *Hum. Mov. Sci.* **2015**, *43*, 183–200. [CrossRef]
52. Afonso, L.C.; Rosa, G.H.; Pereira, C.R.; Weber, S.A.; Hook, C.; Albuquerque, V.H.C.; Papa, J.P. A recurrence plot-based approach for Parkinson's disease identification. *Future Gener. Comput. Syst.* **2019**, *94*, 282–292.
53. Ammour, A.; Aouraghe, I.; Khaissidi, G.; Mrabti, M.; Aboulem, G.; Belahsen, F. A new semi-supervised approach for characterizing the Arabic on-line handwriting of Parkinson's disease patients. *Comput. Methods Programs Biomed.* **2020**, *183*, 104979. [CrossRef] [PubMed]
54. Murphy, K.P. *Machine Learning: A Probabilistic Perspective*; MIT Press: Cambridge, MA, USA, 2012.
55. Ribeiro, L.C.; Afonso, L.C.; Papa, J.P. Bag of Samplings for computer-assisted Parkinson's disease diagnosis based on Recurrent Neural Networks. *Comput. Biol. Med.* **2019**, *115*, 103477. [CrossRef] [PubMed]
56. James, G.; Witten, D.; Hastie, T.; Tibshirani, R. *An Introduction to Statistical Learning*; Springer: Berlin, Germany, 2013; Volume 112.
57. Phillips, J.; Stelmach, G.E.; Teasdale, N. What can indices of handwriting quality tell us about Parkinsonian handwriting? *Hum. Mov. Sci.* **1991**, *10*, 301–314. [CrossRef]
58. Teulings, H.L.; Stelmach, G.E. Control of stroke size, peak acceleration, and stroke duration in Parkinsonian handwriting. *Hum. Mov. Sci.* **1991**, *10*, 315–334. [CrossRef]
59. Contreras-Vidal, J.L.; Stelmach, G.E. A neural model of basal ganglia-thalamocortical relations in normal and parkinsonian movement. *Biol. Cybern.* **1995**, *73*, 467–476. [CrossRef]

60. Van Gemmert, A.; Teulings, H.L.; Contreras-Vidal, J.L.; Stelmach, G. Parkinson's disease and the control of size and speed in handwriting. *Neuropsychologia* **1999**, *37*, 685–694. [CrossRef]
61. Fucetola, R.; Smith, M.C. Distorted visual feedback effects on drawing in Parkinson's disease. *Acta Psychol.* **1997**, *95*, 255–266. [CrossRef]
62. Oliveira, R.M.; Gurd, J.M.; Nixon, P.; Marshall, J.C.; Passingham, R.E. Micrographia in Parkinson's disease: The effect of providing external cues. *J. Neurol. Neurosurg. Psychiatry* **1997**, *63*, 429–433. [CrossRef]
63. Van Gemmert, A.W.; Teulings, H.L.; Stelmach, G.E. The influence of mental and motor load on handwriting movements in Parkinsonian patients. *Acta Psychol.* **1998**, *100*, 161–175. [CrossRef]
64. Swinnen, S.P.; Steyvers, M.; Van Den Bergh, L.; Stelmach, G.E. Motor learning and Parkinson's disease: Refinement of within-limb and between-limb coordination as a result of practice. *Behav. Brain Res.* **2000**, *111*, 45–59. [CrossRef]
65. Van Gemmert, A.W.; Teulings, H.L.; Stelmach, G.E. Parkinsonian patients reduce their stroke size with increased processing demands. *Brain Cogn.* **2001**, *47*, 504–512. [CrossRef] [PubMed]
66. Teulings, H.; Contreras-Vidal, J.L.; Stelmach, G.; Adler, C.H. Adaptation of handwriting size under distorted visual feedback in patients with Parkinson's disease and elderly and young controls. *J. Neurol. Neurosurg. Psychiatry* **2002**, *72*, 315–324. [CrossRef] [PubMed]
67. Van Gemmert, A.; Adler, C.H.; Stelmach, G. Parkinson's disease patients undershoot target size in handwriting and similar tasks. *J. Neurol. Neurosurg. Psychiatry* **2003**, *74*, 1502–1508. [CrossRef] [PubMed]
68. Caligiuri, M.P.; Teulings, H.L.; Filoteo, J.V.; Song, D.; Lohr, J.B. Quantitative measurement of handwriting in the assessment of drug-induced parkinsonism. *Hum. Mov. Sci.* **2006**, *25*, 510–522. [CrossRef] [PubMed]
69. Ponsen, M.M.; Daffertshofer, A.; Wolters, E.C.; Beek, P.J.; Berendse, H.W. Impairment of complex upper limb motor function in de novo Parkinson's disease. *Park. Relat. Disord.* **2008**, *14*, 199–204. [CrossRef] [PubMed]
70. Broderick, M.P.; Van Gemmert, A.W.; Shill, H.A.; Stelmach, G.E. Hypometria and bradykinesia during drawing movements in individuals with Parkinson's disease. *Exp. Brain Res.* **2009**, *197*, 223–233. [CrossRef]
71. Dounskaia, N.; Van Gemmert, A.W.; Leis, B.C.; Stelmach, G.E. Biased wrist and finger coordination in Parkinsonian patients during performance of graphical tasks. *Neuropsychologia* **2009**, *47*, 2504–2514. [CrossRef]
72. Gangadhar, G.; Joseph, D.; Srinivasan, A.; Subramanian, D.; Shivakeshavan, R.; Shobana, N.; Chakravarthy, V. A computational model of Parkinsonian handwriting that highlights the role of the indirect pathway in the basal ganglia. *Hum. Mov. Sci.* **2009**, *28*, 602–618. [CrossRef]
73. Ma, H.I.; Hwang, W.J.; Chang, S.H.; Wang, T.Y. Progressive micrographia shown in horizontal, but not vertical, writing in Parkinson's disease. *Behav. Neurol.* **2013**, *27*, 169–174. [CrossRef]
74. Broeder, S.; Nackaerts, E.; Nieuwboer, A.; Smits-Engelsman, B.C.; Swinnen, S.P.; Heremans, E. The effects of dual tasking on handwriting in patients with Parkinson's disease. *Neuroscience* **2014**, *263*, 193–202. [CrossRef] [PubMed]
75. Senatore, R.; Marcelli, A. A paradigm for emulating the early learning stage of handwriting: Performance comparison between healthy controls and Parkinson's disease patients in drawing loop shapes. *Hum. Mov. Sci.* **2019**, *65*, 89–101. [CrossRef] [PubMed]
76. Contreras-Vidal, J.L.; Poluha, P.; Teulings, H.L.; Stelmach, G.E. Neural dynamics of short and medium-term motor control effects of levodopa therapy in Parkinson's disease. *Artif. Intell. Med.* **1998**, *13*, 57–79. [CrossRef]
77. Poluha, P.; Teulings, H.L.; Brookshire, R. Handwriting and speech changes across the levodopa cycle in Parkinson's disease. *Acta Psychol.* **1998**, *100*, 71–84. [CrossRef]
78. Siebner, H.R.; Ceballos-Baumann, A.; Standhardt, H.; Auer, C.; Conrad, B.; Alesch, F. Changes in handwriting resulting from bilateral high-frequency stimulation of the subthalamic nucleus in Parkinson's disease. *Mov. Disord. Off. J. Mov. Disord. Soc.* **1999**, *14*, 964–971. [CrossRef]
79. Cobbah, W.; Fairhurst, M.C. Computer analysis of handwriting dynamics during dopamimetic tests in Parkinson's disease. In Proceedings of the 26th Euromicro Conference, EUROMICRO 2000, Informatics: Inventing the Future, Maastricht, The Netherlands, 5–7 Septemeber 2000; Volume 2, pp. 414–418.
80. Boylan, L.; Pullman, S.; Lisanby, S.; Spicknall, K.; Sackeim, H. Repetitive transcranial magnetic stimulation to SMA worsens complex movements in Parkinson's disease. *Clin. Neurophysiol.* **2001**, *112*, 259–264. [CrossRef]
81. Lange, K.W.; Mecklinger, L.; Walitza, S.; Becker, G.; Gerlach, M.; Naumann, M.; Tucha, O. Brain dopamine and kinematics of graphomotor functions. *Hum. Mov. Sci.* **2006**, *25*, 492–509. [CrossRef]

82. Randhawa, B.K.; Farley, B.G.; Boyd, L.A. Repetitive transcranial magnetic stimulation improves handwriting in Parkinson's disease. *Park. Dis.* **2013**, *2013*, 1–9. [CrossRef]
83. Smits, E.J.; Tolonen, A.J.; Cluitmans, L.; Van Gils, M.; Zietsma, R.C.; Borgemeester, R.W.; van Laar, T.; Maurits, N.M. Graphical tasks to measure upper limb function in patients with Parkinson's disease: Validity and response to dopaminergic medication. *IEEE J. Biomed. Health Inform.* **2015**, *21*, 283–289. [CrossRef]
84. Danna, J.; Velay, J.L.; Eusebio, A.; Véron-Delor, L.; Witjas, T.; Azulay, J.P.; Pinto, S. Digitalized spiral drawing in Parkinson's disease: A tool for evaluating beyond the written trace. *Hum. Mov. Sci.* **2019**, *65*, 80–88. [CrossRef]
85. Drotár, P.; Mekyska, J.; Smékal, Z.; Rektorová, I.; Masarová, L.; Faundez-Zanuy, M. Prediction potential of different handwriting tasks for diagnosis of Parkinson's. In Proceedings of the 2013 E-Health and Bioengineering Conference (EHB), Piscataway, NJ, USA, 21–23 November 2013; pp. 1–4.
86. Drotár, P.; Mekyska, J.; Rektorová, I.; Masarová, L.; Smékal, Z.; Faundez-Zanuy, M. A new modality for quantitative evaluation of Parkinson's disease: In-air movement. In Proceedings of the 13th IEEE International Conference on BioInformatics and BioEngineering, Chania, Greece, 10–13 November 2013; pp. 1–4.
87. Drotár, P.; Mekyska, J.; Rektorová, I.; Masarová, L.; Smékal, Z.; Faundez-Zanuy, M. Analysis of in-air movement in handwriting: A novel marker for Parkinson's disease. *Comput. Methods Programs Biomed.* **2014**, *117*, 405–411. [CrossRef] [PubMed]
88. Drotár, P.; Mekyska, J.; Smékal, Z.; Rektorová, I.; Masarová, L.; Faundez-Zanuy, M. Contribution of different handwriting modalities to differential diagnosis of Parkinson's disease. In Proceedings of the IEEE International Symposium on Medical Measurements and Applications (MeMeA) Proceedings, Torino, Italy, 7–9 May 2015; pp. 344–348.
89. Moetesum, M.; Siddiqi, I.; Vincent, N.; Cloppet, F. Assessing visual attributes of handwriting for prediction of neurological disorders–A case study on Parkinson's disease. *Pattern Recognit. Lett.* **2019**, *121*, 19–27. [CrossRef]
90. Pereira, C.R.; Pereira, D.R.; Rosa, G.H.; Albuquerque, V.H.; Weber, S.A.; Hook, C.; Papa, J.P. Handwritten dynamics assessment through convolutional neural networks: An application to Parkinson's disease identification. *Artif. Intell. Med.* **2018**, *87*, 67–77. [CrossRef] [PubMed]
91. Kotsavasiloglou, C.; Kostikis, N.; Hristu-Varsakelis, D.; Arnaoutoglou, M. Machine learning-based classification of simple drawing movements in Parkinson's disease. *Biomed. Signal Process. Control.* **2017**, *31*, 174–180. [CrossRef]
92. Zham, P.; Kumar, D.K.; Dabnichki, P.; Poosapadi Arjunan, S.; Raghav, S. Distinguishing different stages of Parkinson's disease using composite index of speed and pen-pressure of sketching a spiral. *Front. Neurol.* **2017**, *8*, 435. [CrossRef] [PubMed]
93. Zham, P.; Arjunan, S.P.; Raghav, S.; Kumar, D.K. Efficacy of guided spiral drawing in the classification of Parkinson's disease. *IEEE J. Biomed. Health Inform.* **2017**, *22*, 1648–1652. [CrossRef]
94. Gallicchio, C.; Micheli, A.; Pedrelli, L. Deep Echo State Networks for Diagnosis of Parkinson's Disease. *arXiv* **2018**, arXiv:1802.06708.
95. Mucha, J.; Mekyska, J.; Galaz, Z.; Faundez-Zanuy, M.; Lopez-de Ipina, K.; Zvoncak, V.; Kiska, T.; Smekal, Z.; Brabenec, L.; Rektorova, I. Identification and Monitoring of Parkinson's Disease Dysgraphia Based on Fractional-Order Derivatives of Online Handwriting. *Appl. Sci.* **2018**, *8*, 2566. [CrossRef]
96. Impedovo, D. Velocity-based signal features for the assessment of Parkinsonian handwriting. *IEEE Signal Process. Lett.* **2019**, *26*, 632–636. [CrossRef]
97. Jerkovic, V.M.; Kojic, V.; Miskovic, N.D.; Djukic, T.; Kostic, V.S.; Popovic, M.B. Analysis of on-surface and in-air movement in handwriting of subjects with Parkinson's disease and atypical parkinsonism. *Biomed. Eng. Tech.* **2019**, *64*, 187–194. [CrossRef]
98. Loconsole, C.; Cascarano, G.D.; Brunetti, A.; Trotta, G.F.; Losavio, G.; Bevilacqua, V.; Di Sciascio, E. A model-free technique based on computer vision and sEMG for classification in Parkinson's disease by using computer-assisted handwriting analysis. *Pattern Recognit. Lett.* **2019**, *121*, 28–36. [CrossRef]
99. Rios-Urrego, C.; Vásquez-Correa, J.; Vargas-Bonilla, J.; Nöth, E.; Lopera, F.; Orozco-Arroyave, J. Analysis and evaluation of handwriting in patients with Parkinson's disease using kinematic, geometrical, and non-linear features. *Comput. Methods Programs Biomed.* **2019**, *173*, 43–52. [CrossRef] [PubMed]

100. Yan, J.H.; Rountree, S.; Massman, P.; Doody, R.S.; Li, H. Alzheimer's disease and mild cognitive impairment deteriorate fine movement control. *J. Psychiatr. Res.* **2008**, *42*, 1203–1212. [CrossRef] [PubMed]
101. Faundez-Zanuy, M.; Sesa-Nogueras, E.; Roure-Alcobé, J.; Garré-Olmo, J.; Lopez-de Ipiña, K.; Solé-Casals, J. Online drawings for dementia diagnose: In-air and pressure information analysis. In Proceedings of the XIII Mediterranean Conference on Medical and Biological Engineering and Computing 2013, Seville, Spain, 25–28 September 2014; Springer: Berlin, Germany, 2014; pp. 567–570.
102. Yu, N.Y.; Chang, S.H. Kinematic analyses of graphomotor functions in individuals with Alzheimer's disease and amnestic mild cognitive impairment. *J. Med Biol. Eng.* **2016**, *36*, 334–343. [CrossRef]
103. Renier, M.; Gnoato, F.; Tessari, A.; Formilan, M.; Busonera, F.; Albanese, P.; Sartori, G.; Cester, A. A correlational study between signature, writing abilities and decision-making capacity among people with initial cognitive impairment. *Aging Clin. Exp. Res.* **2016**, *28*, 505–511. [CrossRef]
104. Kawa, J.; Bednorz, A.; Stępień, P.; Derejczyk, J.; Bugdol, M. Spatial and dynamical handwriting analysis in mild cognitive impairment. *Comput. Biol. Med.* **2017**, *82*, 21–28. [CrossRef]
105. Müller, S.; Preische, O.; Heymann, P.; Elbing, U.; Laske, C. Increased Diagnostic Accuracy of Digital vs. Conventional Clock Drawing Test for Discrimination of Patients in the Early Course of Alzheimer's Disease from Cognitively Healthy Individuals. *Front. Aging Neurosci.* **2017**, *9*, 101. [CrossRef]
106. Müller, S.; Herde, L.; Preische, O.; Zeller, A.; Heymann, P.; Robens, S.; Elbing, U.; Laske, C. Diagnostic value of digital clock drawing test in comparison with CERAD neuropsychological battery total score for discrimination of patients in the early course of Alzheimer's disease from healthy individuals. *Sci. Rep.* **2019**, *9*, 3543. [CrossRef]
107. El-Yacoubi, M.A.; Garcia-Salicetti, S.; Kahindo, C.; Rigaud, A.S.; Cristancho-Lacroix, V. From aging to early-stage Alzheimer's: Uncovering handwriting multimodal behaviors by semi-supervised learning and sequential representation learning. *Pattern Recognit.* **2019**, *86*, 112–133. [CrossRef]
108. Kahindo, C.; El-Yacoubi, M.A.; Garcia-Salicetti, S.; Rigaud, A.S.; Cristancho-Lacroix, V. Characterizing early-stage Alzheimer through spatiotemporal dynamics of handwriting. *IEEE Signal Process. Lett.* **2018**, *25*, 1136–1140. [CrossRef]
109. Ghaderyan, P.; Abbasi, A.; Saber, S. A new algorithm for kinematic analysis of handwriting data; towards a reliable handwriting-based tool for early detection of Alzheimer's disease. *Expert Syst. Appl.* **2018**, *114*, 428–440. [CrossRef]
110. Ishikawa, T.; Nemoto, M.; Nemoto, K.; Takeuchi, T.; Numata, Y.; Watanabe, R.; Tsukada, E.; Ota, M.; Higashi, S.; Arai, T.; et al. Handwriting Features of Multiple Drawing Tests for Early Detection of Alzheimer's Disease: A Preliminary Result. *Stud. Health Technol. Inform.* **2019**, *264*, 168–172. [PubMed]
111. Lella, E.; Amoroso, N.; Lombardi, A.; Maggipinto, T.; Tangaro, S.; Bellotti, R. Communicability disruption in Alzheimer's disease connectivity networks. *J. Complex Netw.* **2018**, *7*, 83–100. [CrossRef]
112. Casalino, G.; Castellano, G.; Pasquadibisceglie, V.; Zaza, G. Contact-Less Real-Time Monitoring of Cardiovascular Risk Using Video Imaging and Fuzzy Inference Rules. *Information* **2019**, *10*, 9. [CrossRef]
113. Senatore, R.; Della Cioppa, A.; Marcelli, A. Automatic Diagnosis of Neurodegenerative Diseases: An Evolutionary Approach for Facing the Interpretability Problem. *Information* **2019**, *10*, 30. [CrossRef]
114. Parziale, A.; Della Cioppa, A.; Senatore, R.; Marcelli, A. A Decision Tree for Automatic Diagnosis of Parkinson's Disease from Offline Drawing Samples: Experiments and Findings. In Proceedings of the International Conference on Image Analysis and Processing, Trento, Italy, 9–13 September 2019; Springer: Berlin, Germany, 2019; pp. 196–206.
115. Luria, G.; Kahana, A.; Rosenblum, S. Detection of deception via handwriting behaviors using a computerized tool: Toward an evaluation of malingering. *Cogn. Comput.* **2014**, *6*, 849–855. [CrossRef]
116. Rosenblum, S.; Luria, G. Applying a handwriting measurement model for capturing cognitive load implications through complex figure drawing. *Cogn. Comput.* **2016**, *8*, 69–77. [CrossRef]
117. Badarna, M.; Shimshoni, I.; Luria, G.; Rosenblum, S. The Importance of Pen Motion Pattern Groups for Semi-Automatic Classification of Handwriting into Mental Workload Classes. *Cogn. Comput.* **2018**, *10*, 215–227. [CrossRef]
118. Likforman-Sulem, L.; Esposito, A.; Faundez-Zanuy, M.; Clémençon, S.; Cordasco, G. EMOTHAW: A novel database for emotional state recognition from handwriting and drawing. *IEEE Trans. Hum. Mach. Syst.* **2017**, *47*, 273–284. [CrossRef]

119. Smekal, Z.; Mekyska, J.; Rektorova, I.; Faundez-Zanuy, M. Analysis of neurological disorders based on digital processing of speech and handwritten text. In Proceedings of the International Symposium on Signals, Circuits and Systems, Iasi, Romania, 11–12 July 2013; pp. 1–6.
120. Griol, D.; Callejas, Z. Mobile conversational agents for context-aware care applications. *Cogn. Comput.* **2016**, *8*, 336–356. [CrossRef]

*applied sciences*

MDPI

*Article*

# Automatic Detection of a Standard Line for Brain Magnetic Resonance Imaging Using Deep Learning

**Hiroyuki Sugimori [1,*] and Masashi Kawakami [2]**

1 Faculty of Health Sciences, Hokkaido University, Sapporo 060-0812, Japan
2 Graduate School of Biomedical Science and Engineering, School of Medicine, Hokkaido University, Sapporo 060-0812, Japan; masashikawakami@eis.hokudai.ac.jp
* Correspondence: sugimori@hs.hokudai.ac.jp; Tel.: +81-11-706-3410

Received: 29 August 2019; Accepted: 9 September 2019; Published: 13 September 2019

**Abstract:** Recently, deep learning technology has been applied to medical images. This study aimed to create a detector able to automatically detect an anatomical structure presented in a brain magnetic resonance imaging (MRI) scan to draw a standard line. A total of 1200 brain sagittal MRI scans were used for training and validation. Two sizes of regions of interest (ROIs) were drawn on each anatomical structure measuring 64 × 64 pixels and 32 × 32 pixels, respectively. Data augmentation was applied to these ROIs. The faster region-based convolutional neural network was used as the network model for training. The detectors created were validated to evaluate the precision of detection. Anatomical structures detected by the model created were processed to draw the standard line. The average precision of anatomical detection, detection rate of the standard line, and accuracy rate of achieving a correct drawing were evaluated. For the 64 × 64-pixel ROI, the mean average precision achieved a result of 0.76 ± 0.04, which was higher than the outcome achieved with the 32 × 32-pixel ROI. Moreover, the detection and accuracy rates of the angle of difference at 10 degrees for the orbitomeatal line were 93.3 ± 5.2 and 76.7 ± 11.0, respectively. The automatic detection of a reference line for brain MRI can help technologists improve this examination.

**Keywords:** object detection; standard line for brain; faster R-CNN; medical image analysis; magnetic resonance imaging

---

## 1. Introduction

The convolutional neural network (CNN) [1] can be trained to extract image features via multiple layers. There are so many types of CNN models [2–4] to choose from and new methods being published and discussed. The object detection technique with CNN [5–8] can detect the locations of regions of interest by distinguishing them from the background, although the image classification with CNN can be classified to specific categories as the whole of an image. Recently, deep learning technology has been adopted in many areas, including image classification [2–4], object detection [5–8], and image segmentation [9]. These deep learning technologies have also been applied to medical images. Examples include computed tomography image classification [10,11], feature extraction [12,13] and automatic detection of lung tumors [14,15], and automatic detection of breast tumors on X-ray images [16,17]. These machine-aided diagnostic techniques have supported the efforts of radiologists to achieve more accurate diagnoses and tumor detection. However, no medical image acquisition technique incorporating deep learning is currently available at this time. Magnetic resonance imaging (MRI) of the brain is one of the most common image acquisitions performed in the hospital. The acquisition of brain MRI scans involves obtaining an arbitrary cross section without radiation exposure. However, determining the ideal angle of the sections is necessary to be able to acquire an arbitrary cross section easily. Currently, some standard lines for brain MRI exist, such as the orbitomeatal line (OM-line) [18] and the anterior commissure–posterior commissure line (AC-PC line) [19]. The use of these standard lines facilitates the location of specific anatomical structures by the technologists manually. The model-based detection [20] for the AC and PC have been reported to dealing with detections of anatomies. However, the anatomies for the OM line could not be supported in this technique. Therefore, the automatic detection of the standard line using a deep learning technique would be useful for technologists seeking to acquire brain MRI scans. The purpose of the present study was to detect standard lines automatically for brain MRI using a deep learning technique.

## 2. Materials and Methods

### *2.1. Subjects and MRI Scans*

The study included 1200 patients (585 males and 615 females, mean age ± standard deviation (SD): 55.8 ± 20.1 years) who were subjected to an MRI examination of the brain between September and November 2016 at Hokkaido University Hospital. All MRI images were obtained using two 1.5-tesla (T) MRI scanners (the Achieva A-series from Philips Healthcare, Best, the Netherlands, or the MAGNETOM Avanto from Siemens Healthcare, Erlangen, Germany) and three 3-T MRI scanners (the GE Discovery MR 750 w from GE Healthcare, Chicago, IL, USA, the TRILLIUM OVAL from Hitachi, Tokyo, Japan, or the Achieva TX from Philips Healthcare, Best, the Netherlands). The direction of the slice in this study incorporated the median sagittal plane. This study was approved by the ethics committee of Hokkaido University Hospital.

### *2.2. Datasets and Preprocessing of Images*

The MRI scans were retrieved from the picture archiving and communication system. To convert the images for use by the training database, they were converted from the Digital Imaging and Communications in Medicine (DICOM) format to the Joint Photographic Experts Group (JPEG) format using a dedicated DICOM software (XTREK view, J-MAC SYSTEM Inc., Sapporo, Japan). The window width and level of the DICOM images were used to preset values in the DICOM tag. The DICOM images were converted to JPEG images with a size of 512 × 512 pixels. The size of the image was unified for inputting as the same size in the software because the middle sagittal images in this study were obtained by several different MRI scanners. JPEG files were loaded into the in-house MATLAB software program (The MathWorks, Inc., Natick, MA, USA). The software was used to draw regions of interest (ROIs) with sizes of 64 × 64 pixels and 32 × 32 pixels at the center of specific structures. The ROIs were drawn according to the anatomies of the root of the nose (Nose), inferior border of the pons (Pons), the AC, and the PC (Figure 1). The ROI data were outputted as a text file, which included the object name, coordinates, and size of each ROI. The dataset was divided into six subsets to complete the six-fold cross-validation. Two hundred images were included in each subset, with a total of five subsets (1000 images) used for training and the other five subsets used for validation (Figure 2). Data augmentation [21,22] was performed involving 1000 images for the improvement of the training. The training images were applied as a data augmentation dataset to the image rotation, which involved angles from −30 to 30 degrees in 3-degree steps (Figure 3).

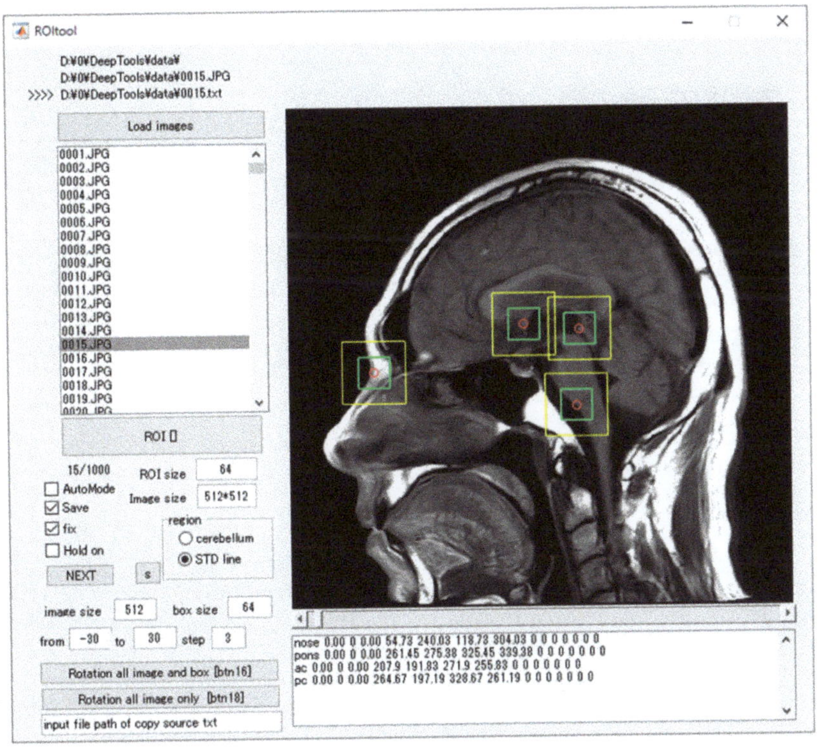


**Figure 1.** The software for training outlined ROIs over the noteworthy anatomical characteristics. The red circle is the center of the target anatomy, the yellow bounding boxes are the ROIs measuring 64 × 64 pixels, and the green bounding boxes are the ROIs measuring 32 × 32 pixels.

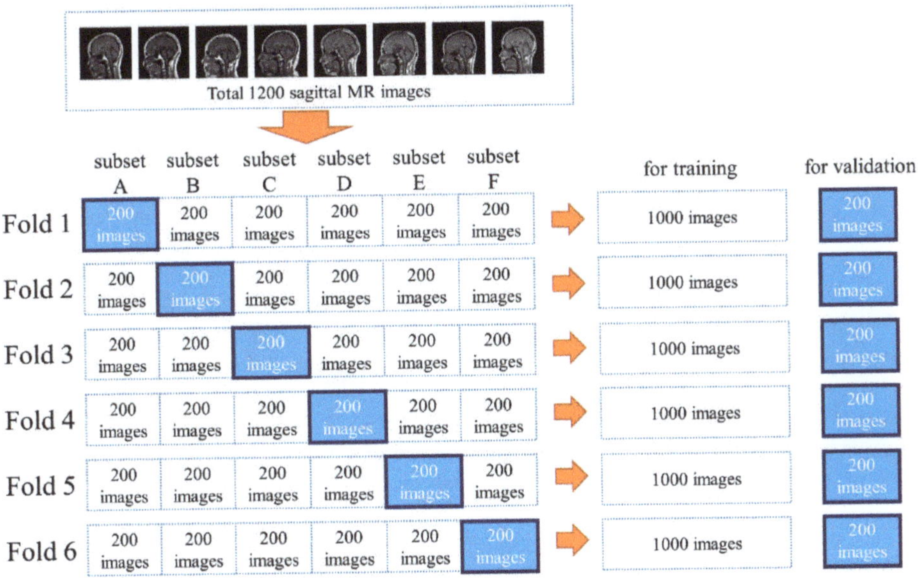


**Figure 2.** A total of 1200 images were divided into six subsets to complete the six-fold cross-validation.

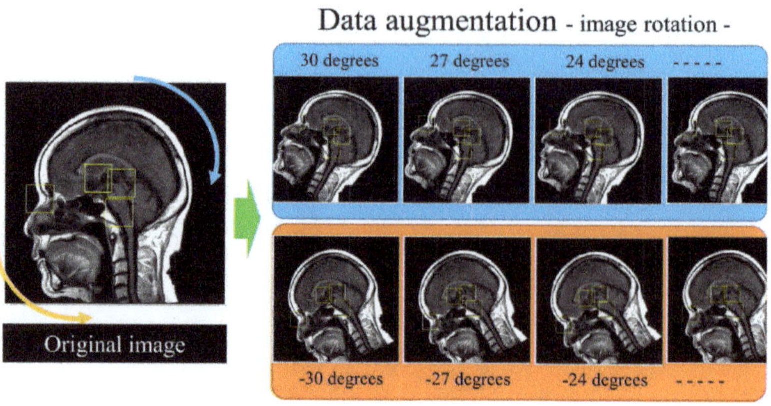


**Figure 3.** The procedure for data augmentation of the training datasets. The original images and supervised bounding boxes were rotated from −30 to 30 degrees in 3-degree steps.

### 2.3. Training of the Images for Model Creation

The software for the deep learning technique was developed via in-house MATLAB software, and a deep learning–optimized machine with a Nvidia GeForce GTX 1080 Ti graphics card (Nvidia Corporation, Santa Clara, CA, USA), with 11.34 tera floating point operations per second (TFLOPS) of single-precision performance, 484 GB/s of memory bandwidth, and 11 GB of memory per board was used. The image training was performed using the faster region-based convolutional neural network (R-CNN) [8] with the Computer Vision System Toolbox on the MATLAB software. Training was divided into four steps inside the MATLAB software. During the first two steps, the region proposal and detection networks were created, while the latter two steps were performed together to train the networks for detection. The hyper-parameters of the training models were as follows: Maximum training epochs, 10; initial learning rates, 0.00001 (first two steps) and 0.000001 (latter two steps); and mini batch size, 1. The stochastic gradient descent with momentum (SGDM) was used for optimization with an initial learning rate. The momentum set to 0.9 and L2 regulation set to 0.0001. Image training was performed six times according to the training datasets in Figure 2.

### 2.4. Evaluation of the Created Models and a Standard Line for Brain MRI

The predicted bounding boxes were incorporated into the MATLAB software to show only one box representing the area with the highest confidence for each anatomy. The detection of different anatomies was evaluated using the average precision (AP) and the mean average precision (mAP) [23]. The ROIs measuring 64 × 64 pixels and 32 × 32 pixels, respectively, were evaluated separately. The higher mAP among the two ROI sizes was selected to evaluate for comparison with the standard lines of the brain imaging, that is, the OM-line and the AC-PC line. The delineation of a standard line for brain MRI was calculated using the MATLAB software (Figure 4). The standard lines were calculated by using the central coordinate of the predicted bounding boxes. If there was no detection of the bounding boxes, which is a necessary requirement to draw the standard line, the line was not represented in the software. The software also had a function of calculating the angle of the predicted line that was formed at an angle relative to the horizontal direction of the image. The number of the predicted lines was calculated as a line-detection rate. The accuracy rates of the delineation for the standard line were calculated to obtain the difference between the original and prediction angles among the detected lines. The processing speeds by calculating the computation time per image were also measured. The angles were evaluated from 0.5 to 10.0 degrees of the absolute value of the angles. All the results were represented as means and SDs according to the number of six-fold datasets.

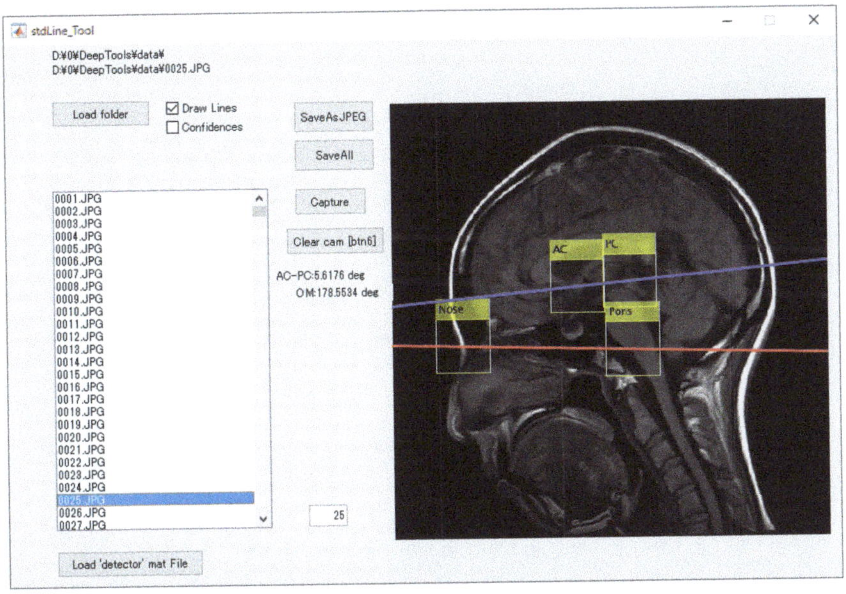


**Figure 4.** The detection software based on the created model and the delineation of the standard line for brain magnetic resonance imaging (MRI). The yellow bounding boxes represent automatically detected anatomy, the red line is the calculated orbitomeatal (OM)-line, and the blue line is the calculated anterior commissure–posterior commissure (AC-PC) line.

## 3. Results

### 3.1. *The Detection of Anatomies*

Tables 1 and 2 show the AP and mAP for the ROI sizes of 64 × 64 pixels and 32 × 32 pixels, respectively. The mAP for the ROIs measuring 64 × 64 pixels was 0.76 ± 0.04, which was higher than those for the ROIs measuring 32 × 32 pixels. Separately, for the ROIs measuring 64 × 64 pixels, the AP of the Nose (0.85 ± 0.07), Pons (0.78 ± 0.11), PC (0.74 ± 0.05), and AC (0.68 ± 0.04) were calculated in the higher order. There was no AP value over 0.5 for the ROIs measuring 32 × 32 pixels in each anatomy.

**Table 1.** Average precision (AP) and mean average precision (mAP) for the ROIs measuring 64 × 64 pixels.

| | **Fold 1** | **Fold 2** | **Fold 3** | **Fold 4** | **Fold 5** | **Fold 6** | **Mean ± SD** |
|---|---|---|---|---|---|---|---|
| Nose | 0.80 | 0.88 | 0.87 | 0.76 | 0.80 | 0.96 | 0.85 ± 0.07 |
| Pons | 0.78 | 0.82 | 0.79 | 0.80 | 0.59 | 0.92 | 0.78 ± 0.11 |
| AC | 0.71 | 0.67 | 0.64 | 0.70 | 0.71 | 0.62 | 0.68 ± 0.04 |
| PC | 0.80 | 0.77 | 0.75 | 0.67 | 0.72 | 0.74 | 0.74 ± 0.05 |
| mAP | 0.77 | 0.79 | 0.76 | 0.73 | 0.71 | 0.81 | 0.76 ± 0.04 |

**Table 2.** AP and mAP for the ROIs measuring 32 × 32 pixels.

| | **Fold 1** | **Fold 2** | **Fold 3** | **Fold 4** | **Fold 5** | **Fold 6** | **Mean ± SD** |
|---|---|---|---|---|---|---|---|
| Nose | 0.33 | 0.46 | 0.62 | 0.43 | 0.47 | 0.54 | 0.48 ± 0.10 |
| Pons | 0.01 | 0.03 | 0.27 | 0.70 | 0.39 | 0.19 | 0.27 ± 0.26 |
| AC | 0.32 | 0.28 | 0.26 | 0.03 | 0.44 | 0.22 | 0.26 ± 0.14 |
| PC | 0.11 | 0.27 | 0.26 | 0.55 | 0.53 | 0.15 | 0.31 ± 0.19 |
| mAP | 0.19 | 0.26 | 0.35 | 0.43 | 0.46 | 0.28 | 0.33 ± 0.10 |

### 3.2. Line-Detection Rates and Accuracies

Table 3 shows the line-detection rates for the brain MRI scans. The detection rate of the OM-line was higher than the AC-PC line. Figure 5 shows the accuracy rates of the delineation for the standard line. The angles of difference at 3, 5, and 10 degrees for the OM-line between the original and predicted lines were 57.5 ± 10.9, 69.8 ± 11.2, and 76.7 ± 11.0, respectively. Additionally, the angles of differences at 3, 5, and 10 degrees for the AC-PC line between the original and predicted lines were 21.7 ± 4.6, 31.6 ± 4.3, and 47.1 ± 5.0, respectively. Table 4 shows the processing speeds by calculating the computation time per image. The computation time per images was 0.11 ± 0.01 s.

**Table 3.** Detection rates of the standard line for brain MRI.

| | Fold 1 | Fold 2 | Fold 3 | Fold 4 | Fold 5 | Fold 6 | Mean ± SD |
|---|---|---|---|---|---|---|---|
| OM line | 99.5 | 86.5 | 89.0 | 94.5 | 98.5 | 92.0 | 93.3 ± 5.2 |
| AC–PC line | 89.5 | 95.0 | 80.5 | 88.5 | 86.5 | 86.5 | 87.8 ± 4.7 |

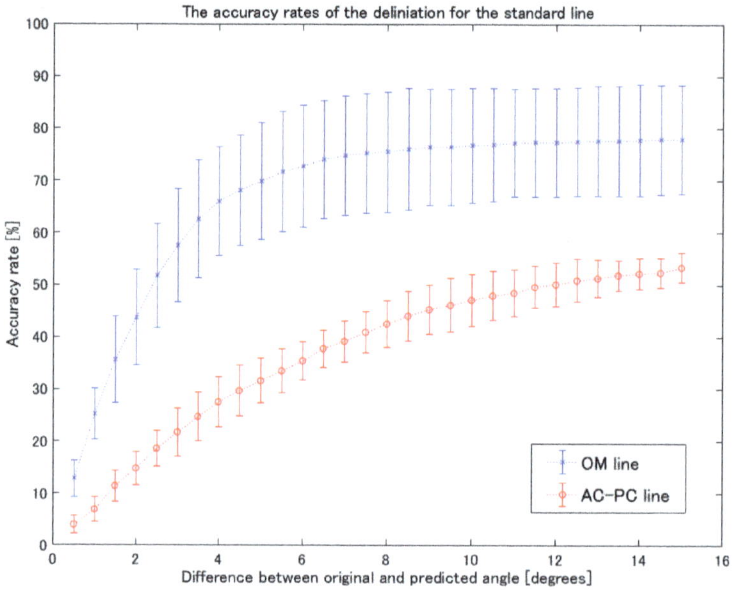


**Figure 5.** The accuracy rates of delineation for the standard line. The blue asterisks indicate the mean accuracy rates of the OM-line. The red circles show the mean accuracy rates of the AC-PC line. The error bars show the standard deviation.

**Table 4.** The processing speeds of calculating the standard line.

| | Fold 1 | Fold 2 | Fold 3 | Fold 4 | Fold 5 | Fold 6 | Mean ± SD |
|---|---|---|---|---|---|---|---|
| Computation time per image [sec] | 0.12 | 0.12 | 0.10 | 0.11 | 0.11 | 0.10 | 0.11 ± 0.01 |

## 4. Discussion

This research was performed to attain detection of the specific anatomy for the delineation of two standard MRI brain lines. With regard to detecting the brain anatomies, the ROI size of 64 × 64 pixels showed a higher accuracy of detection than the 32 × 32 pixels did. This result indicates that the region including peripheral anatomies was effective for training to improve the accuracy of detection because most medical images represent the anatomy as a slice section rather than a volume. Moreover, it is not necessary to extract the anatomical structures because human anatomies are essentially all the same when no malformations are present. Since the object detection ability in self-driving cars need the ability to detect both cars and pedestrians from all directions, the required image features needed to be trained while involving the object itself, rather than the object plus its surroundings. With regard to the delineation of the standard lines for brain MRI, there were some discrepancies between the detection and accuracy rates. The delineation of the standard line was defined by two pairs of points of the Nose and Pons or the AC and PC, respectively (Appendix A). For this reason, the accuracy rates of the standard line were worse than the detection rates because the AP was different for each anatomical part. The difference of angles between the OM-line and the AC-PC line was reported as 12.6 degrees in another report [24]. In comparison, the accuracy rate of the standard line of our results showing less than 12 degrees was suggested to be adequate for the detection of the standard line for brain MRI. Since the angles of difference at 10 degrees on the OM-line and the AC-PC line between the original and predicted lines were 76.7 ± 11.0 and 47.1 ± 5.0, the delineation of the OM-line presented higher reliability for detection. The cause of the reduced accuracy rate of the AC-PC line was the lower detection accuracy on the AC because the anatomy of the AC was described as a tiny point on the sagittal image.

The limitations of the present study are as follows. First, image training was performed only using the faster R-CNN in this study. Many kinds of network models exist for object detection using deep learning. The faster R-CNN was shown to have higher mAP in several papers [6,25], which suggested that the mAP was one of the key indicators for the comparison of models. Though the weak point of the R-CNN was shown as the response speed of detections [25], the speed of response was less important during the acquisition of brain MRI. This study was not conducted to detect the dynamics of the anatomy. Second, the present study was only focusing on the deep learning technique. Though the model-based detection [20] was also one of the techniques for the detection of anatomies, the processing time of deep learning techniques has been improving [6–8]. We could present the processing time for the detection and delineation of the standard line within one second. However, the comparison of both techniques under the same condition should be taken into account in future research because model-based detection is also one of the robust techniques. Third, the number of training images in this study was 1200, including the images used for validation of the training. Another study [11] showed that the number of training images affected the accuracy of deep learning. Therefore, a larger number of images would improve the detection rates of the anatomy and the accuracy rates of the standard line. In this study, we focused on the detection of specific anatomical points for the delineation of the standard line of brain MRI. These results can be applied to other regions of the body and to detect tumors. The methods and results of this study will be useful for the improvement of the accuracy and will contribute to the improvement of medical image analysis, although this study focused specifically on the acquisition of brain MRI. Automatic detection for a standard line for brain MRI can help technologists improve brain MRI scans.

## 5. Conclusions

This study achieved the automatic detection of a standard line for brain MRI using a deep learning technique. It was found that the delineation of the standard line for brain MRI achieved a high accuracy rate on the OM-line. The use of the technique in this study can help technologists improve brain MRI examinations.

**Author Contributions:** H.S. proposed the idea and contributed to data acquisition, data analysis, algorithm construction, and writing and editing of the article. M.K. performed data analysis.

**Funding:** This research received no external funding.

**Acknowledgments:** The authors thank the laboratory students Shota Sato, Kazuki Narita, and Taiga Aoki for their help.

**Conflicts of Interest:** The authors declare that no conflicts of interest exist.

## Appendix A

Figure A1 shows an example of the real time detection of the standard line for brain MRI. The software created a model for the detection of anatomies. The standard line for brain MRI was calculated in real time. This software can be also incorporated into MRI scanner consoles.

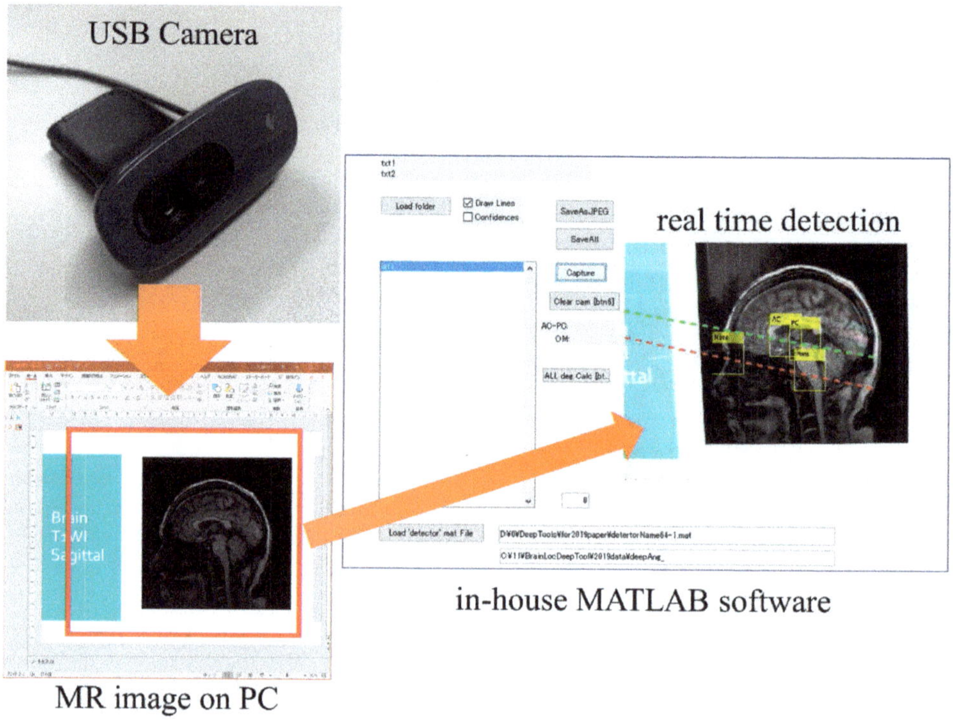


**Figure A1.** Example of the real time detection of the standard line for brain MRI. The red and green dotted lines indicate the OM-line and the AC-PC line, respectively.

## References

1. LeCun, Y.; Bottou, L.; Bengio, Y.; Haffner, P. Gradient-Based Learning Applied to Document Recognition. *Proc. IEEE* **1998**, *86*, 2278–2324. [CrossRef]
2. He, K.; Zhang, X.; Ren, S.; Sun, J. Deep Residual Learning for Image Recognition. In Proceedings of the IEEE Conference on Computer Vision and Pattern Recognition, Las Vegas, NV, USA, 27–30 June 2016; pp. 770–778.
3. Krizhevsky, A.; Sutskever, I.; Hinton, G.E. Imagenet Classification with Deep Convolutional Neural Networks. In *Advances in Neural Information Processing Systems*; The MIT Press: Cambridge, MA, USA, 2012; pp. 1097–1105.
4. Szegedy, C.; Liu, W.; Jia, Y.; Sermanet, P.; Reed, S.; Anguelov, D.; Erhan, D.; Vanhoucke, V.; Rabinovich, A. Going deeper with convolutions. In Proceedings of the IEEE Conference on Computer Vision and Pattern Recognition, Boston, MA, USA, 7–12 June 2015; pp. 1–9.
5. Girshick, R.; Donahue, J.; Darrell, T.; Malik, J.; Malik, J. Rich Feature Hierarchies for Accurate Object Detection and Semantic Segmentation. In Proceedings of the IEEE Conference on Computer Vision and Pattern Recognition, Columbus, OH, USA, 23–28 June 2014; pp. 580–587.
6. Liu, W.; Anguelov, D.; Erhan, D.; Szegedy, C.; Reed, S.; Fu, C.Y.; Berg, A.C. SSD: Single Shot MultiBox Detector. In *European Conference on Computer Vision*; Springer: Berlin/Heidelberg, Germany, 2016; pp. 21–37.
7. Redmon, J.; Farhadi, A. YOLO9000: Better, Faster, Stronger. In Proceedings of the IEEE Conference on Computer Vision and Pattern Recognition (CVPR), Honolulu, HI, USA, 21–26 July 2017; pp. 6517–6525.

8. Ren, S.; He, K.; Girshick, R.; Sun, J. Faster R-CNN: Towards Real-Time Object Detection with Region Proposal Networks. *IEEE Trans. Pattern Anal. Mach. Intell.* **2017**, *39*, 1137–1149. [CrossRef] [PubMed]
9. Ronneberger, O.; Fischer, P.; Brox, T. U-Net: Convolutional Networks for Biomedical Image Segmentation. *Exp. Algorithms* **2015**, *9351*, 234–241.
10. Sugimori, H. Evaluating the Overall Accuracy of Additional Learning and Automatic Classification System for CT Images. *Appl. Sci.* **2019**, *9*, 682. [CrossRef]
11. Sugimori, H. Classification of Computed Tomography Images in Different Slice Positions Using Deep Learning. *J. Health Eng.* **2018**, *2018*, 1–9. [CrossRef] [PubMed]
12. Lai, Z.; Deng, H. Medical Image Classification Based on Deep Features Extracted by Deep Model and Statistic Feature Fusion with Multilayer Perceptron. *Comput. Intell. Neurosci.* **2018**, *2018*, 1–13. [CrossRef] [PubMed]
13. Yang, A.M.; Yang, X.L.; Wu, W.R.; Liu, H.X.; Zhuansun, Y.X. Research on Feature Extraction of Tumor Image Based on Convolutional Neural Network. *IEEE Access* **2019**, *7*, 24204–24213. [CrossRef]
14. Masood, A.; Sheng, B.; Li, P.; Hou, X.; Wei, X.; Qin, J.; Feng, D. Computer-Assisted Decision Support System in Pulmonary Cancer detection and stage classification on CT images. *J. Biomed. Inform.* **2018**, *79*, 117–128. [CrossRef] [PubMed]
15. Zhao, X.; Liu, L.; Qi, S.; Teng, Y.; Li, J.; Qian, W. Agile convolutional neural network for pulmonary nodule classification using CT images. *Int. J. Comput. Assist. Radiol. Surg.* **2018**, *13*, 585–595. [CrossRef] [PubMed]
16. Gardezi, S.J.S.; Elazab, A.; Lei, B.; Wang, T. Breast Cancer Detection and Diagnosis Using Mammographic Data: Systematic Review. *J. Med. Internet Res.* **2019**, *21*, e14464. [CrossRef] [PubMed]
17. Agarwal, R.; Diaz, O.; Lladó, X.; Yap, M.H.; Martí, R. Automatic mass detection in mammograms using deep convolutional neural networks. *J. Med. Imaging* **2019**, *6*, 031409. [CrossRef]
18. Yeoman, L.J.; Howarth, L.; Britten, A.; Cotterill, A.; Adam, E.J. Gantry angulation in brain CT: dosage implications, effect on posterior fossa artifacts, and current international practice. *Radiology* **1992**, *184*, 113–116. [CrossRef] [PubMed]
19. Talairach, J.; Tournoux, P. *Co-Planar Stereotaxic Atlas of the Human Brain*; Thieme Medical Publishers: New York, NY, USA, 1988; p. 122.
20. Ardekani, B.A.; Bachman, A.H. Model-based Automatic Detection of the Anterior and Posterior Commissures on MRI Scans. *NeuroImage* **2009**, *46*, 677–682. [CrossRef] [PubMed]
21. Mash, R.; Borghetti, B.; Pecarina, J. Improved Aircraft Recognition for Aerial Refueling Through Data Augmentation in Convolutional Neural Networks. In Proceedings of the Advances in Visual Computing: 12th International Symposium, ISVC 2016, Las Vegas, NV, USA, 12–14 December 2016; Springer: Berlin/Heidelberg, Germany, 2016; Volume 10072, pp. 113–122.
22. Taylor, L.; Nitschke, G. Improving Deep Learning using Generic Data Augmentation. *arXiv* **2017**, arXiv:170806020.
23. Everingham, M.; Van Gool, L.; Williams, C.K.I.; Winn, J.; Zisserman, A. The Pascal Visual Object Classes (VOC) Challenge. *Int. J. Comput. Vis.* **2010**, *88*, 303–338. [CrossRef]
24. Kim, Y.; Ahn, K.; Chung, Y.; Kim, B.-S. A New Reference Line for the Brain CT: The Tuberculum Sellae-Occipital Protuberance Line is Parallel to the Anterior/Posterior Commissure Line. *Am. J. Neuroradiol.* **2009**, *30*, 1704–1708. [CrossRef] [PubMed]
25. Huang, J.; Rathod, V.; Sun, C.; Zhu, M.; Korattikara, A.; Fathi, A.; Fischer, I.; Wojna, Z.; Song, Y.; Guadarrama, S.; et al. Speed/accuracy trade-offs for modern convolutional object detectors. *arXiv* **2017**, arXiv:161110012.

*applied sciences*

MDPI

*Communication*

# A Prototype of a Portable Gas Analyzer for Exhaled Acetone Detection

**Jakub Sorocki and Artur Rydosz ***

Department of Electronics, AGH University of Science and Technology, 30-059 Krakow, Poland
* Correspondence: artur.rydosz@agh.edu.pl; Tel.: +48-12-617-25-94

Received: 1 June 2019; Accepted: 25 June 2019; Published: 27 June 2019

**Featured Application: Exhaled breath analysis with special emphasis on diabetes. Patients with diabetes (mostly type-1) tend to have higher acetone levels in their breath than healthy people, therefore, the exhaled acetone is considered to be one of the biomarkers of this disease.**

**Abstract:** The paper presents the development of a portable gas analyzer prototype for exhaled acetone detection, employing an application-suited gas sensor array and 3D printing technology. The device provides the functionality to monitor exhaled acetone levels, which could be used as a potential tool for non-invasive diabetes monitoring. The relationship between exhaled acetone concentrations and glucose in blood is confirmed in the literature, including research carried out by the authors. The design process is presented including a general consideration for the sensor array construction, which is the core for sensing gases, as well as requirements for the measurement chamber it is to be placed in. Moreover, the mechanical design of the 3D-printed housing is discussed to ensure the ergonomics of use as a hand-held device while keeping the hardware integrity. Also, the processing hardware is discussed to provide sufficient computing power to handle the stand-alone operation while being energy efficient, enabling long battery-powered operation. Finally, calibration and measurement, as well as the analyzer operation, are shown, validating the proposed class of exhaled acetone-detection capable meters.

**Keywords:** gas sensors; exhaled acetone detection; diabetes; portable breath analyzer

## 1. Introduction

Exhaled human breath consists of several different compounds, including volatile organic compounds (VOCs), which are continuously generated in the human body and are partially emitted via exhaled breath, through the skin, and by urine and feces [1–3]. VOCs are mainly in the ppm–ppt (part per million to part per trillion) range, thus, laboratory methods are used for their detection in breath, such as GC-MS (gas chromatography–mass spectrometry) [4,5], PTR-MS (proton transfer reaction–mass spectrometry) [6,7], IMS-MS (ion-mobility spectrometry–mass spectrometry) [8,9], and SIFT-MS (selected ion flow tube–mass spectrometry) [10,11]. Over the last 40 years, almost 3500 different VOCs have been detected in the human breath [12], and a single breath consists of around 500 various VOCs [13]. The biomarkers present in the exhaled breath are used to indicate several diseases, including lung cancer [14,15], asthma [16,17], chronic obstructive pulmonary disease [18,19], breast cancer [20,21], diabetes [22,23], etc. The total number of diseases that can be detected or controlled by exhaled breath analysis is still unknown [24]. Exhaled human breath analysis has been developing for many years with the utilization of several different methods and techniques, for example, exhaled nitric oxide (FeNO) is currently used in clinical practice. However, it took 12 years from when the first report [25] about an increased level of NO in bronchial asthma was published until the first medical use of FeNO analysis in routine clinical practice. Another example is exhaled breath condensate (EBC) measurements. EBC is a promising source of biomarkers of lung disease; however, it is not a biomarker, but rather a matrix in which biomarkers may be identified [26]. The first papers concerning possible medical applications for EBC were published in early 1980, but a publication boom started in 2006, and the number of papers has increased in each subsequent year. After the American Thoracic Society (ATS) and the European Respiratory Society (ERS) developed guidelines for EBC collection and measurement, such measurements started to be commonly applied in asthma and COPD (Chronic obstructive pulmonary disease) diagnoses. One of the global diseases that is the cause of millions of deaths per year is diabetes. This is a chronic disease that occurs either when the pancreas does not produce enough insulin or when the body cannot effectively use the insulin it produces. Insulin is a hormone that regulates blood sugar. Hyperglycemia, or raised blood sugar, is a common effect of uncontrolled diabetes and, over time, leads to serious damage to many of the body's systems, especially the nerves and blood vessels. Based on actual data provided by the World Health Organization (WHO), 422 million adults have diabetes, and it is projected that diabetes will be the seventh leading cause of death by 2030 [27]. The conventional method for glucose monitoring is based on blood glucose measurements, and it has been used in clinical practice for at least 50 years. Moreover, it has been observed that patients with diabetes tend to have higher acetone levels in their breath than non-diabetics [28]. Due to its increased levels, acetone can be regarded as a biomarker of this disease. The number of people with diabetes increases every year. The exhaled acetone is usually within the range of 0.2–0.9 ppm for non-diabetics, and in the range of 1.25–2.5 ppm for people with diabetes [29]. Some references show that the acetone level can increase up to 25 ppm for type-1 diabetes [30]. The development of gas sensors technology has enabled the possibility to fabricate acetone sensors with sensitivities in the ppm range. The latest achievements in this field are reviewed and discussed in [31–33].

In this paper, we present for the first time, to the best of our knowledge, a prototype of a portable gas analyzer for exhaled acetone detection. By taking advantage of gas sensor technology on one hand and the design freedom offered by the 3D printing technology on the other, a dedicated device with the potential for non-invasive continuous glucose level monitoring is developed. In contrast to the commercially available devices [34–36], the proposed device is highly-selective and highly-sensitive due to its multi-sensor setup and processing. The presented design process includes a general proposal for the sensor array construction, the properties for the gas-sensing layer, and the requirements for the dedicated measurement chamber the gas-sensing layer is to be placed in. Moreover, the mechanical design of the 3D-printed housing is discussed to ensure the ergonomics of its use as a hand-held device while maintaining hardware integrity. Additionally, the processing hardware is discussed to provide sufficient computing power to handle the stand-alone operation while being energy efficient, thus enabling long battery-powered operation. Finally, calibration and measurement techniques as well as the analyzer operation are shown, validating the proposed exhaled acetone-detection meter.

## 2. Development of the Portable Gas Analyzer

This section presents and discusses the development and construction of the portable breath analyzer. The device is constructed upon an electrochemical gas sensor array placed in a dedicated measurement chamber. When the array is exposed to the exhaled breath sample, the resistance of each sensor is dependent upon three factors: gas concentration, humidity, and temperature. In addition to acetone, which is the most important biomarker, other gases, including ethanol, can also be found in exhaled breath. The use of many sensors with different sensitivities to gases in the exhaled air provides excessive information that, after some processing, enables the system to precisely determine the concentration of acetone despite other gases being present. The prototype analyzer was assembled using a custom 3D-printed housing and dedicated PCBs (printed circuit boards) with control electronics. Detailed descriptions of the individual components are presented in the following subsections.

### 2.1. Gas Sensor Array

The prototype is currently patent pending in the European Patent Office as well as specified within a non-disclosure agreement between the designers and investors; therefore, the presented description gives a general overview. The proposed gas sensor array was realized in an arrangement where a sensor that is highly sensitive to acetone is accompanied by a set of sensors sensitive to other exhaled breath constituents with little to no sensitivity to acetone. In the presented study, the gas sensor array is based on four various sensors named S1–S4. Such a configuration allows for the reduction of uncertainty and for an improvement in the selectivity of the exhaled acetone detection. In the presence of a detectable gas, the conductivity of each of the sensors increases depending on the concentration of gas in the air, thus, the sensor acts as a resistor of variable resistance. A simple electrical circuit can translate the change in conductivity into an output signal that corresponds to the gas concentration. The sensor requires two voltage inputs: heater voltage $V_H$ and circuit voltage $V_{CC}$. The heater voltage is applied to the integrated heater in order to maintain the sensing element at a specific temperature that is optimal for sensing. Circuit voltage is applied to allow measurement of voltage $V_{RL}$ across a load resistor $R_L$, which is connected in series with the sensor; this creates a voltage divider, and the voltage can be adapted to the signal processing circuit. The voltage $V_{RL}$ increases when the resistance of the sensor $R_S$ decreases in proportion to the gas concentration. The resistance $R_S$ can be calculated from the formula:

$$R_S = R_L\left(\frac{V_{CC}}{V_{RL}} - 1\right) \tag{1}$$

It should be noted that the above relation leads to uneven measurement resolution as $V_{RL}$ is measured in quantized increments of $V_{\text{LSB}}$ (Analog Digital Converter Least Significant Bit in terms of voltage) while the calculated sensor resistance is an inverse function. Therefore, special care must be taken when selecting $R_L$ to yield a maximum measurement resolution that is centered around the desired range of detectable concentrations.

Sensors used to construct the sensing array have four main unfavorable properties that should be taken into account: First, sensors require a relatively long pre-heating time using the built-in heater, extending the required time before the first measurement from 1 h to 7 days to burn out any pollution and stabilize the working temperature. Secondly, the sensors' responses are relatively slow, and an appropriate time between measurements is required for sensors to return to their original state. Thirdly, sensors are sensitive to changes in the temperature and humidity of the gas volume being tested, for example, a slight blast of cold air causes a change in temperature and humidity, and thus a change in the resistance $R_S$. Lastly, like most of these types of sensors, the main sensor is not only selective for acetone but also for other gases, including ethanol, which means that the measurement of acetone content in the exhaled air is influenced by activities such as, for example, drinking alcohol. The influence of these phenomena can be minimized by means of appropriate heating and calibration of the system to ensure appropriate measurement conditions, in other words, isolating of the sensor from undesirable gases as well as correcting the measurements with respect to the humidity and temperature of the gas volume being tested.

### 2.2. Gas Chamber Design

Operation of the device is based on the measurement of a sample of exhaled breath blown into the measurement chamber of the device through a removable mouthpiece. To achieve an accurate reading of the gas concentration in the exhaled air, the sensor needs time to set and reach the maximum value. This means that the sensor must be exposed to the gas sample to be tested in an isolated environment for a given time. The mechanical construction of such a chamber should, therefore, meet several requirements.

First of all, it is good if the mouthpiece is equipped with a one-way valve that does not allow the air volume to escape, ensures tightness, minimizes the presence of other gases with which the sensor could react, and ensures that the air blown in will have relatively constant pressure.

Secondly, the sensors array should not be located in the direct path of the air blown through the mouthpiece as it could lead to overloading of the sensors and would require a long recovery time. However, the mouthpiece should be at a slightly tilted angle from the perpendicular position to direct the gas stream through reflection from the back wall to the sensor array area.

Thirdly, the tightness of the chamber during the measurement is important, but after the measurement is completed and before proceeding to the next measurement, the chamber must be easily ventilated. This is required so the sensor can return to its original state, and the previously tested volume of air does not affect the next measurement.

Finally, breath is relatively warm and humid, which affects the sensors' response and may lead to a false measurement of gas concentrations. Therefore, a hygro-thermometer is needed to correct for that. To obtain a relatively accurate measurement of the humidity and temperature of the exhaled air sample, a free flow of air must be ensured. Moreover, the meter should not be located too close to the sensors as the integrated heaters heat and dry the air around sensors' housings nor should it be located in a direct path of the air blown through the mouthpiece as the increased pressure may alter the readout.

The gas chamber in the developed analyzer was designed by taking into account the above requirements. The gas chamber is a dedicated compartment in the device housing. The intake is designed in a way that accepts a disposable breathalyzer mouthpiece featuring a one-way valve for blowing the exhaled breath. On the other hand, a small grate opening located perpendicularly to the direct sample path is introduced to expose the chamber to ambient air. Such construction allows the sensing array to be exposed to the breath sample long enough for sensors to produce a stable response while enabling the chamber to be ventilated slowly through convection; blowing in the sample increases pressure in the chamber while the sample is slowly dried and heated due to heaters within the sensor structure. An opening forces pressure equalization to the ambient pressure, thus leading to the sample being vented out. In this way, sensor array can return to its initial state. All of the TGS sensors are attached to a dedicated PCB that connects to the motherboard through an opening in the divider. Moreover, the PCB hosts a daughterboard with a digital humidity and temperature sensor, Sensirion SHT21 in a DFN (Dual Flat No Leads) package, that provides calibrated, linearized signals in digital, $I^2C$ format with the accuracy of ±2% RH and ±0.3 °C with minimal power consumption. A drawing of the chamber is provided in Figure 1. The total volume of the chamber equals 48.4 mm × 30 mm × 20.4 mm (29.6 mL).

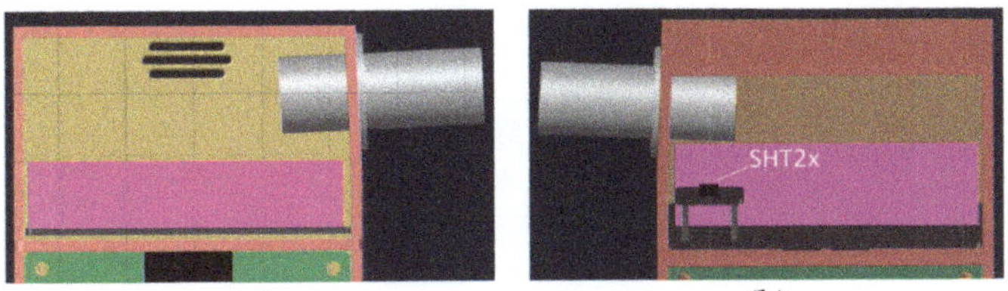


(a) (b)

**Figure 1.** Drawing of the designed gas chamber as a dedicated compartment in the device housing (in salmon). View of the mouthpiece (in light gray) and ventilation grate (in beige) (**a**). The SHT21 is soldered to a raised daughter board (in dark gray) allowing for indirect airflow as well as thermal separation from the gas sensor array (in violet) printed circuit boards (PCB) (**b**).

### 2.3. *3D Printable Housing Design*

The housing of the device was designed in a way to accommodate the gas chamber described in Section 2.2 as well as the motherboard PCB described in Section 2.4. As the developed analyzer is a portable device, its size and ergonomics are of great importance. Therefore, a 3D printing technology was employed offering a high degree of design freedom. The design was an iterative process. The first iteration was evaluated in terms of ergonomics of use, in other words, the legibility of the display, the positioning of the buttons with regard to the possibility of one-hand operation, and the grip in the hand. When the design met the above requirements, the second iteration was carried out to position I/O ports (including openings in the housing), ensuring the convenience of use and avoiding interference when the analyzer is held. The final design is shown in Figure 2. The housing is comprised of two parts: front and back pieces. The front piece has the user interface panel and the back piece has two separated compartments: the gas chamber and the electronics compartment. To close the housing, a set of latches on the back panels snap into indents in the front piece allowing for easy and tool-free access. In the electronics compartment, a shallow indentation the size of the display was added for ease of alignment and mounting along with properly spaced mounting posts for the motherboard PCB. The housing size was designed so that there is sufficient space for the electronics and sensors while providing a compact body of the device. The overall external dimensions of the device are 95.2 mm × 23.7 mm × 51.6 mm.

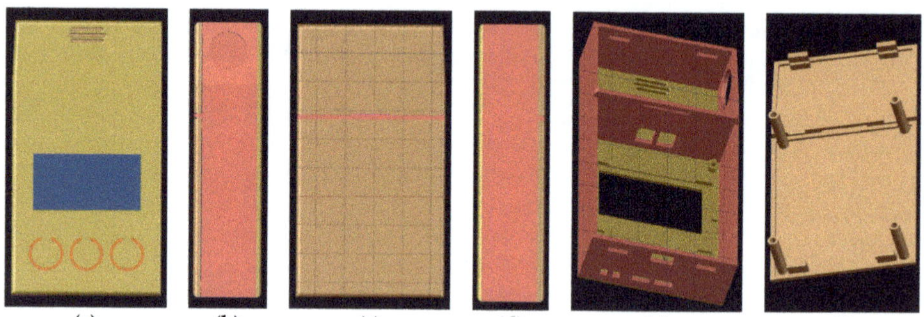

(a) (b) (c) (d) (e) (f)

**Figure 2.** 3D view and projections of the developed housing for the portable gas analyzer designed using CAD software. The housing is comprised of two parts: a front piece with user interface elements (**a**), mouthpice port (**b**), USB port and μSD card slots and back panels (**c**,**d**) allowing easy assembly (clips and slots) and access inside the device (**e**,**f**). Total dimensions are 95.2 mm × 23.7 mm × 51.6 mm.

### 2.4. Hardware Design

The designed device provides three main functionalities: signals measurement; data analysis; and communication and interaction with the user and data presentation. Control electronics were designed accordingly to provide appropriate hardware resources. As discussed earlier, the device's housing needed to be divided into two compartments: one is the gas chamber and the other accommodates the control electronics. Therefore, two separate PCBs were designed, the layout of both are shown in Figure 3. The sensor array PCB located in the gas chamber hosts all the sensors for which heater supply, circuit supply, and signal outputs are routed via a pin header connector to the motherboard PCB located in the second compartment, which contains the control electronics of the analyzer. The header pinout is described in Table 1.

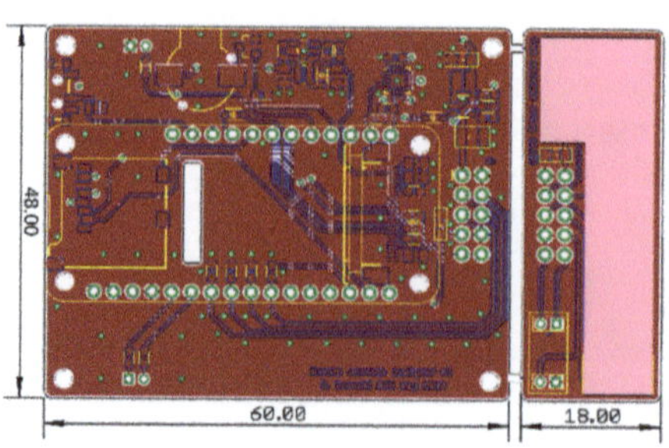


**Figure 3.** The layout of the motherboard and daughterboard PCBs. Dimensions in mm. Sensing array's layout visibility is limited due to the patent pending.

**Table 1.** List of the signals routed through the mother–daughter board the pin-header connector. Sensor signals are arranged to provide shielding from the digital bus by putting the power supply lines in-between.

| | J1 | | |
|---|---|---|---|
| Heaters +5 $V_{DC}$ supply | 1 | 2 | GND |
| Sensor 2 signal output | 3 | 4 | Sensor 1 signal output |
| Sensor 3 signal output | 5 | 6 | Sensor 3 signal output |
| SHT +3.3 $V_{DC}$ supp | 7 | 8 | GND |
| I2C bus SCL | 9 | 10 | I2C bus SCA |

The main module on the motherboard PCB is the Adafruit Feather M0 Bluefruit LE development board as it is an excellent compromise between computing power, number, type of peripherals, size, and price. The board is powerful enough to handle the stand-alone operation of the analyzer, including the future implementation of a neural network (raw measurement, pre-processing, feature extraction, pattern recognition, classification, and decision making). Importantly, due to compatibility with Arduino Zero, the firmware can be written in C++ using the Arduino Development Environment which includes a wide range of libraries. The heart of the Feather M0 board is a 32-bit ATSAMD21G18 microcontroller with an ARM Cortex M0 core, clocked at 48 MHz, supporting logic in the 3.3 V standard. The chip has 256K FLASH memory and 32K RAM. It is equipped with built-in USB support giving it the ability to program and debug USB-to-serial without the need for an additional chip. The board is equipped with a dedicated connector for 3.7 V lithium-polymer batteries and a built-in battery charging system via a micro USB connector and available battery voltage monitoring functionality. In addition, it is possible to power it directly through the micro USB connector. The board has an automatic selector of the power source depending on whether it is powered by a battery or via the USB connector. In addition, the board is equipped with a Bluetooth Low Energy communication module. The user interface hardware is comprised of a 1.5′ monochrome OLED display (SSD1305 driver) with a resolution of 128 × 64 pixels, a buzzer, and three tack-switches: up, OK, and down. Alternatively, a command-line interface is available. Analog signals from the sensor array are measured as a voltage drop across a 0.1% tolerant load resistor $R_{Lx}$ located on the motherboard using a 12-bit microcontroller built-in ADC ($V_{CC}$ = $V_{ADCref}$ = +3.3 $V_{DC}$). To ensure a stable heater supply $V_H$ to sensors, a high-efficiency step-up DC-DC converter was used providing +5 $V_{DC}$ and up to 250 $mA_{DC}$ from either the Li-Po battery or the USB line. It is also possible to disable the converter and thus implement the sleep mode. In addition, there is a μSD card slot on the board that allows local acquisition of measurement data, a USB interface for data acquisition from the PC level, and a BLE (Bluetooth Low Energy) modem for wireless acquisition.

Each of the components was chosen to take into account its power consumption as the analyzer is to be battery powered. Total power drawn by the device was estimated to be ~400 mA at 3.3 $V_{DC}$ where the majority of the power is consumed by the sensors' heaters (~380 mA). Therefore, a mid-size Li-po battery having 1500 mAh capacity would allow more than 3.5 h of operation. However, the array can be pre-heated overnight using a wall adapter and then using appropriate sensor-stabilization cycles, a full day of operation is possible.

### 2.5. Prototype Manufacturing and Assembly

Designed printed circuit boards were made by an external company offering production services of prototype series. The housing was printed in the in-house workshop using a Prusa Research Prusa i3 MK3 3D printer out of PET-G filament (Polyethylene terephthalate glycol-modifies). The printer uses a FFF (fused filament fabrication) process which in combination with the used filament leads to low-cost prototyping while providing satisfactory print resolution and mechanical strength. Finally, the components were assembled and the prototype of the device was tested for correctness. Photographs of the manufactured analyzer are shown in Figure 4.

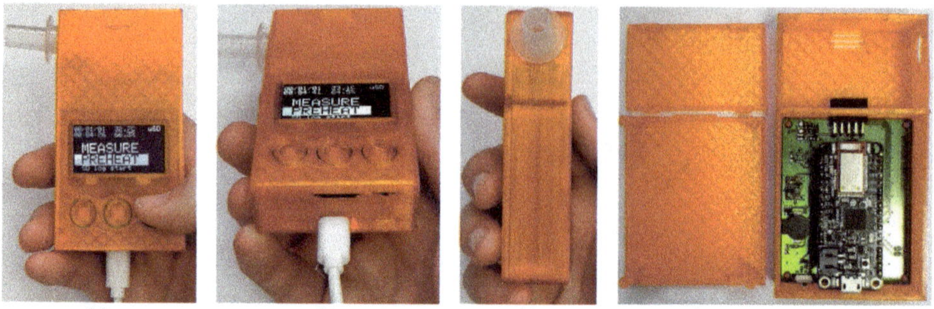


(a) (b) (c) (d) (e)

**Figure 4.** A photograph of the manufactured analyzer. Front panel (**a**), bottom view with connectors (**b**), side panel with a disposable mouthpiece (**c**), the gas chamber (**d**) and the electronics compartment (**e**).

## 3. Results and Discussion

In this section, calibration and measurements using the developed portable gas analyzer for exhaled acetone detection are presented and discussed. The calibration procedure is shown allowing for the determination of the reference (normalization) resistance $R_{0x}$ for each of the sensors in the array. Moreover, the influence of temperature and humidity of the gas sample on the response of the sensor is considered and appropriate corrections are proposed. Finally, the procedure to determine gas concentration based on the measured and corrected sensor's response is presented.

### *3.1. Sensors Reference Resistance Calibration*

Before proceeding with actual measurements, it is necessary to calibrate the sensors, in other words, determine the value of the reference resistance $R_{0x}$. The procedure is described for an example of the main sensor, but it also applies for the other sensors used. The main sensor datasheet provides its typical sensitivity characteristics, in other words, the change of its resistance $R_S$ for different gas types in the measurable range of concentrations. The characteristic, however, is normalized to a reference resistance $R_0$ being the sensor resistance in 300 ppm ethanol at 20 °C and 65% RH Knowing this value, it is possible to calculate the dependence between sensor resistance and the gas concentration for all other gas concentrations. Further, in the datasheet, it is stated that $R_0$ is in the range of 1–10 kΩ, which is a relatively wide range, therefore, it is impossible to take a specific value from this range as $R_0$. On the other hand, calibration using a reference gas sample having a proper concentration in a specific temperature and humidity would require a laboratory and reference measuring device. Such a calibration method is hardly possible to implement and very expensive for a portable analyzer. Therefore, it was important to find an alternative calibration technique.

The proposed approach relies on the fact that the sensor's resistance ratio $R_0/R_S$ is constant over the measurable concentration range when exposed to air. Reference resistance can be calculated as:

$$R_0 = \frac{R_{Sair}}{c_{r0}} \tag{2}$$

where $c_{r0}$ is a constant coefficient estimated based on the data provided in the datasheet while $R_{Sair}$ is the sensor resistance in air corrected for temperature and humidity (see Section 3.2). The remaining sensors in the array can be calibrated in the same way. Such calibration is easy to implement and convenient from the user's perspective. The disadvantage is higher measurement error when related to one using a dedicated calibration gas sample, however, it is sufficient for non-clinical applications. The measurement error associated with the described technique has three main sources: estimation of the $c_{r0}$ coefficient based on graphical data and the sensor's manufacturing tolerance; cleanness of the air when measured: and temperature and humidity of the calibration gas. The first source can hardly be minimized and is considered as a systematic error. The impact of the former two can be minimized when sensors are placed in an enclosed environment and pre-heated using embedded heaters before measurements to allow the sensor to stabilize its response and to stabilize temperature and humidity in the gas chamber. In addition, any possible impurities or debris on the sensing element surface can be burned-out or evaporated at each sensor.

### 3.2. Correction for Temperature and Humidity

Having calibrated the system, the actual measurements can be carried out. However, since the response of the sensor is not only a function of a detectable gas concentration but also temperature and humidity, it is necessary to correct the measured resistance $R_S$ for the last two parameters. This dependence exists regardless of the gas being measured and its concentration. Unfortunately, the required data is presented in datasheets only in a graphical form, which hinders analysis and is a source of systematic error. To allow for calculating the correction, a linear polynomial regression of two variables was used to model the temperature–humidity dependence out of the collected data points. A linear function describes accurately enough the relation while being very computationally inexpensive. The general formula is:

$$corr(t,h) = c_{00} + c_{10}h + c_{01}t \tag{3}$$

where $t$ and $h$ are the measured temperatures in °C and humidity in % at the time the sensor's resistance is measured while $c_{xx}$ are constant coefficients estimated for a given sensor.

### 3.3. Gas Concentration Calculation

Having measured the raw sensor array response on the gas sample and corrected it for humidity and temperature, the gas concentration can be established. In order to determine the analytical relationship between the resistance ratio $R_S/R_0$ and the concentration of a given gas in the tested air volume, the sensitivity characteristics presented by the manufacturer in the datasheet expressed in log-log scale was analyzed. Therefore, the following variable is is introduced to simplify calculations:

$$r = log_{10}\left(\frac{R_S}{R_0}\cdot \mathrm{corr}(t,h)\right) \tag{4}$$

Unfortunately, the required data is presented only in a graphical form, which hinders analysis and is a source of systematic error. Moreover, those relations are provided under the assumption that the sensor is exposed only to one particular gas. Square polynomial regression was used to derive the relation out of the collected data points. A quadratic function describes accurately enough the relation while being fairly computationally inexpensive. The general formula is:

$$gas(\mathrm{r}) = p_0 + p_1 r + p_2\, r^2 [ppm] \tag{5}$$

where $p_x$ are constant coefficients estimated based on the data provided in the datasheet. The polynomial relation coefficients are established based on data for a give a measurable range of gas concentrations, therefore, outside that range, it may not hold. However, since it is a continuous function and the relation appears to be monotonic, Equation (5) can be used to estimate the concentration outside the above-specified range.

### 3.4. Portable Analyzer Operation and Measurements

The developed portable analyzer allows for breath acetone detection. The information from the main sensor is calculated to the acetone concentration while the information from the accompanying sensors is taken to check the validity of the measurement, in other words, to determine if other gases are deviating the response or not. The use of the device is intuitive and easy. The user menu and available functionalities presented on the screen (see Figure 5) are directly related to the state in which the device currently is as the firmware was written using a finite-state machine. When the device boots, the welcome screen appears briefly showing the information on the hardware and firmware version that is currently used/loaded into the device. The device then automatically changes to the main screen. The user can select either to go directly to measurements or to preheat the sensor array. At the moment of entering the measurement state, the device self-calibrates the array (measures each sensor' resistance in air and calculates the reference resistance). If preheating is selected, the user can define the preheating time or the device can preheat for as long as the sensor array response determines, the end of which in each case is signaled by the buzzer. When in the measure state, the array is sampled continuously to monitor its stability. To perform measurement of the exhaled breath, the user needs to blow constantly through the mouthpiece for a given time, the end of which is signaled by the buzzer. Then the readouts are processed and the acetone concentration is displayed along with the information if the concentration is within or outside sensor's range as well as if it is valid and the maximum concentration measured. Before proceeding to the next measurement, the gas chamber needs to be properly ventilated and the sensor array needs to return to its default state. When the analyzer is ready, appropriate information is displayed and the device can be used again.

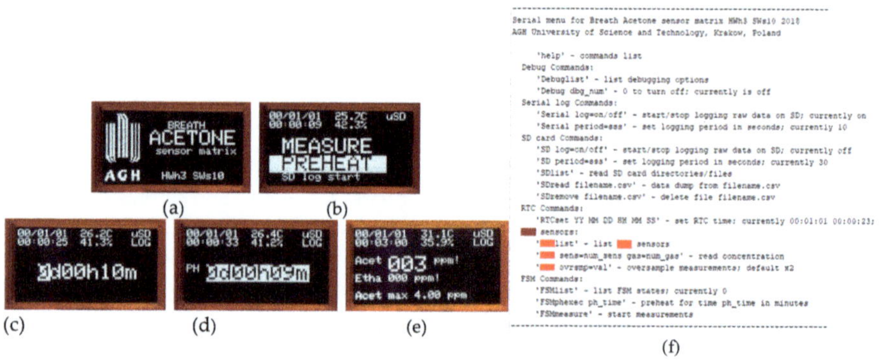


**Figure 5.** User interface screens of the developed portable gas analyzer. Splash screen (**a**), main menu (**b**), preheat setting (**c**), and executing (**d**) measurement screen presenting the acetone concentration and the raw data from each sensor (**e**). Alternatively, the analyzer can be controlled through a serial interface (**f**).

### 3.5. Gas-Sensing Setup Characterization

The developed sensing setup composed of a set of four different sensors located in the gas chamber of the device was characterized by conducting a series of measurements under exposure to diabetes biomarkers such as acetone and ethanol. Figure 6 shows the gas-sensing characteristics under exposure to various acetone concentration (20–200 ppm) of each of the gas sensor *S1*–*S4* within the array. For a better perspective, all curves are shown on the same scale. As can be observed, the sensors exhibit a response to acetone, however, with a largely differentiated sensitivity as expected. Moreover, the relatively short recovery time is to be dealt with in the devices firmware. The sharp lines are the artifacts related to measurement conditions that switched the effect of the gas-dosing system.

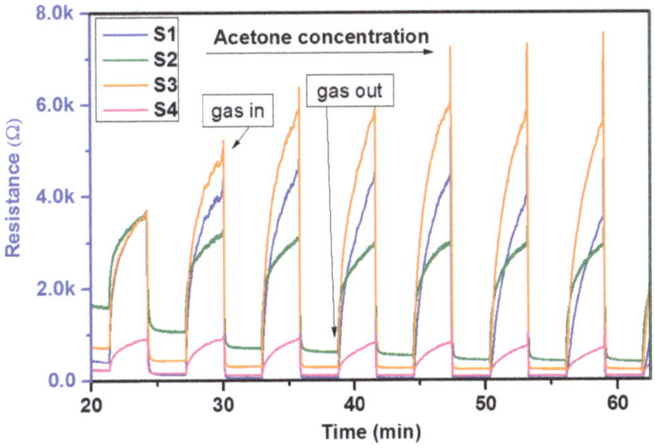


**Figure 6.** The gas-sensing characteristics of each particular gas sensor in the gas-sensing chamber under exposure to acetone in the 20–200 ppm range. Optimal operating temperature (given by the manufacturer) was ensured and 50% relative humidity was stabilized.

Despite the S1 sensor being the one dedicated for acetone sensing due to its high sensitivity, there were others acting in a similar fashion. Figure 7 shows the gas-sensing responses of sensor S3, which was non-dedicated to acetone detection, however, as can be observed under exposure to acetone, it reacted with resistance changes. This cross-sensitivity effect is a disadvantage of metal-oxide based sensors and has to be taken into account within the firmware.

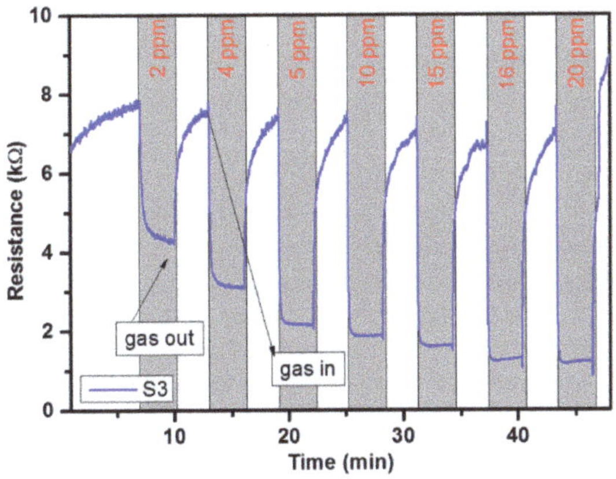


**Figure 7.** The resistance changes of sensor S3 (non-dedicated to acetone detection) in a gas sensor array under exposure to 2–20 ppm of acetone. Optimal operating temperature (given by the manufacturer) was ensured and 50% relative humidity was stabilized.

Finally, Figure 8 shows the calibration curves of all four sensors in the gas sensor array in the 2–20 ppm range of acetone. The sensors S1–S3 reacted with acetone in the measurement range and their response can be fitted using an exponential function with high $R^2$ values of 0.91, 0.95, and 0.95, respectively. On the other hand, Sensor S4 exhibited a constant response (~5.5), which was related to the humidity level. Sensors S1 and S2 are dedicated to acetone measurement while S3 and S4 are not. Despite that, sensor S3 exhibits a cross-sensitivity issue.

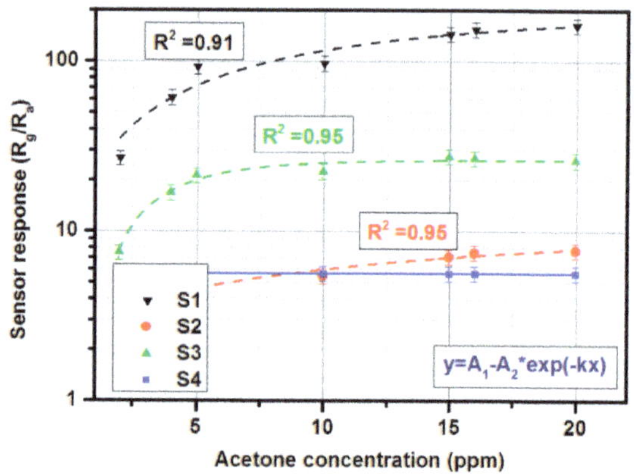


**Figure 8.** The calibration curves of all four sensors (S1–S4) in the gas sensor array under exposure to 2–20 ppm of acetone. Optimal operating temperature (given by the manufacturer) was ensured and 50% relative humidity was stabilized.

It is important to note, however, that for the purpose of measurements the developed sensing array is to be considered as a whole, in other words, its multi-dimensional response characteristics can be characterized and calibrated into the firmware. Further processing blocks such as a neural network are to be trained beforehand and fed in with the measurement data. The key point is to select sensors in such a way that they exhibit variable (including high) sensitivity on the diabetes markers as well as on other breath constituents. Such a combination along with proper signal processing enables measurements of low acetone concentrations with a high level of confidence.

## 4. Conclusions and Further Perspectives

We developed a portable gas analyzer prototype for exhaled acetone detection employing an application-suited gas sensor array. The device was shown to be capable of serving as a tool for non-invasive diabetes monitoring with the additional advantage of having a hand-held form factor and battery powered operation. The prototype is currently patent pending in the European Patent Office and is specified within a non-disclosure agreement between the designers and investors, therefore, the presented description gives only a general overview of the system. However, all components have been widely discused, including the gas sensor array, gas chamber design, 3D printable housing design, hardware and software design, and correction for temperature and humidity. Finally, the calibration and measurement technique as well as the analyzer operation was shown, validating the proposed type of exhaled acetone-detection meters. After the certification process, the prototype will be validated in clinical tests in the University Hospital in Krakow, Poland.

**Author Contributions:** A.R. conceptualization; J.S. and A.R. methodology; J.S. software; A.R. validation; A.R. and J.S. investigations; J.S. and A.R. writing—original draft preparation and writing—review and editing; A.R. supervision; A.R. project administration.

**Funding:** The work was financially supported by the National Centre for Research and Development under LIDER/252/L-6/NCBR/2015.

**Conflicts of Interest:** The authors declare no conflict of interest.

## References


1. De Lacy Costello, B.; Amann, A.; Al-Kateb, H.; Flynn, C.; Filipiak, W.; Khalid, T.; Osborne, D.; Ratcliffe, N.M. A review of the volatiles from the healthy human body. *J. Breath Res.* **2014**, *8*, 014001. [CrossRef] [PubMed]
2. Phillips, M.; Byrnes, R.; Cataneo, R.N.; Chaturvedi, A.; Kaplan, P.D.; Libardoni, M.; Mehta, V.; Mundada, M.; Patel1, U.; Ramakrishna, N.; et al. Detection of volatile biomarkers of therapeutic radiation in breath. *J. Breath Res.* **2013**, *7*. [CrossRef] [PubMed]
3. Al-Kateb, H.; de Lacy Costello, B.; Ratcliffe, N. An investigation of volatile organic compounds from the saliva of healthy individuals using headspace-trap/GC-MS. *J. Breath Res.* **2013**, *7*, 036004. [CrossRef] [PubMed]
4. Deng, C.; Zhang, J.; Yu, X.; Zhang, W.; Zhang, X. Determination of acetone in human breath by gas chromatography-mass spectrometry and solid-phase microextraction with on-fiber derivatization. *J. Chromatogr. B* **2004**, *810*, 269–275. [CrossRef]
5. Lubes, G.; Goodarzi, M. GC-MS based metabolomics used for the identification of cancer volatile organic compounds as biomarkers. *J. Phar. Biomed. Anal.* **2018**, *147*, 313–322. [CrossRef] [PubMed]
6. Fedrigo, M.; Hoeschen, Ch.; Oeh, U. Multidimensional statistical analysis of PTR-MS breath samples: A test study on irradiation detection. *Int. J. Mass Spectrom.* **2010**, *295*, 13–20. [CrossRef]
7. Righettoni, M.; Schmid, A.; Amann, A.; Pratsinis, S.E. Correlations between blood glucose and breath components from portable gas sensors and PTR-TOF-MS. *Breath Res.* **2013**, *7*, 037110. [CrossRef] [PubMed]
8. Lapthorn, C.; Pullen, F.; Chowdhry, B.Z. Ion mobility spectrometry-mass spectrometry (IMS-MS) of small molecules: Separating and assigning structures to ions. *Mass Spectrom. Rev.* **2013**, *32*, 43–71. [CrossRef] [PubMed]
9. Szymanska, E.; Tinnevelt, G.H.; Brodrick, E.; Williams, M.; Davies, A.N.; van Manen, H.-J.; Buydens, L.M.C. The increasing conclusiveness of clinical breath analysis by improved baseline correction of the multi-capillary column—Ion mobility spectrometry (MCC-IMS) data. *J. Pharm. Biomed. Anal.* **2016**, *127*, 170175. [CrossRef] [PubMed]
10. Michalcikova, R.B.; Dryahina, K.; Spanel, P. SIFT-MS quantification of several breath biomarkers of inflammatory bowel disease, IBD.; A detailed study of the ion chemistry. *Int. J. Mass Spect.* **2016**, *396*, 35–41. [CrossRef]
11. Smith, D.; Spanel, P. Direct, rapid quantitative analyses of BVOCs using SIFT-MS and PTR-MS obviating sample collection. *TrAC* **2011**, *30*, 945–959. [CrossRef]
12. Schmidt, K.; Podmore, I. Current challenges in volatile organic compounds analysis as potential biomarkers of cancer. *J. Biomark.* **2015**, *2015*, 981458. [CrossRef] [PubMed]
13. Smolinska, A.; Klaassen, E.M.M.; Dallinga, J.W.; van de Kant, K.D.G.; Jobis, Q.; Moonen, E.J.C.; van Schayck, Q.P.; Dompeling, E.; van Schooten, F.J. Profiling of volatile organic compounds in exhaled breath as a strategy to find early predictive signatures of asthma in children. *PLoS ONE* **2014**, *9*, e95668. [CrossRef] [PubMed]
14. Saalberg, Y.; Wolff, M. VOC breath biomarkers in lung cancer. *Clin. Chim. Acta* **2016**, *459*, 5–9. [CrossRef] [PubMed]
15. Chang, J.-E.; Lee, D.-S.; Ban, S.-W.; Oh, J.; Jung, M.Y.; Kim, S.-H.; Park, S.J.; Persaud, K.; Jheon, S. Analysis of volatile organic compounds in exhaled human breath for lung cancer diagnosis using a sensor system. *Sens. Actuators B* **2018**, *255*, 800–807. [CrossRef]
16. Bos, L.D.; Sterk, P.J.; Fowler, S.J. Breathomics in the setting of asthma and chronic obstructive pulmonary disease. *J. Allergy Clin. Immunol.* **2016**, *138*, 970–976. [CrossRef] [PubMed]
17. Hanania, N.A.; Pharm, M.M.; Jain, N. Measurement of fractional exhaled nitric oxide in real-world clinical practice alters asthma treatment decisions. *Ann. Allergy Asthma Immunol.* **2018**, *120*, 414–418. [CrossRef] [PubMed]
18. Anders, C.; Rømhild, D.J.; Ingrid, T.; Jørgen, V.; Jan, B. A systematic review of breath analysis and detection of volatile organic compounds in COPD. *J. Breath Res.* **2016**, *10*, 034002.
19. Bregy, L.; Nussbaumer-Ochsner, Y.; Sinues, P.M.-L.; Garcia-Gomez, D.; Suter, Y.; Gaisl, T.; Stebler, N.; Gaugg, M.T.; Kohler, M.; Zenobi, R. Real-time mass spectrometric identification of metabolites characteristic of chronic obstructive pulmonary disease in exhaled breath. *Clinic. Mass Spectr.* **2018**, *7*, 29–35. [CrossRef]
20. Herman-Saffar, O.; Boger, Z.; Libson, S.; Lieberman, D.; Gonen, R.; Zeiri, Y. Early non-invasive detection of breast cancer using exhaled breath and urine analysis. *Comput. Biol. Med.* **2018**, *96*, 227–232. [CrossRef]

21. LI, H.; Peng, Y.; Duan, Y. Diagnosis of breast cancer based on breath analysis: An emerging method. *Crit. Rev. Oncol. Hemat.* **2013**, *87*, 28–40. [CrossRef] [PubMed]
22. Minh, T.D.Ch.; Blake, D.R.; Galassetti, P.R. The clinical potential of exhaled breath analysis for diabetes mellitus. *J. Diab. Res.* **2012**, *97*, 195–205. [CrossRef]
23. Karyakin, A.A.; Nikulina, S.V.; Vokhmyanina, D.V.; Karyakina, E.E.; Anaev, E.K.H.; Chuchalin, A.G. Non-invasive monitoring of diabetes through analysis of the exhaled breath condensate (aerosol). *Electrochem. Commun.* **2017**, *83*, 81–84. [CrossRef]
24. Amann, A.; Smith, D. *Volatile Biomarkers: Non-Invasive Diagnosis in Physiology and Medicine*; Elsevier: Edinburgh, UK, 2013.
25. Piacentini, G.L.; Cattazzo, E.; Tezz, G.; Peroni, D.G. Exhaled nitric oxide in pediatrics: What is new for ptracrive purposes and clinical research in children? *J. Breath Res.* **2012**, *6*, 027103. [CrossRef] [PubMed]
26. Warwick, G.; Kotluar, E.; Chow, S.; Thomas, P.S.; Yate, D.H. Exhaled breath condensate in pulmonary arterial hypertension. *J. Breath Res.* **2012**, *6*, 036006. [CrossRef]
27. Diabetes—World Health Organization. Available online: http://www.who.int/diabetes/en/ (accessed on 27 October 2018).
28. Rydosz, A. A negative correlation between blood glucose and acetone measured in healthy and type-1 diabetes mellitus patient breath. *J. Diabetes Sci. Technol.* **2015**, *9*, 881–884. [CrossRef]
29. Rydosz, A. Micropreconcentrator in LTCC technology with mass spectrometry for the detection of acetone in healthy and type-1 diabetes mellitus patient breath. *Metabolites* **2014**, *4*, 921–931. [CrossRef]
30. Rydosz, A. Sensors for enhanced detection of acetone as a potential tool for noninvasive diabetes monitoring. *Sensors* **2018**, *18*, 2298. [CrossRef]
31. Saasa, V.; Malwela, T.; Beukes, M.; Mokgotho, M.; Liu, Ch.-P.; Mwakikunga, B. Sensing technologies for detection of acetone in human breath for diabetes diagnosis and monitoring. *Diagnostics* **2018**, *8*, 12. [CrossRef]
32. Todd, C.; Salvetti, P.; Naylor, K.; Albatat, M. Towards non-invasive extraction and determination of blood glucose levels. *Bioengineering* **2017**, *4*, 82. [CrossRef]
33. Wang, Ch.; Sahay, P. Breath analysis using laser spectroscopic techniques: Breath biomarkers, spectral fingerprints, and detection limits. *Sensors* **2009**, *9*, 8230–8262. [CrossRef] [PubMed]
34. KETONIX®Breath Ketone Analyzer. Available online: https://www.ketonix.com/ (accessed on 29 May 2019).
35. KetoIQ Keto Breathalyzer. Available online: https://www.amazon.com/dp/B07BKTNWF7/?tag=ketoconcern-20 (accessed on 29 May 2019).
36. Ketone Breath Analyzer. Available online: https://www.amazon.com/dp/B07CKW8HZQ/?tag=ketoconcern-20 (accessed on 29 May 2019).

MDPI
St. Alban-Anlage 66
4052 Basel
Switzerland
www.mdpi.com

*Applied Sciences* Editorial Office
E-mail: applsci@mdpi.com
www.mdpi.com/journal/applsci

www.ingramcontent.com/pod-product-compliance
Lightning Source LLC
Chambersburg PA
CBHW061446210926
51850CB00008B/1628
* 9 7 8 3 7 2 5 8 1 2 9 9 8 *